SCHAUM'S OUTLINE OF

THEORY AND PROBLEMS

of

PHYSICS
for Engineering and Science

•

by

DARE A. WELLS, Ph.D.

Emeritus Professor of Physics
University of Cincinnati

and

HAROLD S. SLUSHER, D.Sc., Ph.D.

Assistant Professor of Physics
University of Texas at El Paso

•

SCHAUM'S OUTLINE SERIES
McGRAW-HILL BOOK COMPANY

New York St. Louis San Francisco Auckland Bogotá Hamburg London
Madrid Mexico Milan Montreal New Delhi Panama Paris
São Paulo Singapore Sydney Tokyo Toronto

DARE A. WELLS is Emeritus Professor of Physics at the University of Cincinnati, Cincinnati, Ohio. Professor Wells received his Ph.D. from the University of Cincinnati. His published research includes some twenty papers in the fields of spectroscopy, small oscillations, Lagrangian dynamics, and the treatment of electromechanical systems by Lagrangian methods. He is the originator of a general form of "P-function" for the determination of generalized dissipative forces. Professor Wells is also the author of the advanced Schaum's Outline, *Lagrangian Dynamics*.

HAROLD S. SLUSHER is Assistant Professor of Physics at the University of Texas at El Paso and Professor of Planetary Sciences at the Graduate School of the Institute for Creation Research, San Diego, California. Professor Slusher holds a D.Sc. from Indiana Christian University and a Ph.D. from Columbia Pacific University. His published research includes monographs in the fields of cosmogony, cosmology, and geochronology.

Schaum's Outline of Theory and Problems of
PHYSICS FOR ENGINEERING AND SCIENCE

6 7 8 9 10 11 12 13 14 15 SH SH 8 9 8

ISBN 0-07-069254-8

Editing Supervisor, Marthe Grice
Production Manager, Nick Monti

Library of Congress Cataloging in Publication Data

Wells, Dare A., 1899-
    Schaum's outline of theory and problems of
physics for engineering and science.

    (Schaum's outline series)
    Includes index.
    1. Physics. 2. Physics--Problems, exercises,
etc. I. Slusher, Harold Schultz. II. Title.
QC21.2.W44 1983   530   82-4715
ISBN 0-07-069254-8

Preface

The fundamental principles of physics, in conjunction with certain branches of mathematics, constitute the all-important building blocks of every physical science as well as all branches of engineering. The primary purpose of this text is to help the student of science and engineering to obtain, without an undue expenditure of time and effort, a thorough understanding and appreciation of the basic principles and mathematical methods of the subject.

The guiding principle followed in the preparation of the book rests on the firm belief that a specific example, solved in detail, is the ultimate means of making clear physical principles and mathematical procedures; that the solved example is a truly effective way of "explaining the explanation" of the classroom text and/or lecture; and finally that it is an important and noteworthy means of arousing interest in this (sometimes not-so-easy) basic science. Moreover, a great many students have expressed this opinion.

Hence each chapter beyond the first (which is a summary of certain important background material) begins with a brief statement of physical principles and corresponding mathematical relations, such as are found in standard texts. This is then followed by an ample collection of carefully selected and graded examples, with a step-by-step solution of each. Finally, for use as an effective self-examination of the student's understanding of topics treated, a number of specific problems, with answers, are included.

We wish to express our sincere gratitude to the thousands of former students whose interest and lack of interest, difficulties and successes have contributed greatly to the motivation and content of this undertaking.

DARE A. WELLS
HAROLD S. SLUSHER

Contents

CONTENTS

CONTENTS

CONTENTS

Chapter 1

Review of Background Material
Vector Methods, Units,
Dimensional Analysis

1.1 SCALARS AND VECTORS

Quantities such as time, mass, density, work, temperature, etc., which have magnitude but no direction, are referred to as *scalars*. They are denoted in lightface type as A, B, m, t, ρ, Q, etc.

Quantities such as velocity, acceleration, force, electric field, etc., which have direction as well as magnitude, are referred to as *vectors*. They are denoted in boldface type as **A**, **B**, **F**, **E**, etc. Three numbers are required to specify a vector, while only one is required for a scalar.

1.2 GRAPHICAL REPRESENTATION OF VECTORS

Any vector—as force **F**, Fig. 1-1(a), or velocity **v**, Fig. 1-1(b)—can be represented by a straight line. The length of the line, measured in any convenient units (inches, centimeters, etc.), represents the vector's magnitude and the angles it makes with, for example, rectangular axes, X, Y, Z, the vector's direction.

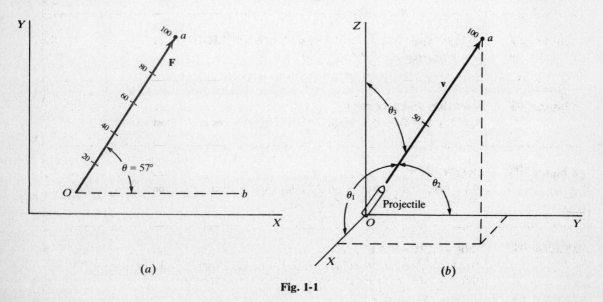

(a) (b)

Fig. 1-1

EXAMPLE 1.1. The line Oa, Fig. 1-1(a), represents a force of 100 N acting on point O. Here the line is taken in the XY plane at an angle of 57° with Ob. Note that both magnitude and direction of **F** are represented by the line Oa.

The space line Oa, Fig. 1-1(b), represents the velocity **v** of a projectile moving at 100 m/s. The length of Oa (100 units) indicates the magnitude of **v** and θ_1, θ_2, θ_3 give its direction. Note that if values of $\cos \theta_1$ and $\cos \theta_2$ are given, θ_3 follows from

$$\cos \theta_3 = \pm\sqrt{1 - \cos^2 \theta_1 - \cos^2 \theta_2}$$

1

Graphical Addition

In Fig. 1-2, forces $\mathbf{F}_1$ and $\mathbf{F}_2$ are acting on point O. Magnitudes and directions are drawn to scale as shown. Now to "add" these vectors (that is, to find a single vector entirely equivalent to the two) we complete the parallelogram (dashed lines) and draw in the diagonal. This line, measured in the same units as $\mathbf{F}_1$ and $\mathbf{F}_2$, represents in both *magnitude and direction* the "vector sum" $\mathbf{R}$, written symbolically as

$$\mathbf{R} = \mathbf{F}_1 + \mathbf{F}_2$$

EXAMPLE 1.2. Suppose a nail driven in a board at O, Fig. 1-2. By pulling on cords tied to the nail we exert forces of 75 and 100 N in the directions of Oa and Ob. The nail does not sense the existence of two forces, but instead it "feels" a single force $\mathbf{R}$, the magnitude of which is (by rough measurements) $R = 152$ N, acting at an angle $\alpha \approx 35°$. Of course, given $\mathbf{F}_1$, $\mathbf{F}_2$ and θ, R and α can be computed. But here we are concerned with graphical methods only.

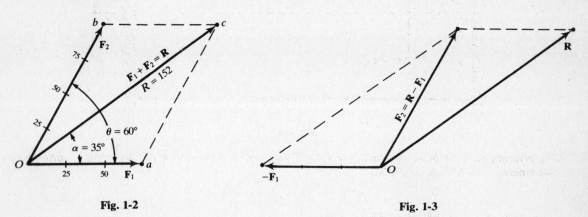

Fig. 1-2   Fig. 1-3

EXAMPLE 1.3. Suppose that the nail in Example 1.2 is replaced by a small object (free to move) having mass $m = 0.2$ kg. At the instant the forces are applied, what is its acceleration $\mathbf{a}$?

The net force and the acceleration are related by $\mathbf{R} = m\mathbf{a}$. Thus the magnitude of $\mathbf{a}$ is

$$a = \frac{152}{0.2} = 760 \text{ m/s}^2$$

and its direction is that of $\mathbf{R}$.

EXAMPLE 1.4. If an airplane moves with a velocity of 152 m/s in the direction of Oc, Fig. 1-2, this is equivalent to moving simultaneously in the directions of Oa and Ob with velocities of 75 and 100 m/s, respectively.

Graphical Subtraction

Subtraction of a vector means that it is reversed in direction and added just as above.

EXAMPLE 1.5. Given $\mathbf{R}$ and $\mathbf{F}_1$ in Fig. 1-2, find $\mathbf{F}_2$.

From $\mathbf{R} = \mathbf{F}_1 + \mathbf{F}_2$, we write $\mathbf{F}_2 = \mathbf{R} - \mathbf{F}_1$. As indicated in Fig. 1-3, $\mathbf{F}_1$ is reversed in direction and added to $\mathbf{R}$ by completing the parallelogram to give $\mathbf{F}_2$.

EXAMPLE 1.6. A boat is to cross a river along the straight line AB, Fig. 1-4. The water has a velocity of 4 m/s downstream, as indicated. On still water the boat travels at speed $v_2 = 6$ m/s. What is its speed v_3 along AB? In what direction must the boat be steered (what is α?), and what is the time from A to B? (*Note:* We are here given the magnitude and direction of $\mathbf{v}_1$, the magnitude of $\mathbf{v}_2$, the direction of $\mathbf{v}_3$, and the magnitude and direction of segment AB; we are to find α and the magnitude of $\mathbf{v}_3$.)

A graphical solution may be obtained as follows. Draw: (1) river and line AB to any convenient scale; (2) a circle of radius 6 units, with center at some point O of AB; (3) Oa 4 units long (this is $-\mathbf{v}_1$); ab, parallel to AB

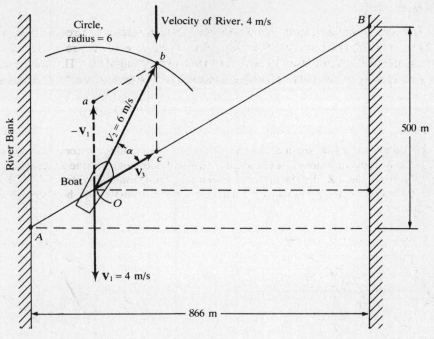

Fig. 1-4

and intersecting circle at b; (4) bc parallel to Oa. Then, angle α and $v_3 = \overline{Oc}$ are determined. From rough measurements, $v_3 = 3.0$ m/s and $\alpha = 35°$. Since

$$\overline{AB} = \sqrt{(500)^2 + (866)^2} = 1000 \text{ m}$$

the time from A to B is $1000/3.0 = 333$ s $= 5.6$ min.

1.3 COMPONENTS OF VECTORS

In Fig. 1-5 the dashed perpendiculars from P to X and Y determine the magnitudes and directions of the vector components $\mathbf{F}_x$ and $\mathbf{F}_y$ of vector $\mathbf{F}$. The *signed* magnitudes of these components, which are scalar quantities, are written as F_x, F_y. Note that in Fig. 1-2 $\mathbf{F}_1$ and $\mathbf{F}_2$ are the vector components of $\mathbf{R}$ taken along the oblique lines Oa and Ob respectively. In Fig. 1-6, $\mathbf{F}_x$, $\mathbf{F}_y$, $\mathbf{F}_z$ are the rectangular vector components of $\mathbf{F}$; scalar components are written as F_x, F_y, F_z.

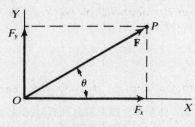

Fig. 1-5

Computation of the Magnitudes of Components

In Fig. 1-5, it is clear that

$$F_x = F \cos \theta \qquad F_y = F \sin \theta$$

In Fig. 1-6, the components of $\mathbf{F}$ are given by

$$F_x = F \cos \theta_1 \qquad F_y = F \cos \theta_2 \qquad F_z = F \cos \theta_3$$

Or for convenience, writing $\cos\theta_1 = \ell$, $\cos\theta_2 = m$, $\cos\theta_3 = n$,

$$F_x = F\ell \qquad F_y = Fm \qquad F_z = Fn$$

Letters ℓ, m, and n are referred to as the *direction cosines* of **F**. As can be shown,

$$\ell^2 + m^2 + n^2 = 1$$

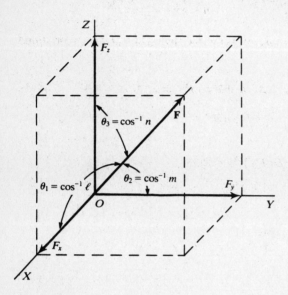

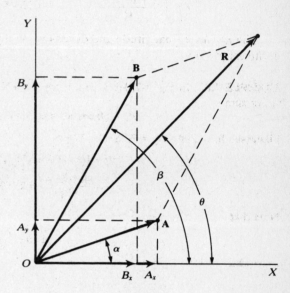

Fig. 1-6   **Fig. 1-7**

EXAMPLE 1.7. (a) Let **F**, Fig. 1-5, have a magnitude of 300 N and $\theta = 30°$. Then

$$F_x = 300\cos 30° = 259.8\text{ N} \qquad F_y = 300\sin 30° = 150\text{ N}$$

(b) Suppose that $F = 300$ N and $\theta = 145°$ (**F** is here in the second quadrant).

$$F_x = 300\cos 145° = (300)(-0.8192) = -245.75\text{ N (in the negative direction of }X)$$
$$F_y = 300\sin 145° = (300)(+0.5736) = 172.07\text{ N}$$

EXAMPLE 1.8. In Fig. 1-6, let **F** represent a force of 200 N. Let $\theta_1 = 60°$, $\theta_2 = 40°$. Then,

$$\ell = 0.5 \qquad m = 0.766 \qquad n = (1 - \ell^2 - m^2)^{1/2} = 0.404$$

(assuming that F_z is positive; otherwise, $n = -0.404$), and the rectangular components of **F** are:

$$F_x = (200)(0.5) = 100\text{ N} \qquad F_y = 153.2\text{ N} \qquad F_z = 80.8\text{ N}$$

As a check, $(100^2 + 153.2^2 + 80.8^2)^{1/2} \approx 200$. Note that $\theta_3 = 66.17°$.

Addition by Components

Adding **A** and **B**, Fig. 1-7, we write $\mathbf{A} + \mathbf{B} = \mathbf{R}$. **R** is referred to as the *resultant* or *vector sum* of **A** and **B**.

The components of **A** and **B** are $A_x = A\cos\alpha$, $A_y = A\sin\alpha$, $B_x = B\cos\beta$, $B_y = B\sin\beta$. **A** and **B** can now be replaced by these components, and **R** is a vector having rectangular components

$$R_x = A_x + B_x \qquad R_y = A_y + B_y$$

Since R_x and R_y are at right angles

$$R = (R_x^2 + R_y^2)^{1/2} = [(A_x + B_x)^2 + (A_y + B_y)^2]^{1/2}$$

The direction cosines of **R** are given by

$$\ell = \cos\theta = \frac{A_x + B_x}{R} \qquad m = \sin\theta = \frac{A_y + B_y}{R} \qquad n = 0$$

Let us find the vector sum of, say, three vectors, $\mathbf{F}_1$, $\mathbf{F}_2$, $\mathbf{F}_3$, drawn from O. Following the above procedure, the magnitude of the resultant is given by

$$R = [(F_{1x} + F_{2x} + F_{3x})^2 + (F_{1y} + F_{2y} + F_{3y})^2 + (F_{1z} + F_{2z} + F_{3z})^2]^{1/2}$$

where F_{1x} is the X component of $\mathbf{F}_1$, etc. The direction cosines of $\mathbf{R}$ are given by

$$\ell = \frac{F_{1x} + F_{2x} + F_{3x}}{R} \qquad m = \frac{F_{1y} + F_{2y} + F_{3y}}{R} \qquad n = \frac{F_{1z} + F_{2z} + F_{3z}}{R}$$

The *resultant* (magnitude and direction) of any number of vectors drawn from O can be obtained in the same way.

EXAMPLE 1.9. In Fig. 1-7, let $\mathbf{A}$ be a force of 50 N at $\alpha = 20°$ and $\mathbf{B}$ a force of 80 N at $\beta = 60°$. Find the vector sum.

$$A_x = 50 \cos 20° = 46.98 \text{ N} \qquad A_y = 50 \sin 20° = 17.1 \text{ N}$$

Likewise, $B_x = 40$ N, $B_y = 69.28$ N. Thus,

$$R = [(46.98 + 40)^2 + (17.1 + 69.28)^2]^{1/2} = 122.6 \text{ N}$$

$$\ell = \frac{46.98 + 40}{122.6} = 0.709 \qquad m = 0.705 \qquad n = 0$$

Note that

$$\tan \theta = \frac{17.1 + 69.28}{46.98 + 40} \approx 1$$

from which $\theta \approx 45°$.

1.4 UNIT VECTORS

Any vector $\mathbf{F}$ may be written as

$$\mathbf{F} = F\mathbf{e}$$

where F is the magnitude of $\mathbf{F}$ and where $\mathbf{e}$ is a *unit vector* (a vector whose magnitude is 1) in the direction of $\mathbf{F}$. That is, the magnitude of $\mathbf{F}$ is indicated by F and its direction is that of $\mathbf{e}$. $\mathbf{F}$ carries units (e.g. N, m/s), F carries the same units; $\mathbf{e}$ is a dimensionless vector.

Unit Vectors Along Rectangular Axes

In Fig. 1-6, let us introduce unit vectors $\mathbf{i}$, $\mathbf{j}$, $\mathbf{k}$ along X, Y, Z, respectively. Then the *vector* components of $\mathbf{F}$ can be written as $F_x\mathbf{i}$, $F_y\mathbf{j}$, $F_z\mathbf{k}$. Since $\mathbf{F}$ is the resultant of its vector components, we obtain the very important expression

$$\mathbf{F} = F_x\mathbf{i} + F_y\mathbf{j} + F_z\mathbf{k}$$

In this expression, $F_x = F \cos \theta_1 = F\ell$, etc., as previously shown; and, as before, the magnitude and direction (direction cosines, that is) are obtained as:

$$F = (F_x^2 + F_y^2 + F_z^2)^{1/2}$$

$$\ell = \frac{F_x}{F} \qquad m = \frac{F_y}{F} \qquad n = \frac{F_z}{F}$$

EXAMPLE 1.10. Referring to Example 1.8 and Fig. 1-6, where

$$F_x = 100 \text{ N} \qquad F_y = 153.2 \text{ N} \qquad F_z = 80.8 \text{ N}$$

vector $\mathbf{F}$ can be written as

$$\mathbf{F} = 100\mathbf{i} + 153.2\mathbf{j} + 80.8\mathbf{k}$$

with magnitude $F = (100^2 + 153.2^2 + 81^2)^{1/2} = 200$ N and direction

$$\ell = \frac{100}{200} = 0.5 \qquad m = 0.766 \qquad n = 0.404$$

Strictly, we should have written

$$\mathbf{F} = (100\text{ N})\mathbf{i} - (153.2\text{ N})\mathbf{j} + (80.8\text{ N})\mathbf{k} \qquad \text{or} \qquad \mathbf{F} = 100\mathbf{i} + 153.2\mathbf{j} + 80.8\mathbf{k} \quad \text{N}$$

but, for simplicity, we commonly omit the units when expressing a vector in terms of its components.

EXAMPLE 1.11. The rectangular components of an acceleration vector $\mathbf{a}$ are $a_x = 6$, $a_y = 4$, $a_z = 9$ m/s^2. Hence, in vector notation,

$$\mathbf{a} = 6\mathbf{i} + 4\mathbf{j} + 9\mathbf{k}$$

The magnitude of $\mathbf{a}$ is $a = (6^2 + 4^2 + 9^2)^{1/2} = 11.53$ m/s^2, and the direction cosines of $\mathbf{a}$ are

$$\ell = \frac{6}{11.53} \qquad m = \frac{4}{11.53} \qquad n = \frac{9}{11.53}$$

Vector Expression for a Line Segment

The straight line ab, Fig. 1-8, is determined by points P_1 and P_2. Regarding the line segment from P_1 to P_2 as a vector $\mathbf{s}$, we can write

$$\mathbf{s} = (x_2 - x_1)\mathbf{i} + (y_2 + y_1)\mathbf{j} + (z_2 - z_1)\mathbf{k}$$

with magnitude

$$s = [(x_2 - x_1)^2 + (y_2 - y_1)^2 + (z_2 - z_1)^2]^{1/2}$$

and direction

$$\ell = \frac{x_2 - x_1}{s} \qquad m = \frac{y_2 - y_1}{s} \qquad n = \frac{z_2 - z_1}{s}$$

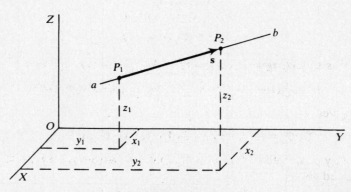

Fig. 1-8

An important special case of this is the so-called *radius vector* $\mathbf{r}$, the directed segment from the origin O to a point $P(x, y, z)$.

$$\mathbf{r} = x\mathbf{i} + y\mathbf{j} + z\mathbf{k}$$

with $r = (x^2 + y^2 + z^2)^{1/2}$ and

$$\ell = \frac{x}{r} \qquad m = \frac{y}{r} \qquad n = \frac{z}{r}$$

1.5 MULTIPLICATION OF VECTORS

Three types of multiplication must be considered. The method and the utility of each will become evident from the various geometrical and physical examples given below.

Multiplication of a Vector by a Scalar

A vector $\mathbf{F}$ can be multiplied by a scalar b. The quantity $b\mathbf{F}$ is a vector having magnitude $|b|\,F$ (the absolute value of b times the magnitude of $\mathbf{F}$); the direction of $b\mathbf{F}$ is that of $\mathbf{F}$ or $-\mathbf{F}$, depending on whether b is positive or negative.

EXAMPLE 1.12. Consider the velocity vector

$$\mathbf{v} = 16\mathbf{i} + 30\mathbf{j} + 24\mathbf{k} \quad \text{m/s}$$

with $v = (16^2 + 30^2 + 24^2)^{1/2} = 41.62$ m/s, direction given by $\ell = 16/41.62$, etc.

Now let us multiply $\mathbf{v}$ by 10: $10\mathbf{v} = 160\mathbf{i} + 300\mathbf{j} + 240\mathbf{k} \equiv \mathbf{v}_1$.   Then

$$v_1 = [(160)^2 + (300)^2 + (240)^2]^{1/2} = (10)(41.62) = 10v$$

and the direction cosines of $\mathbf{v}_1$ are

$$\ell_1 = \frac{160}{(10)(41.62)} = \frac{16}{41.62} = \ell \qquad m_1 = m \qquad n_1 = n$$

which shows that $\mathbf{v}_1$ has the direction of $\mathbf{v}$.

The Scalar or Dot Product

The *dot product* of any two vectors, as $\mathbf{F}_1$ and $\mathbf{F}_2$, Fig. 1-2, is written as $\mathbf{F}_1 \cdot \mathbf{F}_2$ and is defined as the product of their magnitudes and the cosine of the included angle.   That is,

$$\mathbf{F}_1 \cdot \mathbf{F}_2 = F_1 F_2 \cos \theta$$

which is a scalar quantity.   In Fig. 1-2, $F_1 = 75$, $F_2 = 100$, $\theta = 60°$.   Thus,

$$\mathbf{F}_1 \cdot \mathbf{F}_2 = (75)(100)(0.5) = 3750$$

Dot products of the unit vectors along X, Y, Z.   Since $\mathbf{i}, \mathbf{j}, \mathbf{k}$ are mutually perpendicular and of unit magnitude, the definition of the dot product gives

$$\mathbf{i} \cdot \mathbf{i} = \mathbf{j} \cdot \mathbf{j} = \mathbf{k} \cdot \mathbf{k} = 1$$
$$\mathbf{i} \cdot \mathbf{j} = \mathbf{i} \cdot \mathbf{k} = \mathbf{j} \cdot \mathbf{k} = 0$$

Dot product in terms of rectangular components.   Write any two vectors as

$$\mathbf{F}_1 = F_{1x}\mathbf{i} + F_{1y}\mathbf{j} + F_{1z}\mathbf{k} \qquad \mathbf{F}_2 = F_{2x}\mathbf{i} + F_{2y}\mathbf{j} + F_{3y}\mathbf{k}$$

Their dot product is given by

$$\mathbf{F}_1 \cdot \mathbf{F}_2 = (F_{1x}\mathbf{i} + F_{1y}\mathbf{j} + F_{1z}\mathbf{k}) \cdot (F_{2x}\mathbf{i} + F_{2y}\mathbf{j} + F_{2z}\mathbf{k})$$

The right-hand side may be simplified by assuming that the distributive law holds and employing the values of $\mathbf{i} \cdot \mathbf{i}$, etc., found above.

$$\mathbf{F}_1 \cdot \mathbf{F}_2 = F_{1x}F_{2x} + F_{1y}F_{2y} + F_{1z}F_{2z}$$

To show that $\mathbf{F}_1 \cdot \mathbf{F}_2$ is just the quantity $F_1 F_2 \cos \theta$, where θ is the angle between $\mathbf{F}_1$ and $\mathbf{F}_2$, multiply and divide the right side through by $F_1 F_2$, giving

$$\mathbf{F}_1 \cdot \mathbf{F}_2 = F_1 F_2 \left(\frac{F_{1x}}{F_1} \frac{F_{2x}}{F_2} + \frac{F_{1y}}{F_1} \frac{F_{2y}}{F_2} + \frac{F_{1z}}{F_1} \frac{F_{2z}}{F_2} \right) = F_1 F_2 (\ell_1 \ell_2 + m_1 m_2 + n_1 n_2)$$

Now, the familiar addition formula in two dimensions,

$$\cos \theta = \cos (\theta_1 - \theta_2) = \cos \theta_1 \cos \theta_2 + \sin \theta_1 \sin \theta_2 = \ell_1 \ell_2 + m_1 m_2$$

extends to three dimensions as $\cos \theta = \ell_1 \ell_2 + m_1 m_2 + n_1 n_2$.   Hence, the above becomes $F_1 F_2 \cos \theta$, and therefore this method of multiplication is in accord with the definition of the dot product.

EXAMPLE 1.13.   Let $\mathbf{F}_1 = 10\mathbf{i} - 15\mathbf{j} - 20\mathbf{k}$, $\mathbf{F}_2 = 6\mathbf{i} + 8\mathbf{j} - 12\mathbf{k}$.

$$\mathbf{F}_1 \cdot \mathbf{F}_2 = (10)(6) + (-15)(8) + (-20)(-12) = 180$$

Now note that $F_1 = (10^2 + 15^2 + 20^2)^{1/2} = 26.93$, $F_2 = 15.62$.   Hence, the angle θ between $\mathbf{F}_1$ and $\mathbf{F}_2$ is given by

$$\cos \theta = \frac{\mathbf{F}_1 \cdot \mathbf{F}_2}{F_1 F_2} = \frac{180}{(26.93)(15.62)} = 0.4279 \qquad \theta = 64.66°$$

Of course, the same value can be obtained from $\cos \theta = \ell_1 \ell_2 + m_1 m_2 + n_1 n_2$.

Projection of any vector along a straight line.   The projection of vector $\mathbf{A} = (A_x, A_y, A_z)$ along the line determined by the radius vector $\mathbf{r} = (x, y, z)$ is $A_r = A \cos \theta$, where θ is the angle between $\mathbf{r}$ and $\mathbf{A}$.   From the definition of the dot product,

$$\mathbf{A} \cdot \mathbf{r} = (Ar \cos \theta) = A_r r$$

Hence,

$$A_r = \frac{1}{r} \mathbf{A} \cdot \mathbf{r} = \frac{1}{r}(A_x \mathbf{i} + A_y \mathbf{j} + A_z \mathbf{k}) \cdot (x\mathbf{i} + y\mathbf{j} + z\mathbf{k}) = A_x \frac{x}{r} + A_y \frac{y}{r} + A_z \frac{z}{r}$$

$$= A_x \ell + A_y m + A_z n$$

where ℓ, m, n are the direction cosines of line considered.   The expression for A_r remains valid even when the line does not pass through the origin.

EXAMPLE 1.14.   Find the projection of $\mathbf{A} = 10\mathbf{i} + 8\mathbf{j} - 6\mathbf{k}$ along $\mathbf{r} = 5\mathbf{i} + 6\mathbf{j} + 9\mathbf{k}$.
   Here $r = (5^2 + 6^2 + 9^2)^{1/2} = 11.92$ and

$$A_r = A_x \ell + A_y m + A_z n = 10\left(\frac{5}{11.92}\right) + 8\left(\frac{6}{11.92}\right) - 6\left(\frac{9}{11.92}\right) = \frac{44}{11.92} = 3.69$$

The Vector or Cross Product

The *cross product* of two vectors, as $\mathbf{F}_1$ and $\mathbf{F}_2$, Fig. 1-9, written as $\mathbf{F} = \mathbf{F}_1 \times \mathbf{F}_2$, is defined as a vector $\mathbf{F}$ having a magnitude

$$F = F_1 F_2 \sin \theta$$

and a direction which is the direction of advance of a right-hand screw when turned from $\mathbf{F}_1$ to $\mathbf{F}_2$ through angle θ, it being assumed that the axis of the screw is normal to the plane determined by $\mathbf{F}_1$ and $\mathbf{F}_2$ (the *right-hand screw rule*).   Or, if the curled fingers of the right-hand point from $\mathbf{F}_1$ to $\mathbf{F}_2$, the extended thumb points in the direction of $\mathbf{F}$ (*right-hand rule*).

Note that, in accord with the right-hand screw rule, $\mathbf{F}_1 \times \mathbf{F}_2 = -(\mathbf{F}_2 \times \mathbf{F}_1)$.

Cross products of the unit vectors.   Since $\mathbf{i}$, $\mathbf{j}$, $\mathbf{k}$ are mutually perpendicular and of unit magnitude, it follows from the definition of the cross product that

$$\mathbf{i} \times \mathbf{i} = \mathbf{j} \times \mathbf{j} = \mathbf{k} \times \mathbf{k} = 0$$

$$\mathbf{i} \times \mathbf{j} = \mathbf{k} \qquad \mathbf{j} \times \mathbf{k} = \mathbf{i} \qquad \mathbf{k} \times \mathbf{i} = \mathbf{j}$$

$$\mathbf{j} \times \mathbf{i} = -\mathbf{k} \qquad \mathbf{k} \times \mathbf{j} = -\mathbf{i} \qquad \mathbf{i} \times \mathbf{k} = -\mathbf{j}$$

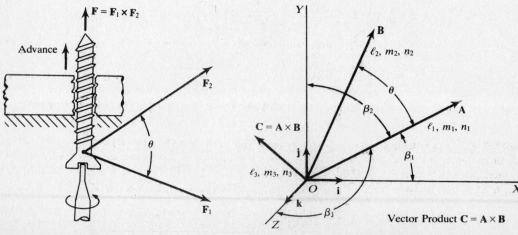

Fig. 1-9   Fig. 1-10

Cross products in terms of rectangular components.   Given two vectors, as in Fig. 1-10,

$$\mathbf{A} = A_x\mathbf{i} + A_y\mathbf{j} + A_z\mathbf{k} \qquad \mathbf{B} = B_x\mathbf{i} + B_y\mathbf{j} + B_z\mathbf{k}$$

their cross product is

$$\mathbf{C} = \mathbf{A} \times \mathbf{B} = (A_x\mathbf{i} + A_y\mathbf{j} + A_z\mathbf{k}) \times (B_x\mathbf{i} + B_y\mathbf{j} + B_z\mathbf{k})$$

Applying the distributive law to the right-hand side and using the values of $\mathbf{i} \times \mathbf{i}$, etc., found above, we obtain

$$\mathbf{C} = \mathbf{A} \times \mathbf{B} = (A_yB_z - A_zB_y)\mathbf{i} + (A_zB_x - A_xB_z)\mathbf{j} + (A_xB_y - A_yB_x)\mathbf{k}$$

Equivalently, $\mathbf{A} \times \mathbf{B}$ may be expressed as a determinant,

$$\mathbf{C} = \mathbf{A} \times \mathbf{B} = \begin{vmatrix} \mathbf{i} & \mathbf{j} & \mathbf{k} \\ A_x & A_y & A_z \\ B_x & B_y & B_z \end{vmatrix}$$

as may be verified by expanding the determinant with respect to the first row.   Note that the X, Y, Z components of $\mathbf{C}$ are

$$C_x = A_yB_z - A_zB_y \qquad C_y = (A_zB_x - A_xB_z) \qquad C_z = A_xB_y - A_yB_x$$

Hence the magnitude of $\mathbf{C}$ is $C = (C_x^2 + C_y^2 + C_z^2)^{1/2}$ and its direction cosines are

$$\ell = \frac{C_x}{C} \qquad m = \frac{C_y}{C} \qquad n = \frac{C_z}{C}$$

Vector $\mathbf{C}$ is, of course, normal to the plane of vectors $\mathbf{A}$ and $\mathbf{B}$.

EXAMPLE 1.15.   Assuming that vectors $\mathbf{A}$ and $\mathbf{B}$, Fig. 1-11, are in the XY plane, let us determine the magnitude and direction of $\mathbf{C} = \mathbf{A} \times \mathbf{B}$.

$$C = (200)(100) \sin (55° - 15°) = 20\,000 \sin 40° = 12\,855.75$$

and by the right-hand rule the direction of $\mathbf{C}$ is that of $+Z$.   Vectorially we can write $\mathbf{C} = 12\,855.75\mathbf{k}$.

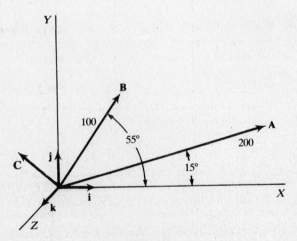

Fig. 1-11

EXAMPLE 1.16.   Referring to Fig. 1-10, let $\mathbf{A} = 20\mathbf{i} - 10\mathbf{j} + 30\mathbf{k}$ and $\mathbf{B} = -6\mathbf{i} + 15\mathbf{j} - 25\mathbf{k}$.   (a) Find the magnitude of $\mathbf{A}$ and $\mathbf{B}$.   (b) Find the direction cosines of $\mathbf{A}$.   (c) Find the vector product $\mathbf{C} = \mathbf{A} \times \mathbf{B}$.   (d) Find the magnitude and direction of $\mathbf{C}$.   (e) Find angle θ between $\mathbf{A}$ and $\mathbf{B}$.   (f) Find the values of the direction cosines ℓ_2, m_2, n_2 of $\mathbf{B}$, and find angles $\alpha_{21}, \alpha_{22}, \alpha_{23}$ between $\mathbf{B}$ and the $X, Y,$ and Z axes, respectively.

(a)
$$A = (20^2 + 10^2 + 30^2)^{1/2} = 37.42 \qquad B = 29.77$$

(b)
$$\ell_1 = \frac{20}{37.42} \qquad m_1 = \frac{-10}{37.42} \qquad n_1 = \frac{30}{37.42}$$

(c) Applying the determinant formula,

$$\mathbf{C} = \begin{vmatrix} \mathbf{i} & \mathbf{j} & \mathbf{k} \\ 20 & -10 & 30 \\ -6 & 15 & -25 \end{vmatrix}$$

$$\mathbf{C} = \mathbf{i}[(-10)(-25) - (15)(30)] - \mathbf{j}[(20)(-25) - (30)(-6)] + \mathbf{k}[(20)(15) - (-10)(-6)]$$
$$= -200\mathbf{i} + 320\mathbf{j} + 240\mathbf{k} = 200(-\mathbf{i} + 1.6\mathbf{j} + 1.2\mathbf{k})$$

(d) The magnitude of $\mathbf{C}$ is

$$C = 200 \, (1^2 + 1.6^2 + 1.2^2)^{1/2} = 447.21$$

The direction cosines are

$$\ell_3 = \frac{-200}{447.21} \qquad m_3 = \frac{320}{447.21} \qquad n_3 = \frac{240}{447.21}$$

Note that $\mathbf{C} = C(\ell_3\mathbf{i} + m_3\mathbf{j} + n_3\mathbf{k})$.

(e)
$$C = AB \sin \theta$$
$$447.21 = (37.42)(29.77) \sin \theta$$
$$\sin \theta = 0.40145$$
$$\theta = 23.67°$$

(f)
$$\mathbf{B} = -6\mathbf{i} + 15\mathbf{j} - 25\mathbf{k} = B(\ell_2\mathbf{i} + m_2\mathbf{j} + n_2\mathbf{k})$$

Thus

$$B\ell_2 = -6 \qquad Bm_2 = 15 \qquad Bn_2 = -25$$
$$B = (6^2 + 15^2 + 25^2)^{1/2} = 29.766$$
$$\ell_2 = -0.2016 \qquad m_2 = 0.5039 \qquad n_2 = -0.8399$$

Corresponding angles are

$$\alpha_{21} = 101.63° \qquad \alpha_{22} = 59.74° \qquad \alpha_{23} = 147.13°$$

1.6 PHYSICAL ENTITIES

Physical entities (quantities) which are of importance in the treatment of the general fields of mechanics, electricity and magnetism are mass, length, time, velocity, acceleration, force, work, energy, electric charge, voltage (and many more).

But of all these, four and only four, *mass, length, time*, and *electric current* (or *charge*, see below) are regarded as independent, *basic* entities. All others are defined by simple relations between the basic ones, and are referred to as *derived* quantities.

Basic Entities

In keeping with modern practice, the International System of Units (SI) is used throughout the text except where otherwise stated. In this system names, symbols and definition of corresponding units are:

Length: The *meter* (m) = length of 1 650 763.73 wavelengths in vacuum of a certain spectral line of krypton-86.

Mass: The *kilogram* (kg) = mass of a particular cylinder of platinum-iridium kept in Sèvres, France.

Time: The *second* (s) = duration of 9 192 631 770.0 periods of oscillation of a certain spectral line of cesium-133.

Electric Current: The *ampere* (A). Consider two fine, very long, parallel wires located one meter apart in vacuum, connected in series and carrying a steady electric current I. Suppose I is adjusted until the magnetic force per meter length on each wire is exactly 2×10^{-7} newton. This value of I is defined as one ampere.

Electric Charge: The *coulomb* (C) is defined as the quantity of charge which passes per second through a cross section of wire in which there is a steady current of one ampere.   It is approximately equal in value to the total charge on 6.2419×10^{18} electrons.

Since coulombs = amperes × seconds, it is clear that amperes and coulombs are *not independent*. Hence there is a choice; either may be treated as independent.   The other must then be regarded as a derived quantity.

Length, mass, time, and electric current (or charge) are often referred to as *physical dimensions*.

For the treatment of matters having to do with temperature, light and luminous intensity and the molecular entity the mole, corresponding independent units are defined in the chapters that follow.

Derived Entities

A derived entity is one defined in terms of two or more of the basic entities.   *Examples*: linear velocity = length/time;   acceleration = length/time^2;   force = (mass × length)/time^2.   These relations are valid regardless of what units are employed.

When specific units are introduced, corresponding *unit-dimensional* (u-d) relations can be written.   For example, using SI units,

Velocity: $\mathbf{u} = \dfrac{d\mathbf{s}}{dt}$; u-d relation, $\mathbf{u} = \left| \dfrac{m}{s} \right|$.

Acceleration: $\mathbf{a} = \dfrac{d\mathbf{u}}{dt}$; u-d relation, $\mathbf{a} = \left| \dfrac{m}{s^2} \right|$.

Force: $\mathbf{F} = M\mathbf{a}$; u-d relation, $\mathbf{F} = \left| \dfrac{kg \cdot m}{s^2} \right| = |N|$.

Work: $W = \displaystyle\int \mathbf{F} \cdot d\mathbf{s}$; u-d relation, $W = |N \cdot m| = \left| \dfrac{kg \cdot m^2}{s^2} \right| = |J|$.

In like manner, starting with the fundamental definition of any derived entity, the corresponding u-d expression can be written at once.

1.7 UNIT DIMENSIONAL ANALYSIS OF PHYSICAL EQUATIONS

A *physical equation* expresses mathematically the relations between physical quantities.   The importance of *dimensional analysis* stems from the fact that *each separate term* in a physical equation must represent the same physical entity; *each must be dimensionally the same.   If this is not the case, the equation is in error.*   And, for a correct solution of a problem, all terms must be expressed in the same basic units throughout.

To make a dimensional check, replace each symbol representing length in meters by m, mass in kilograms by kg, time in seconds by s, velocity in meters per second by m/s, acceleration in meters per second by m/s^2, force in newtons by (kg · m)/s^2, etc.   After reduction of terms, a glance indicates if the equation is dimensionally the same throughout.   If constants occur in the equation, their dimensions must be known from previous considerations and taken into account.   It should be noted that the fact that an equation is dimensionally the same throughout *does not guarantee the correctness of the equation itself*.   Many examples of dimensional analysis are found in the chapters that follow.

Chapter 2

Motion of a Particle Along a Straight Line with Constant Acceleration

2.1 DEFINITIONS OF VELOCITY AND ACCELERATION

Average speed, v_{avg}, is a scalar quantity. A particle which covers a distance s (say from a to b, Fig. 2-1) in time t does so with an average speed given by

$$v_{avg} = \frac{s}{t} \qquad \text{or} \qquad s = v_{avg}t \qquad (2.1)$$

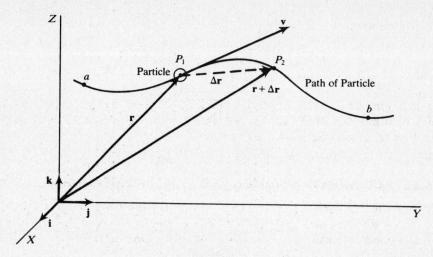

Fig. 2-1

Instantaneous linear velocity, $\mathbf{v}$ (a vector quantity), is defined (see Fig. 2-1) by

$$\mathbf{v} = \lim_{\Delta t \to 0} \frac{\Delta \mathbf{r}}{\Delta t} = \frac{d\mathbf{r}}{dt} \qquad (2.2)$$

Or, since $\mathbf{r} = x\mathbf{i} + y\mathbf{j} + z\mathbf{k}$,

$$\mathbf{v} = \frac{dx}{dt}\mathbf{i} + \frac{dy}{dt}\mathbf{j} + \frac{dz}{dt}\mathbf{k} = \dot{x}\mathbf{i} + \dot{y}\mathbf{j} + \dot{z}\mathbf{k} \qquad (2.3)$$

where x, y, z are rectangular coordinates of the particle at P_1, Fig. 2-1; $\mathbf{i}$, $\mathbf{j}$, $\mathbf{k}$ are unit vectors along X, Y, Z; and, for convenience, dx/dt is written as $\dot{x}$, etc. Observe that $\mathbf{v}$ is tangent to the path at P_1. The units of $\mathbf{v}$ (as well as of v_{avg}) are m/s.

Instantaneous linear acceleration, $\mathbf{a}$ (a vector), is the instantaneous time-rate-of-change of the vector velocity $\mathbf{v}$. Referring to Fig. 2-2, the particle at P_1 has velocity $\mathbf{v}_1$. A short time, Δt, later its velocity at P_2 is $\mathbf{v}_2$. The change in velocity is $\Delta \mathbf{v} = \mathbf{v}_2 - \mathbf{v}_1$, and the instantaneous acceleration at P_1 is

$$\mathbf{a} = \lim_{\Delta t \to 0} \frac{\Delta \mathbf{v}}{\Delta t} = \frac{d\mathbf{v}}{dt} \qquad (2.4)$$

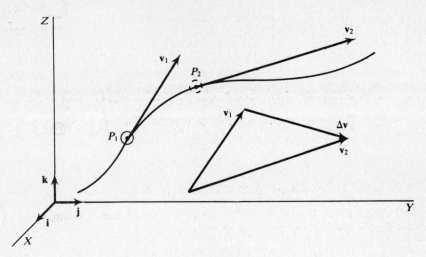

Fig. 2-2

Or, from (2.3), we can write

$$\mathbf{a} = \frac{d^2x}{dt^2}\mathbf{i} + \frac{d^2y}{dt^2}\mathbf{j} + \frac{d^2z}{dt^2}\mathbf{k} = \ddot{x}\mathbf{i} + \ddot{y}\mathbf{j} + \ddot{z}\mathbf{k} \tag{2.5}$$

The units of **a** are m/s².

Equations (2.2) through (2.5) are general expressions for linear velocity and linear acceleration for three-dimensional motion. They are, of course, applicable to special cases, as motion along a line, motion in a plane, motion on the surface of a sphere, etc.

2.2 UNIFORMLY ACCELERATED MOTION ALONG A STRAIGHT LINE

When **a** is constant in magnitude and direction, and when the motion is along the line of **a**, we may set up a positive direction along the line (either the direction of **a** or the direction of −**a**) and work with signed numbers instead of vectors. We then have the relations:

$$v_{\text{avg}} = \frac{1}{2}(v_0 + v) \qquad v = v_0 + at \qquad v^2 = v_0^2 + 2a(s - s_0)$$

$$s = s_0 + v_0 t + \frac{1}{2}at^2 \tag{2.6}$$

where v_{avg} is the average velocity over the time interval 0 to t, and where v_0 and s_0 are the velocity and distance at $t = 0$.

In almost all problems the coordinate axes are chosen such that $s_0 = 0$. Thus, for motion at constant acceleration along X, with the particle initially at the origin, (2.6) takes the form

$$\dot{x}_{\text{avg}} = \frac{1}{2}(\dot{x}_0 + \dot{x}) \qquad \dot{x} = \dot{x}_0 + \ddot{x}t \qquad \dot{x}^2 = \dot{x}_0^2 + 2\ddot{x}x$$

$$x = \dot{x}_0 t + \frac{1}{2}\ddot{x}t^2$$

Gravitational acceleration. Every body falling freely near the surface of the earth has an almost-constant downward acceleration of $g = 9.8$ m/s².

Solved Problems

2.1. The particle shown in Fig. 2-3 moves along X with a constant acceleration of -4 m/s^2. As it passes the origin, moving in the $+$ direction of X, its velocity is 20 m/s. In this problem time t is measured from the moment the particle is first at the origin. (a) At what distance x' and time t' does $v = 0$? (b) At what time is the particle at $x = 15$ m, and what is its velocity at this point? (c) What is the velocity of the particle at $x = +25$ m? at $x = -25$ m? Try finding the velocity of the particle at $x = 55$ m.

(a) Applying $v = v_0 + at$ (or $\dot{x} = \dot{x}_0 + \ddot{x}t$),

$$0 = 20 + (-4)t' \qquad \text{or} \qquad t' = 5 \text{ s}$$

Then

$$x' = \dot{x}_0 t + \frac{1}{2}\ddot{x}t'^2 = (20)(5) + \frac{1}{2}(-4)(5)^2 = 50 \text{ m}$$

Or, from $\dot{x}^2 = \dot{x}_0^2 + 2\ddot{x}x$,

$$0 = (20)^2 + 2(-4)x' \qquad \text{or} \qquad x' = 50 \text{ m}$$

(b)

$$15 = 20t + \frac{1}{2}(-4)t^2 \qquad \text{or} \qquad 2t^2 - 20t + 15 = 0$$

Solving this quadratic,

$$t = \frac{20 \pm \sqrt{(20)^2 - 4(2)(15)}}{4} = \frac{1}{4}(20 \pm 16.7332)$$

Thus $t_1 = 0.8167$ s, $t_2 = 9.1833$ s, where t_1 is the time from the origin to $x = 15$ m and t_2 is the time to go from O out beyond $x = 15$ m and return to that point. At $x = 15$ m,

$$v_1 = 20 - 4(0.8167) = +16.7332 \text{ m/s} \qquad v_2 = 20 - 4(9.1833) = -16.7332 \text{ m/s}$$

Observe that the speeds are equal.

(c) At $x = +25$ m,

$$v^2 = (20)^2 + 2(-4)(25) \qquad \text{or} \qquad v = \pm 14.1421 \text{ m/s}$$

and at $x = -25$ m,

$$v^2 = 20^2 + 2(-4)(-25) \qquad \text{or} \qquad v = -24.4949 \text{ m/s}$$

(Why has the root $v = +24.4949$ m/s been discarded?)

Assuming that $x = 55$ m, $v^2 = 20^2 + 2(-4)(55)$, from which $v = \pm\sqrt{-40}$. The imaginary value of v is to be expected since x is never greater than 50 m.

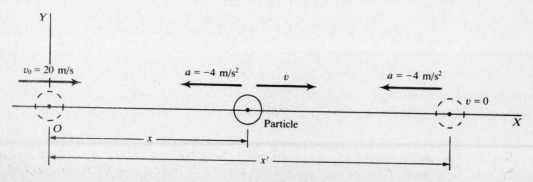

Fig. 2-3

2.2. A rocket-propelled car starts from rest at $x = 0$ and moves in the $+$ direction of X with constant acceleration $\ddot{x} = 5$ m/s^2 for 8 s until the fuel is exhausted. It then continues with constant velocity. What distance does the car cover in 12 s?

The distance from O at the moment fuel is exhausted is

$$x_1 = (0)(8) + \frac{1}{2}(5)(8)^2 = 160 \text{ m}$$

and at this point $v = (2ax_1)^{1/2} = 50.5964$ m/s. Hence the distance covered in 12 s is

$$x_2 = x_1 + v(12 - 8) = 160 + (50.5964)(4) = 362.38 \text{ m}$$

2.3. A ball is thrown vertically upward with a velocity of 20 m/s from the top of a tower having a height of 50 m, Fig. 2-4. On its return it misses the tower and finally strikes the ground. (a) What time t_1 elapses from the instant the ball was thrown until it passes the edge of the tower? What velocity v_1 does it have at this time? (b) What total time t_2 is required for the ball to reach the ground? With what velocity v_2 does it strike the ground?

(a) For the coordinate system shown in Fig. 2-4, $y = v_0 t + \frac{1}{2}at^2$. But at the edge of the roof $y = 0$, and thus

$$0 = 20t_1 + \frac{1}{2}(-9.8)t_1^2$$

from which $t_1 = 0$, indicating the instant at which the ball is released, and also $t_1 = 4.0816$ s, which is the time to go up and return to the edge. Then, from $v = v_0 + at$,

$$v_1 = 20 + (-9.8)(4.0816) = -20 \text{ m/s}$$

which is the negative of the initial velocity.

(b) $$-50 = 20t_2 + \frac{1}{2}(-9.8)t_2^2 \qquad \text{or} \qquad t_2 = 5.8315 \text{ s}$$

$$v_2 = 20 + (-9.8)(5.8315) = -37.15 \text{ m/s}$$

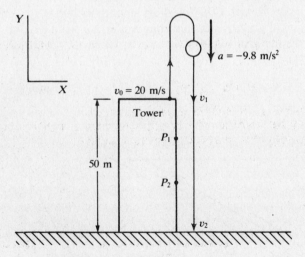

Fig. 2-4

2.4. Refer to Problem 2.3 and Fig. 2-4. (a) What is the maximum height above ground reached by the ball? (b) Points P_1 and P_2 are 15 and 30 m respectively below the top of the tower. What time interval is required for the ball to travel from P_1 to P_2? (c) It is desired that after passing the edge, the ball will reach the ground in 3 s. With what velocity must it be thrown upward from the roof?

(a) Maximum height above ground: $h = y_{max} + 50$. From $v_0^2 + 2ay_{max} = 0$,

$$y_{max} = \frac{-(20)^2}{-2(9.8)} = 20.40 \text{ m}$$

Thus, $h = 70.4082$ m.

(b) If t_1 and t_2 are the times to reach P_1 and P_2 respectively,

$$-15 = 20t_1 - 4.9 t_1^2 \qquad \text{and} \qquad -30 = 20t_2 - 4.9 t_2^2$$

Solving, $t_1 = 4.723$ s, $t_2 = 5.248$ s, and the time from P_1 to P_2 is $t_2 - t_1 = 0.519$ s.

(c) If v_i is the desired initial velocity, then $-v_i$ is the velocity upon passing the edge (why?). Then, applying

$$y = v_0 t + \frac{1}{2} at^2$$

to the trip down the tower, we find

$$-50 = (-v_i)(3) - 4.9(3)^2 \qquad \text{or} \qquad v_i = 1.967 \text{ m/s}$$

2.5. A ball, after having fallen from rest under the influence of gravity for 6 s, crashes through a horizontal glass plate, thereby losing 2/3 of its velocity. If it then reaches the ground in 2 s, find the height of the plate above the ground.

From $v = v_0 t + \frac{1}{2} at^2$, the velocity just before striking the glass is

$$v_1 = 0 - 4.9 (6)^2 = -176.4 \text{ m/s}$$

and so the velocity after passing through glass is $(1/3)v_1 = -58.8$ m/s. Thus

$$-h = (-58.8)(2) - 4.9(2)^2 \qquad \text{or} \qquad h = 137.2 \text{ m}$$

2.6. An inclined plane, Fig. 2-5, makes an angle θ with the horizontal. A groove OA cut in the plane makes an angle α with OX. A short smooth cylinder is free to slide down the groove under the influence of gravity, starting from rest at the point (x_0, y_0). Find: (a) its downward acceleration along the groove, (b) the time to reach O, (c) its velocity at O. Let $\theta = 30°$, $x_0 = 3$ m, $y_0 = 4$ m.

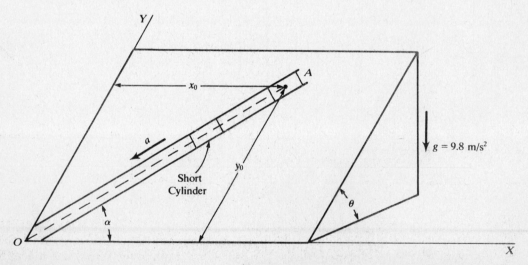

Fig. 2-5

(a) The downward component of **g** parallel to OY is $g \sin \theta$; hence, the downward component along the groove is $a = g \sin \theta \sin \alpha$. Since

$$\sin \alpha = \frac{y_0}{(x_0^2 + y_0^2)^{1/2}} = 0.8 \qquad \text{and} \qquad \sin \theta = 0.5$$

$a = (9.8)(0.5)(0.8) = 3.92 \text{ m/s}^2.$

(b) $$s = v_0 t + \frac{1}{2} a t^2$$

where $s = (x_0^2 + y_0^2)^{1/2} = 5$ m and $v_0 = 0$. Thus

$$s = \frac{1}{2} (3.92) t^2 \qquad \text{or} \qquad t = 1.597 \text{ s}$$

(c) $$v = 0 + (3.92)(1.597) = 6.26 \text{ m/s}$$

2.7.   A bead, Fig. 2-6, is free to slide down a smooth wire tightly stretched between points P_1 and P_2 on a vertical circle of radius R. If the bead starts from rest at P_1, the highest point on the circle, find (a) its velocity v on arriving at P_2, (b) the time to arrive at P_2 and show that this time is the same for any chord drawn from P_1.

(a) The acceleration of the bead down the wire is $g \cos \theta$ and the length of the wire is $2R \cos \theta$. Hence,

$$v^2 = 0^2 + 2(g \cos \theta)(2R \cos \theta) \qquad \text{or} \qquad v = 2(\sqrt{gR}) \cos \theta$$

(b) $$t = \frac{v}{a} = \frac{2\sqrt{gR} \cos \theta}{g \cos \theta} = 2\sqrt{\frac{R}{g}}$$

which is the same regardless of where P_2 is located on the circle.

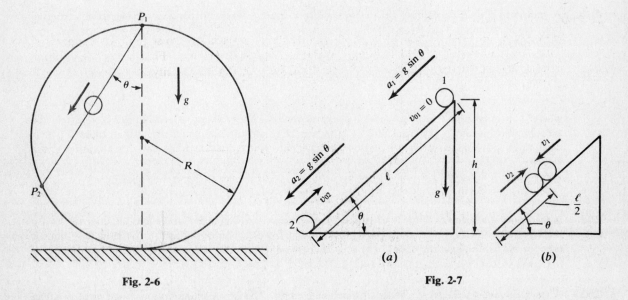

Fig. 2-6   Fig. 2-7

2.8.   Body 1, Fig. 2-7, is released from rest at the top of a smooth inclined plane, and at the same instant body 2 is projected upward from the foot of the plane with such velocity that they meet halfway up the plane.   Determine (a) the velocity of projection, (b) the velocity of each body when they meet.

(a) In the common time t, body 1 travels the distance

$$\frac{\ell}{2} = (0)t + \frac{1}{2}(g \sin \theta)t^2$$

and body 2 travels the distance

$$\frac{\ell}{2} = v_{02}t + \frac{1}{2}(-g \sin \theta)t^2$$

Adding these two equations gives $\ell = v_{02}t$ or $t = \ell/v_{02}$. Substituting this value of t in the first equation and solving for v_{02}, we obtain

$$v_{02} = \sqrt{g\ell \sin \theta} = \sqrt{gh}$$

(b) $v_1^2 = 0^2 + 2(g \sin \theta)\dfrac{\ell}{2}$ or $v_1 = \sqrt{g\ell \sin \theta} = \sqrt{gh}$

$v_2^2 = v_{02}^2 + 2(-g \sin \theta)\dfrac{\ell}{2} = g\ell \sin \theta - g\ell \sin \theta = 0$ or $v_2 = 0$

Supplementary Problems

2.9. A stone is thrown vertically upward with velocity 40 m/s at the edge of a cliff having a height of 110 m. Neglecting air resistance, compute the time required to strike the ground at the base of the cliff. With what velocity does it strike?   *Ans.* 11.93 s; −76.89 m/s

2.10. A proton in a uniform electric field moves along a straight line with constant acceleration. Starting from rest it attains a velocity of 1000 km/s in a distance of 1 cm. (a) What is its acceleration? (b) What time is required to reach the given velocity?   *Ans.* . (a) 5×10^{13} m/s^2; (b) 2×10^{-8} s

2.11. An object is forced to move along the X axis in such a way that its displacement is given by

$$x = 30 + 20t - 15t^2$$

where x is in m and t is in s. (a) Find expressions for the velocity $\dot{x}$ and acceleration $\ddot{x}$. Is the acceleration constant? (b) What are the initial position and the initial velocity of the object? (c) At what time and distance from the origin is the velocity zero? (d) At what time and location is the velocity −50 m/s?
Ans.   (a) $\dot{x} = 20 - 30t$; $\ddot{x} = -30$ m/s^2 = constant   (c) $t = 0.66667$ s, $x = 36.6667$ m
(b) $x_0 = 30$ m, $\dot{x}_0 = 20$ m/s   (d) $t = 2.3333$ s, $x = -5$ m

2.12. A man runs at a speed of 4 m/s to overtake a standing bus. When he is 6 m behind the door (at $t = 0$), the bus moves forward and continues with a constant acceleration of 1.2 m/s^2. (a) How long does it take for the man to gain the door? (b) If in the beginning he is 10 m from the door, will he (running at the same speed) ever catch up?   *Ans.* (a) 4.387 s; (b) no

2.13. A truck is moving forward at a constant speed of 21 m/s. The driver sees a stationary car directly ahead at a distance of 110 m. After a "reaction time" of Δt, he applies the brakes, which gives the truck an acceleration of −3 m/s^2. (a) What is the maximum allowable Δt to avoid a collision, and what distance will the truck have moved before the brakes take hold? (b) Assuming a reaction time of 1.4 s, how far behind the car will the truck stop, and in how many seconds from the time the driver first saw the car?   *Ans.* (a) 1.7381 s, 36.5 m; (b) 7.1 m, 8.4 s

2.14. A ball is released from rest at the edge of a deep ravine. Assume that air resistance gives it an acceleration of $-b\dot{y}$, where y is measured positive downward. (This negative acceleration is proportional to its speed, $\dot{y}$; the positive constant b can be found by experiment.) The ball has a total acceleration of $-b\dot{y} + g$, and so

$$\ddot{y} = -b\dot{y} + g \tag{1}$$

is the differential equation of motion. (a) Show by differentiation and substitution that

$$y = k(e^{-bt} - 1) + (g/b)t \tag{2}$$

is a solution of (1) for an arbitrary value of the constant k and that (2) gives $y = 0$ for $t = 0$. (b) Show from (2) that

$$\dot{y} = -kbe^{-bt} + g/b \tag{3}$$

Since at $t = 0$, $\dot{y} = 0$, prove that $k = g/b^2$. Show from (3) that as $t \to \infty$, $\dot{y} \to g/b$; that is, the velocity reaches a limiting value such that the negative acceleration due to air resistance exactly offsets the positive acceleration of gravity and thus $\ddot{y} = 0$. (c) Assuming that $b = 0.1$ s^{-1}, find the distance fallen and the speed reached after 10 s. (d) Show that after 1 min the ball will have essentially reached its terminal velocity of 98 m/s.   *Ans.* (c) 360.522 m, 61.95 m/s

2.15. A ball is thrown vertically upward from the origin of axes (Y regarded + upward), with initial velocity $\dot{y}_0$. Assuming as in Problem 2.14 an acceleration $-b\dot{y}$ due to air resistance, we write

$$\ddot{y} = -b\dot{y} - g \tag{1}$$

Notice that when $\dot{y}$ changes sign, so does $-b\dot{y}$; hence (1) is valid for the trip down as well as the trip up. (a) Show that

$$y = k(e^{-bt} - 1) - (g/b)t \tag{2}$$

is a solution of (1) for any value of k. (b) Show that $\dot{y} = -kbe^{-bt} - g/b$ and, since $\dot{y} = \dot{y}_0$ at $t = 0$, prove that

$$k = -\frac{1}{b}\left(\dot{y}_0 + \frac{g}{b}\right)$$

(c) Assuming that $b = 0.1$ s^{-1} and $\dot{y}_0 = 50$ m/s, find the height and speed at $t = 3$ s. (d) How long does it take for the ball to attain its maximum height and what is this height? (*Hint:* ln 1.51 = 0.41211.) (e) Show that with no air resistance the ball would reach a maximum height of 127.55 m in 5.10 s. (f) Show, by substituting in (2), that the time to go up and back to earth again is about 8.9 s. *Ans.* (c) 89.59 m, 11.64 m/s; (d) 4.121 s, 96 m

Chapter 3

Motion of a Particle
in a Plane with Constant Acceleration

Relations (2.1) through (2.5) are applicable to the most general type of motion of a particle (or point), whether along a line, in a plane, or in space, and for which the acceleration **a** may or may not be constant. In the special case of motion in a plane at constant acceleration, the vector expressions for velocity and acceleration reduce to:

$$\mathbf{v} = v_x\mathbf{i} + v_y\mathbf{j} = \dot{x}\mathbf{i} + \dot{y}\mathbf{j} \tag{3.1}$$

$$\mathbf{a} = a_x\mathbf{i} + a_y\mathbf{j} = \ddot{x}\mathbf{i} + \ddot{y}\mathbf{j} \tag{3.2}$$

in which $\ddot{x}$ and $\ddot{y}$ are each constant. The magnitudes and directions of these vectors are given by

$$v = (\dot{x}^2 + \dot{y}^2)^{1/2} \qquad \tan\beta = \frac{\dot{y}}{\dot{x}} \tag{3.3}$$

$$a = (\ddot{x}^2 + \ddot{y}^2)^{1/2} \qquad \tan\alpha = \frac{\ddot{y}}{\ddot{x}} \tag{3.4}$$

where β and α are the angles between **v** and X and between **a** and X.

Expressions for velocity **v** and displacement **r** (the position vector of the particle) in terms of time t, as found by integration, are:

$$\mathbf{v} = \mathbf{v}_0 + \mathbf{a}t \tag{3.5}$$

$$\mathbf{r} = \mathbf{r}_0 + \mathbf{v}_0 t + \frac{1}{2}\mathbf{a}t^2 \tag{3.6}$$

in which $\mathbf{v}_0$ and $\mathbf{r}_0$ are the values of **v** and **r** at $t = 0$. The scalar components of (3.5) and (3.6) yield a set of relations of the form (2.6) in each coordinate:

$$
\begin{aligned}
x &= x_0 + \dot{x}_0 t + \frac{1}{2}\ddot{x}t^2 & y &= y_0 + \dot{y}_0 t + \frac{1}{2}\ddot{y}t^2 \\
\dot{x}_{\text{avg}} &= \frac{1}{2}(\dot{x}_0 + \dot{x}) & \dot{y}_{\text{avg}} &= \frac{1}{2}(\dot{y}_0 + \dot{y}) \\
\dot{x} &= \dot{x}_0 + \ddot{x}t & \dot{y} &= \dot{y}_0 + \ddot{y}t \\
\dot{x}^2 &= \dot{x}_0^2 + 2\ddot{x}(x - x_0) & \dot{y}^2 &= \dot{y}_0^2 + 2\ddot{y}(y - y_0)
\end{aligned}
\tag{3.7}
$$

The fact that **a** is constant in magnitude and direction does not imply that the motion is along a straight line. In general, the particle moves along a parabola. This is easiest shown by choosing the coordinate axes such that one of them, say X, is parallel to **a** and such that the particle is at the origin at $t = 0$. Then the first two equations (3.7) become

$$x = \dot{x}_0 t + \frac{1}{2}\ddot{x}t^2 \qquad y = \dot{y}_0 t$$

When t is eliminated between these, the result is

$$\left(y + \frac{\dot{x}_0\dot{y}_0}{\ddot{x}}\right)^2 = \frac{2\dot{y}_0^2}{\ddot{x}}\left(x + \frac{\dot{x}_0^2}{2\ddot{x}}\right) \tag{3.8}$$

which is the equation of a parabola (see Fig. 3-1). In the special case $\dot{y}_0 = 0$, the path is a straight line, the X axis.

20

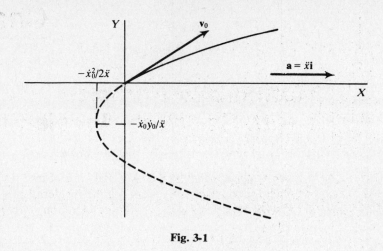

Fig. 3-1

Solved Problems

3.1. A projectile, Fig. 3-2, is fired upward with an initial velocity $v_0 = 200$ m/s at an angle $\theta = 60°$. (a) Find the position and velocity of the projectile 10 s after firing. (b) Find the maximum height h and the time to reach this position. (c) Find the total time of flight and the range R. Derive a general expression for R. (d) Write an equation for the path. (e) When the projectile is at a height $y = 1000$ m, what is its velocity?

First note that

$$x_0 = y_0 = 0 \qquad \dot{x}_0 = v_0 \cos\theta = 200 \cos 60° = 100 \text{ m/s} \qquad \dot{y}_0 = v_0 \sin\theta = 173.2 \text{ m/s}$$

Acceleration of gravity is $g = 9.8$ m/s^2 in the negative direction of Y. Thus $\ddot{x} = 0$, $\ddot{y} = -9.8$ m/s^2.

(a) Applying (3.7),

$$x = (100)(10) + 0 = 1000 \text{ m} \qquad y = (173.2)(10) + \frac{1}{2}(-9.8)(10)^2 = 1242 \text{ m}$$

$$\dot{x} = 100 + 0 = 100 \text{ m/s} \qquad \dot{y} = 173.2 + (-9.8)(10) = 75.2 \text{ m/s}$$

The magnitude of the velocity is $v = [(100)^2 + (75.2)^2]^{1/2} = 125.12$ m/s; the direction is given by

$$\tan\beta = \frac{75.2}{100} \qquad \text{or} \qquad \beta = 36.94°$$

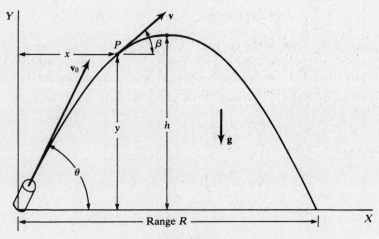

Fig. 3-2

(b) When $y = h$, $\dot{y} = 0 = \dot{y}_0 - gt$. Thus $t = \dot{y}_0/g = 17.67$ s, and

$$h = (173.2)(17.67) - \frac{1}{2}(9.8)(17.67)^2 = 1530.5 \text{ m}$$

(c) On striking the ground, $y = 0$. Thus

$$0 = \dot{y}_0 t - \frac{1}{2}gt^2 \qquad \text{or} \qquad t = 35.35 \text{ s}$$

Then $R = \dot{x}_0 t = (100)(35.35) = 3535$ m.
As above, time of flight is $2\dot{y}_0/g$, and $R = \dot{x}_0(2\dot{y}_0/g)$. But $\dot{x}_0 = v_0 \cos \theta$, $\dot{y}_0 = v_0 \sin \theta$. Thus

$$R = \frac{2v_0^2 \sin \theta \cos \theta}{g} = \frac{v_0^2 \sin 2\theta}{g}$$

(d) Eliminating t from $y = \dot{y}_0 t - \frac{1}{2}gt^2$ by $x = \dot{x}_0 t$ yields

$$y = \dot{y}_0\left(\frac{x}{\dot{x}_0}\right) - \frac{1}{2}g\left(\frac{x^2}{\dot{x}_0^2}\right) \qquad \text{or} \qquad y = (\tan \theta)x - \frac{gx^2}{2v_0^2 \cos^2 \theta}$$

as the equation of the path. Alternatively, the path is given by (3.8), with x and y replaced by $-y$ and x, respectively.

(e) By (3.7), $\dot{x}^2 = \dot{x}_0^2$ and $\dot{y}^2 = \dot{y}_0^2 - 2gy$. Hence $v^2 = \dot{x}^2 + \dot{y}^2 = v_0^2 - 2gy = (200)^2 - 2(9.8)(1000) = 2.04 \times 10^4$ m²/s² or $v = 143$ m/s. The direction of the velocity is given by

$$\tan \beta = \frac{\dot{y}}{\dot{x}} = \frac{\pm(\dot{y}_0^2 - 2gy)^{1/2}}{\dot{x}_0} = \frac{\pm[3 \times 10^4 - 2(9.8)(1000)]^{1/2}}{100} = \pm 1.02$$

or $\beta = \pm 45.6°$. (Why are there two values for the angle?)

3.2. A ball is thrown upward from the top of a 35 m tower, Fig. 3-3, with initial velocity $v_0 = 80$ m/s at an angle $\theta = 25°$. (a) Find the time to reach the ground and the distance R from P to the point of impact. (b) Find the magnitude and direction of the velocity at the moment of impact.

(a) At the point of impact, $y = -35$ m and $x = R$. From

$$y = -35 = (80 \sin 25°)t - \frac{1}{2}(9.8)t^2$$

$t = 7.814$ s. Then $x = R = (80 \cos 25°)(7.814) = 566.55$ m.

(b) At impact, $\dot{y} = 80 \sin 25° - (9.8)(7.814) = -42.77$ m/s and $\dot{x} = \dot{x}_0 = 80 \cos 25° = 72.5$ m/s. Thus $v = (42.77^2 + 72.5^2)^{1/2} = 84.18$ m/s and

$$\tan \beta = \frac{-42.77}{72.5} \qquad \text{or} \qquad \beta = -30.54°$$

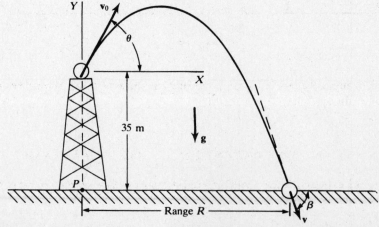

Fig. 3-3

3.3. A projectile, Fig. 3-4, is fired upward with velocity v_0 at an angle θ. (*a*) At what point $P(x, y)$ does it strike the roof of the building, and in what time? (*b*) Find the magnitude and direction of **v** at P. Let $\theta = 35°$, $v_0 = 40$ m/s, $\alpha = 30°$, and $h = 15$ m.

First note that

$$\dot{x}_0 = v_0 \cos 35° = 32.7661 \text{ m/s} \qquad \dot{y}_0 = v_0 \sin 35° = 22.943 \text{ m/s}$$

and, from the equation of the roof,

$$y = h - x \tan \alpha = 15 - (0.57735)x \tag{1}$$

(*a*) Eliminating t from $y = \dot{y}_0 t - 4.9 t^2$ by $x = \dot{x}_0 t$, we have

$$y = \left(\frac{\dot{y}_0}{\dot{x}_0}\right)x - \frac{4.9 x^2}{\dot{x}_0^2} \tag{2}$$

for the path of the projectile. Equating y in (*1*) to y in (*2*) and inserting numerical values,

$$0.004564 x^2 - 1.277558 x + 15 = 0$$

from which $x = 12.28$ m. Then $y = h - (12.28) \tan \alpha = 7.90$ m. The time to strike is given by

$$12.28 = 32.7661 t \qquad \text{or} \qquad t = 0.375 \text{ s}$$

(*b*) At P,

$$\dot{x} = \dot{x}_0 = 32.766 \text{ m/s}$$
$$\dot{y} = \dot{y}_0 - 9.8t = 22.943 - (9.8)(0.375) = 19.268 \text{ m/s}$$

Thus $v = (\dot{x}^2 + \dot{y}^2)^{1/2} = 38.0$ m/s and $\tan \beta = \dot{y}/\dot{x} = 0.588$, or $\beta = 30.46°$ where β is the angle **v** makes with X at P.

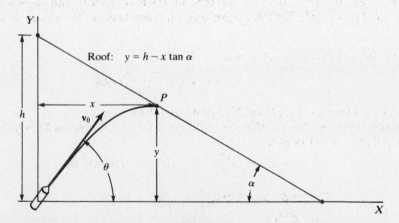

Fig. 3-4

3.4. In Problem 3.3, angle θ can be adjusted. Find the value of θ for which the projectile strikes the roof in a minimum time.

Again

$$y = \dot{y}_0 t - \frac{1}{2} g t^2 = (v_0 \sin \theta)t - \frac{1}{2} g t^2 \qquad \text{and} \qquad y = h - x \tan \alpha$$

Equating these two expressions for y and eliminating x by $x = \dot{x}_0 t = (v_0 \cos \theta)t$, we obtain the following equation for the time of striking:

$$\frac{1}{2} g t^2 - v_0(\cos \theta \tan \alpha + \sin \theta)t + h = 0$$

or, using the addition formula $\sin(\theta + \alpha) = \sin\theta\cos\alpha + \cos\theta\sin\alpha$,

$$\frac{1}{2}gt^2 - \left[\frac{v_0}{\cos\alpha}\sin(\theta + \alpha)\right]t + h = 0 \tag{1}$$

For a minimum t, we must have $dt/d\theta = 0$. Differentiating (1) with respect to θ and setting $dt/d\theta = 0$, we obtain

$$-\left[\frac{v_0}{\cos\alpha}\cos(\theta + \alpha)\right]t_{min} = 0$$

which implies that (since $t_{min} \neq 0$)

$$\cos(\theta + \alpha) = 0 \qquad \text{or} \qquad \theta = 90° - \alpha$$

This result means that the projectile should be aimed in the direction of minimum *distance*, just as though the acceleration of gravity did not exist. However, gravity cannot be ignored in this problem. If we seek to determine the value of t_{min} by substituting $\theta + \alpha = 90°$ into (1) and solving, we obtain

$$t_{min} = \frac{v_0 - \sqrt{v_0^2 - 2gh\cos^2\alpha}}{g\cos\alpha}$$

which is complex if $v_0 < \sqrt{2gh}\cos\alpha$. In other words, if $v_0 < \sqrt{2gh}\cos\alpha$, the projectile never reaches the roof, whatever the value of θ, and the concept of a minimum time becomes meaningless.

3.5.   With reference to Fig. 3-5, the projectile is fired with an initial velocity $v_0 = 30$ m/s at an angle $\theta = 23°$. The truck is moving along X with a constant speed of 15 m/s. At the instant the projectile is fired, the back of the truck is at $x = 45$ m. (a) Find the time for the projectile to strike the back of the truck, if the truck is very tall. (b) What will happen if the truck is only 2 m tall?

(a)   In this case, the projectile hits the back of the truck at the moment of overtaking it, which is the moment at which the distance of the back of the truck,

$$x_1 = 45 + 15t$$

equals the horizontal distance of the projectile,

$$x = (v_0\cos\theta)t = 32.22\,t$$

Thus
$$t = \frac{45}{32.22 - 15} = 2.614 \text{ s}$$

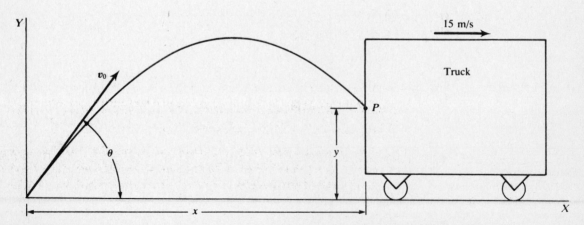

Fig. 3-5

(b) At $t = 2.614$ s, when the projectile overtakes the back of the truck, its height is ($v_0 \sin \theta = 13.67$ m/s):

$$y = (13.67)(2.614) - \frac{1}{2}(9.8)(2.614)^2 = 2.27 \text{ m}$$

i.e. 27 cm above the top of the truck.   Since the projectile travels faster horizontally than does the truck, it is clear that thereafter the projectile remains ahead of the back of the truck, and so never hits the back.

The projectile will reach (for the second time) a height of 2 m in a total time t_2 given by

$$2 = (13.67)t_2 - \frac{1}{2}(9.8)t_2^2 \qquad \text{or} \qquad t_2 = 2.635 \text{ s}$$

that is, $2.635 - 2.614 = 0.021$ s after overtaking the back of the truck.   Thus the projectile hits the top of the truck a distance of

$$(32.22 - 15)(0.021) = 0.36 \text{ m} = 36 \text{ cm}$$

in front of the rear edge.

3.6. Referring to Problem 3.5(a), find a value of v_0, all other conditions remaining the same, for which the projectile hits the truck at $y = 3$ m.

The time taken to overtake the back of the truck is given by

$$45 + 15t = (v_0 \cos \theta)t \qquad \text{or} \qquad t = \frac{45}{v_0 \cos \theta - 15}$$

at which time

$$y = 3 = (v_0 \sin \theta)t - \frac{1}{2}(9.8)t^2 = (v_0 \sin \theta)\left(\frac{45}{v_0 \cos \theta - 15}\right) - \frac{1}{2}(9.8)\left(\frac{45}{v_0 \cos \theta - 15}\right)^2$$

Inserting the numerical values of $\sin \theta$ and $\cos \theta$, we obtain the following quadratic equation for v_0:

$$v_0^2(4.54772) - v_0(60.29936) - 3532.5 = 0$$

Solving, $v_0 = 35.27775$ m/s.

3.7. A particle moving in the XY plane has X and Y components of velocity given by

$$\dot{x} = b_1 + c_1 t \qquad \dot{y} = b_2 + c_2 t \qquad\qquad (1)$$

where x and y are measured in meters and t in seconds.   (a) What are the units and dimensions of the constants b_1 and b_2?   of c_1 and c_2?   (b) Integrate the above relations to obtain x and y as functions of time.   (c) Denoting total acceleration as **a** and total velocity as **v**, find expressions for the magnitude and direction of **a** and of **v**.   (d) Write **v** in terms of the unit vectors.

(a)   Inspection of (1) shows that b_1 and b_2 must represent velocities in meters per second (m/s); unit-dimensionally c_1 and c_2 must be $|$m/s$^2|$, thus accelerations.

(b) $$x = x_0 + b_1 t + \frac{1}{2} c_1 t^2 \qquad y = y_0 + b_2 t + \frac{1}{2} c_2 t^2$$

where x_0, y_0 are the values of x and y at $t = 0$.

(c)   Differentiating (1) with respect to t, $\ddot{x} = c_1$, $\ddot{y} = c_2$.   Then

$$a = (\ddot{x}^2 + \ddot{y}^2)^{1/2} = (c_1^2 + c_2^2)^{1/2} \qquad \tan \alpha = \frac{\ddot{y}}{\ddot{x}} = \frac{c_2}{c_1}$$

where α is the angle **a** makes with X.   Note that **a** is constant in magnitude and direction.   For the velocity,

$$v = (\dot{x}^2 + \dot{y}^2)^{1/2} = [(b_1 + c_1 t)^2 + (b_2 + c_2 t)^2]^{1/2} \qquad \tan \beta = \frac{b_2 + c_2 t}{b_1 + c_1 t}$$

where β is the angle **v** makes with X.

(d) $$\mathbf{v} = (b_1 + c_1 t)\mathbf{i} + (b_2 + c_2 t)\mathbf{j}$$

3.8.   Refer to Fig. 3-6. A projectile is fired from the origin with initial velocity $v_1 = 100$ m/s at an angle $\theta_1 = 30°$. Another is fired at the same instant from a point on X at a distance $x_0 = 60$ m from the origin, with initial velocity $v_2 = 80$ m/s at angle θ_2. It is desired that the two projectiles collide at some point $P(x, y)$. (a) Determine the required value of θ_2. (b) At what time and at what point do they collide? (c) Find the velocity components of each just at impact.

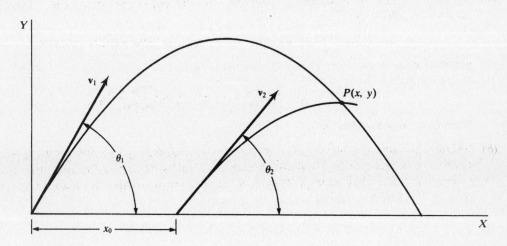

Fig. 3-6

(a)   Let (x_1, y_1) and (x_2, y_2) represent the coordinates of the first and second projectiles, respectively, at any time t. Then

$$y_1 = (v_1 \sin \theta_1)t - 4.9\,t^2 \qquad x_1 = (v_1 \cos \theta_1)t$$
$$y_2 = (v_2 \sin \theta_2)t - 4.9\,t^2 \qquad x_2 = x_0 + (v_2 \cos \theta_2)t$$

In order that they collide, $y_1 = y_2$ (and also $x_1 = x_2$). Thus

$$(v_1 \sin \theta_1)t - 4.9\,t^2 = (v_2 \sin \theta_2)t - 4.9\,t^2$$

from which $v_1 \sin \theta_1 = v_2 \sin \theta_2$ or

$$\sin \theta_2 = \frac{v_1}{v_2} \sin \theta_1 = \frac{100}{80} \sin 30° = 0.625 \qquad \text{or} \qquad \theta_2 = 38.6822°$$

(b)   Since $x_1 = x_2$, $(v_1 \cos \theta_1)t = x_0 + (v_2 \cos \theta_2)t$, from which

$$t = \frac{x_0}{v_1 \cos \theta_1 - v_2 \cos \theta_2} = \frac{60}{100 \cos 30° - 80 \cos 38.6822°} = 2.48421 \text{ s}$$

Then

$$x_1 = (100 \cos 30°)(2.48421) = 215.139 \text{ m} = x_2$$
$$y_1 = (100 \sin 30°)(2.48421) - (4.9)(2.48421)^2 = 93.971 \text{ m} = y_2$$

(c) $\dot{y}_1 = 100 \sin 30° - (9.8)(2.48421) = 25.655$ m/s

$$\dot{x}_1 = 100 \cos 30° = 86.6025 \text{ m/s}$$
$$\dot{y}_2 = 80 \sin 38.6822° - (9.8)(2.48421) = 25.6548 \text{ m/s}$$
$$\dot{x}_2 = 80 \cos 38.6822° = 62.45 \text{ m/s}$$

Actually, $\dot{y}_1 = \dot{y}_2$ (why?); the apparent difference is a roundoff error.

3.9. A ball, B_1, is fired upward from the origin of X, Y with initial velocity $v_1 = 100$ m/s at an angle $\theta_1 = 40°$. After $t_1 = 10$ s, as can easily be shown, the ball is at point $P(x_1, y_1)$, where $x_1 = 766.0444$ m, $y_1 = 152.7876$ m. Some time later, another ball, B_2, is fired upward, also from the origin, with velocity v_2 at angle $\theta_2 = 35°$. (a) Find a value of v_2 such that B_2 will pass through the point $P(x_1, y_1)$. (b) Find when B_2 must be fired in order that the two balls will collide at $P(x_1, y_1)$.

(a) Let (x_1, y_1, t_1) refer to the coordinates and time of B_1 and (x_2, y_2, t_2) to those of B_2. Since B_2 is to pass through $P(x_1, y_1)$,

$$x_2 = (v_2 \cos 35°)t_2 = 766.0444 \qquad y_2 = (v_2 \sin 35°)t_2 - 4.9\, t_2^2 = 152.7876$$

Eliminating t_2,

$$152.7876 = (766.0444)\tan 35° - (4.9)\left(\frac{766.0444}{v_2 \cos 35°}\right)^2$$

from which $v_2 = 105.69313$ m/s.

(b) Inserting the value of v_2 in $x_2 = (v_2 \cos 35°)t_2 = 766.0444$, we find $t_2 = 8.84795$ s. Hence, with $v_2 = 105.69313$ m/s and $\theta_2 = 35°$, B_2 passes through $P(x_1, y_1)$ 8.84795 s after it is fired. But B_1 arrives at this point 10 s after starting. Hence, if the two are to collide, the firing of B_2 must be delayed $10 - 8.84795 = 1.152$ s.

Supplementary Problems

3.10. A ball is thrown upward from a point on the side of a hill which slopes upward uniformly at an angle of $28°$. Initial velocity of ball: $v_0 = 33$ m/s, at an angle $\theta = 65°$ (with respect to the horizontal). At what distance up the slope does the ball strike and in what time?   *Ans.*   72.5 m; 4.59 s

3.11. A projectile is fired with initial velocity $v_0 = 95$ m/s at an angle $\theta = 50°$. After 5 s it strikes the top of a hill. What is the elevation of the hill above the point of firing? At what horizontal distance from the gun does the projectile land?   *Ans.*   241.37 m; 305.32 m

3.12. Rework Problem 3.4 in a coordinate system with axes perpendicular to and parallel to the roof. Show that the condition $v_0^2 \geq 2gh \cos^2 \alpha$ has a simple interpretation in this system.

3.13. The motion of a particle in the XY plane is given by

$$x = 25 + 6t^2 \qquad y = -50 - 20t + 8t^2$$

(a) Find the following initial values: x_0, y_0, $\dot{x}_0$, $\dot{y}_0$, v_0.
(b) Find magnitude and direction of **a**, the acceleration of the particle.
(c) Write an equation for the particle's path (find y as a function of x).
Ans. (a) $x_0 = 25$ m, $y_0 = -50$ m, $\dot{x}_0 = 0$, $\dot{y}_0 = -20$ m/s, $v_0 = 20$ m/s; (b) $a = 20$ m/s^2, $\alpha = 53.13°$ with horizontal; (c) $6y = -500 + 8x - 120[(x - 25)/6]^{1/2}$

3.14. In Fig. 3-7, α-particles from a bit of radioactive material enter through slit S into the space between two large parallel metal plates, A and B, connected to a source of voltage. As a result of the uniform electric field between the plates, each particle has a constant acceleration $a = 4 \times 10^{13}$ m/s^2 normal to and toward B. If $v_0 = 6 \times 10^6$ m/s and $\theta = 45°$, determine h and R.   *Ans.*   22.5 cm; 90 cm

3.15. The arrangement in Fig. 3-8 is the same as that in Fig. 3-7 except that α-particles enter slit S from two sources, A_1 and A_2, at angles θ_1 and θ_2, respectively. v_0 and **a** are the same for both groups. Given that $v_0 = 6 \times 10^6$ m/s, $a = 4 \times 10^{13}$ m/s^2, $\theta_1 = 45° + 1°$, $\theta_2 = 45° - 1°$, show that all particles are "focused" at a single point P. Find the values of R, h_1, and $h_2 - h_1$.
Ans. $R = 89.945$ cm; $h_1 = 23.285$ cm; $h_2 - h_1 = 2.114$ cm

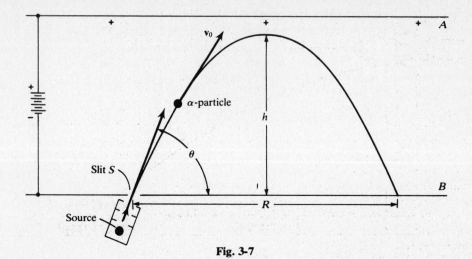

Fig. 3-7

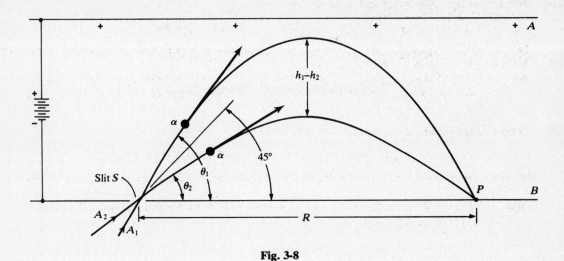

Fig. 3-8

3.16. A ball is thrown upward with initial velocity $v_0 = 15$ m/s at an angle of 30° with the horizontal. The thrower stands near the top of a long hill which slopes downward at an angle of 20°. (a) When does the ball strike the slope? (b) How far down the slope does it strike? (c) At what velocity does it hit? (Specify horizontal and vertical components.)

Ans. (a) 2.495 s after thrown   (c) $\dot{x} = 13.824$ m/s, $\dot{y} = 16.96$ m/s

      (b) 34.50 m, measured down the incline

3.17. A bomber, Fig. 3-9, is flying level at a speed $v_1 = 72$ m/s (about 161 mph), at an elevation of $h = 100$ m. When directly over the origin bomb B is released and strikes the truck T, which is moving along a level road (the X axis) with constant speed v_2. At the instant the bomb is released the truck is at a distance $x_0 = 125$ m from O. Find the value of v_2 and the time of flight of B.

Ans. 44.33 m/s (almost 100 mph); 4.51754 s

3.18. A particle moves in the XY plane along the path given by $y = 10 + 3x + 5x^2$. The X component of velocity, $\dot{x} = 4$ m/s, is constant, and at $t = 0$, $x = x_0 = 6$ m. (a) Write y and x as functions of t. (b) Find y_0 and $\dot{y}_0$. (c) Find $\ddot{y}$ and $\ddot{x}$, the components of acceleration of the particle.

Ans. (a) $y = 208 + 252t + 80t^2$, $x = 4t + 6$; (b) 208 m, 252 m/s; (c) $\ddot{y} = 160$ m/s², $\ddot{x} = 0$

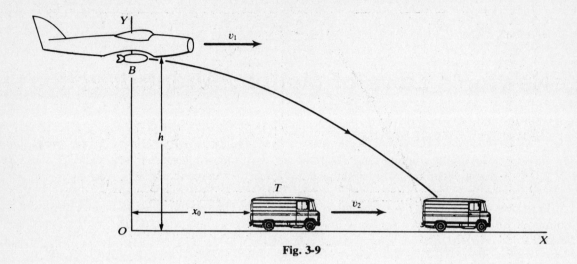

Fig. 3-9

3.19. The motion of a particle in the XY plane is given by

$$x = 10 + 12t - 20t^2 \qquad y = 25 + 15t + 30t^2$$

(a) Find values of x_0, $\dot{x}_0$; y_0, $\dot{y}_0$.   (b) Find the magnitude and direction of $\mathbf{v}_0$.   (c) Find $\ddot{x}$, $\ddot{y}$ and $\mathbf{a}$.   (d) Is the motion along a straight line?

Ans.   (a) $x_0 = 10$, $\dot{x}_0 = 12$, $y_0 = 25$, $\dot{y}_0 = 15$;   (b) $v_0 = (12^2 + 15^2)^{1/2} = 19.21$, $\beta = \arctan(15/12) = 51.34°$;
    (c) $\ddot{x} = -40$, $\ddot{y} = 60$, $a = (40^2 + 60^2)^{1/2} = 72.11$, $\alpha = \arctan(-60/40) = 123.69°$ or $-56.31°$;   (d) no

3.20. Consider the case where the motion of a particle is given by

$$x = 5 + 10t + 17t^2 + 4t^3 \qquad y = 8 + 9t + 20t^2 - 6t^3$$

(a) Find expressions for $\ddot{x}$, $\ddot{y}$.   (b) Is this a case of motion with constant acceleration as in all previous problems?

Ans.   (a) $\ddot{x} = 34 + 24t$, $\ddot{y} = 40 - 36t$;   (b) $\mathbf{a}$ is not constant, for its components are not constant.

Chapter 4

Newton's Laws of Motion: An Introduction

4.1 NEWTON'S LAWS OF MOTION

Law 1: Every body continues in its state of rest or of uniform motion in a straight line unless it is compelled by external, unbalanced forces to change that state.

From this law we have *force* defined as anything that changes or tends to change the state of motion of an object. Also, Newton's first law implicitly defines *inertial coordinate frames* (see Section 4.3).

Law 2: If a body of mass m is subject to various forces and if **a** is its acceleration as observed in an inertial coordinate frame, then

$$\sum \mathbf{F} = m\mathbf{a}$$

where $\sum \mathbf{F}$ is the vector sum of *all* the forces acting on the body.

For the special case when the resultant force is zero, Newton's second law gives $\mathbf{a} = 0$, which implies that the body's velocity is constant in magnitude and direction.

Law 3: If body 1 exerts a force $\mathbf{F}_2$ on body 2 and the latter exerts a force $\mathbf{F}_1$ on the former, then, regardless of what other forces may be acting on the two bodies, these forces are equal and opposite:

$$\mathbf{F}_1 = -\mathbf{F}_2$$

According to Newton's third law, no force occurs by itself. Action and reaction forces never balance out, because they are exerted on *different* bodies.

4.2 MASS AND WEIGHT

The property a body has of resisting any change in its state of rest or of uniform motion in a straight line is called *inertia*. The inertia of a body is related to what can be loosely thought of as the "amount of matter" it contains. A quantitative measure of inertia is *mass*.

The *weight* of a body is the gravitational *force* exerted on the body. A body of mass m has a weight $\mathbf{w} = m\mathbf{g}$ at a location where the gravitational acceleration is $\mathbf{g}$.

4.3 REFERENCE FRAMES

There exist certain reference frames, called *inertial* frames, relative to which any particle has a constant velocity vector when free from all external forces.

EXAMPLE 4.1. The reference frame attached to the "fixed stars" is generally taken as an inertial frame. Any other *frame is inertial if and only if its velocity with respect to this particular frame is constant.*

In a noninertial frame, i.e. a frame that has an acceleration $\mathbf{a}_0$ relative to every inertial frame, the equation of motion of a particle is

$$m \frac{d\mathbf{v}'}{dt} = \sum \mathbf{F} - m\mathbf{a}_0 \tag{4.1}$$

where $d\mathbf{v}'/dt$ is the acceleration of the particle in this system, m is its mass, and $\Sigma \mathbf{F}$ is the total force acting on it.

EXAMPLE 4.2. It is seen from (4.1) that Newton's second law will formally hold in a noninertial frame if the term $-m\mathbf{a}_0$ is interpreted as a force on the particle (called an *inertial force* because it is proportional to the mass or inertia, m), which is added to the actual forces.

4.4 PROCEDURE FOR CALCULATING FORCES AND ACCELERATIONS

(1) Draw a reasonably careful picture of the situation.

(2) Isolate the object in question.

(3) Draw all forces acting on this object, roughly indicating their magnitudes and directions.

(4) Find the resultant force, **R**. (The forces will all be acting at a point if we have idealized the object to a point with mass.)

(5) Select an inertial reference frame. (For nearly all practical problems, a frame fixed in the earth may be considered inertial.)

(6) Apply $\mathbf{R} = m\mathbf{a}$. Here $\mathbf{a}$ is relative to the chosen inertial frame.

Solved Problems

4.1. A constant force acts on a particle of mass 20 kg for 5 s, causing it to reach a velocity of 50 m/s from rest. Find (*a*) the force, (*b*) the acceleration this force would give a particle of mass 2000 kg, (*c*) the distance through which the particle of mass 2000 kg would move while being accelerated from rest to 44 m/s.

(*a*) Choose the X axis to lie along the constant force. Then the force is

$$F_x = ma_x = m\frac{\Delta v_x}{t} = 20\left(\frac{50-0}{5}\right) = 200 \text{ N}$$

(*b*)
$$a_x = \frac{F_x}{m} = \frac{200}{2000} = 0.1 \text{ m/s}^2$$

(*c*) From $v_x^2 = v_0^2 + 2ax$,

$$x = \frac{v_x^2 - v_0^2}{2a} = \frac{(44)^2 - 0}{2(0.1)} = 9680 \text{ m}$$

4.2. A coconut of mass 0.5 kg falls from a tree 10 m high and comes to rest after penetrating 0.1 m of sand. Find the force of resistance, F_r, of the sand, assuming it is constant.

First, find the velocity of the coconut just before hitting the sand (Fig. 4-1):

$$v_y^2 = 0^2 + 2a(y - y_0) = 2(-9.8)(0 - 10) = 196 \text{ m}^2/\text{s}^2$$

The acceleration of the coconut while moving through the sand has the constant value a_1, given by

$$0^2 = v_y^2 + 2a_1(s - s_0) \qquad \text{or} \qquad a_1 = -\frac{v_y^2}{2(s - s_0)} = -\frac{196}{2(-0.1 - 0)} = 980 \text{ m/s}^2$$

Then $\Sigma F_y = F_r - mg = ma_1$, or

$$F_r = m(g + a_1) = (0.5)(9.8 + 980) = 494.9 \text{ N}$$

The force is positive, since it opposes the motion, which is in the $-Y$ direction.

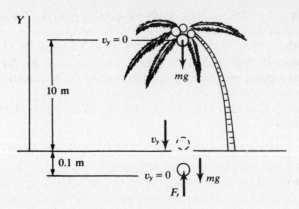

Fig. 4-1

4.3. Two carts are arranged to roll on a straight horizontal track.   At least one cart is set in motion so that a collision occurs.   As a result of the collision (interaction), the velocity of cart A changes as shown in Fig. 4-2.   If the duration of the interaction was 0.02 s and m_A is 0.5 kg, find the average force of interaction between A and B.

$$\bar{F}_A = m_A \frac{\Delta v_A}{\Delta t} = (0.5)\frac{-0.1 - 0.5}{0.02} = -15 \text{ N}$$

The minus sign indicates that the force *on* A (exerted *by* B) is to the left.   The force $\bar{F}_B$ can be stated as $\bar{F}_B = +15$ N, from Newton's third law of motion.   These are *average* forces acting during the time interval Δt.

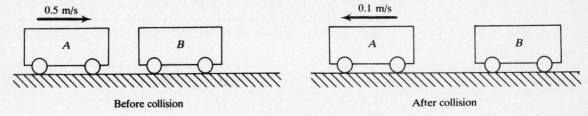

Before collision   After collision

Fig. 4-2

4.4. Blocks A and B, with masses 4 kg and 6 kg, respectively, are in contact on the smooth horizontal surface shown in Fig. 4-3.   An external force of 20 N is exerted on block A.   (*a*) What is the acceleration of the blocks?   (*b*) What is the force of block A on block B?   of block B on block A?

(*a*) $$a = \frac{\Sigma F}{m} = \frac{F}{m_A + m_B} = \frac{20}{4 + 6} = 2 \text{ m/s}^2$$

(*b*) $$F_{A \text{ on } B} = m_B a = 6(2) = 12 \text{ N, to the right}$$

(*c*) By Newton's third law, $F_{B \text{ on } A} = 12$ N, to the left.

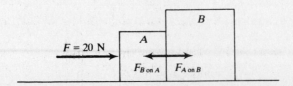

Fig. 4-3

4.5. A box of mass 4 kg at rest on a horizontal frictionless surface is attached to a rope that passes over a frictionless pulley having a very small mass (see Fig. 4-4). The box is initially 4 m horizontally from the pulley, and the rope initially makes an angle of 30° with respect to the horizontal. A constant pull of 56 N is applied to the rope by two men. At what point relative to its starting point will the box just start to leave the surface?

The box will leave the surface when the vertical component of the pull in the rope is just greater than the weight of the box.

$$T \sin \theta = mg \qquad \text{or} \qquad \sin \theta = \frac{mg}{T} = \frac{4(9.8)}{56} = 0.7$$

and $\theta = 44.4°$. Now the height of the pulley is $h = 4 \tan 30° = (4/\sqrt{3})$ m. Therefore,

$$\ell = h \cot 44.4° = \frac{4}{\sqrt{3}} (1.0212) = 2.4 \text{ m}$$

and the distance from the starting point is $d = 4 - 2.4 = 1.6$ m.

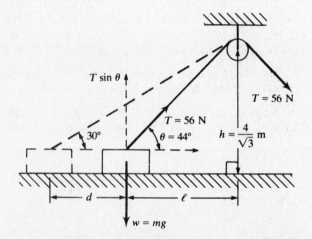

Fig. 4-4

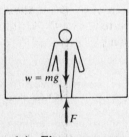

(a) Elevator at rest

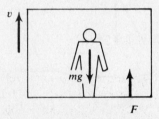

(b) Elevator moving with constant velocity

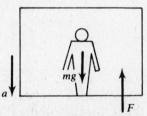

(d) Elevator moving with downward acceleration ($a < g$)

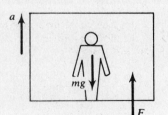

(c) Elevator moving with upward acceleration

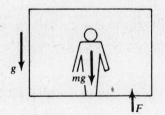

(e) Elevator in free fall

Fig. 4-5

4.6. Find the force F exerted by the elevator floor on the man's feet for the situations shown in Fig. 4-5. In these situations the positive direction is taken as that of $\mathbf{a}$, or as *up* if $\mathbf{a} = 0$.

In each case we apply Newton's second law to the man, the reference frame being the inertial frame of the earth.

(a) $\mathbf{a} = 0$: $\Sigma \mathbf{F} = m\mathbf{a}$, $F - mg = m(0)$, $F = mg$.

(b) $\mathbf{a} = 0$: same as (a).

(c) $\Sigma \mathbf{F} = m\mathbf{a}$, $F - mg = ma$, $F = m(g + a)$.

(d) $\Sigma \mathbf{F} = m\mathbf{a}$, $mg - F = ma$, $F = m(g - a)$. (What would happen if $a > g$?)

(e) $\Sigma \mathbf{F} = m\mathbf{a}$, $mg - F = mg$, $F = 0$.

4.7. A bird, in level flight at constant acceleration $\mathbf{a}_0$ relative to the ground frame X, Y (Fig. 4-6), lets fall a worm from its beak. What is the path of the worm, as seen by the bird?

In the bird's noninertial coordinate system X', Y' (Fig. 4-6), the equation of motion of the worm is

$$m \frac{d\mathbf{v}'}{dt} = m\mathbf{g} - m\mathbf{a}_0 \qquad \text{or} \qquad \frac{d\mathbf{v}'}{dt} = \mathbf{g} - \mathbf{a}_0 = \text{constant}$$

Thus the acceleration of the worm is constant, and its path is a straight line (supposing that it was dropped from rest). The slope of the line with respect to the horizontal is

$$\tan \theta = \frac{g}{a_0}$$

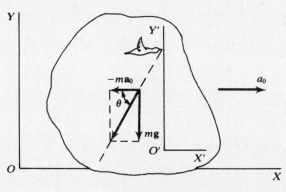

Fig. 4-6

4.8. Refer to Problem 4.7 and Fig. 4-6. (a) Determine the path of the worm as seen from the ground. (b) Verify that the two descriptions of the path are equivalent.

(a) In the ground frame X, Y, the worm has constant acceleration $\ddot{y} = -g$ and an initial velocity $\dot{x}_0 = v_0$, where v_0 is the speed of the bird at the instant the worm is released (call this time $t = 0$). Hence

$$x = x_0 + v_0 t \qquad y = y_0 - \tfrac{1}{2}gt^2 \tag{1}$$

and the path is a parabola.

(b) Let us suppose that at $t = 0$ the two coordinate frames coincide. At time t, O' will have advanced a distance $v_0 t + \tfrac{1}{2}a_0 t^2$ along the X axis, so that the coordinates (x, y) and (x', y') of the worm in the two systems are related by

$$x = x' + (v_0 t + \tfrac{1}{2}a_0 t^2) \qquad y = y' \tag{2}$$

The path in the X', Y' system is obtained by substituting the expressions (2) into (1):

$$x' = v_0 t + \tfrac{1}{2}a_0 t^2 = x_0 + v_0 t \qquad y' = y_0 - \tfrac{1}{2}g t^2$$

or

$$\frac{y' - y_0}{x' - x_0} = \frac{g}{a_0}$$

which is a straight line of slope g/a_0, as found in Problem 4.7.

4.9. A blimp is descending with an acceleration a. How much ballast must be jettisoned for the blimp to rise with the same acceleration a? There is a buoyant force acting upward on the blimp which is equal to the weight of the air displaced by the blimp; assume that the buoyant force is the same in both cases.

From Fig. 4-7, the equations of motion are

descending: $\qquad m_1 g - F_b = m_1 a$

ascending: $\qquad F_b - m_2 g = m_2 a$

Adding gives $(m_1 - m_2)g = (m_1 + m_2)a$. But $m_1 - m_2 = m$, the mass of the discarded ballast. Therefore,

$$mg = [m_1 + (m_1 - m)]a \qquad \text{or} \qquad m = \left(\frac{2a}{g + a}\right)m_1$$

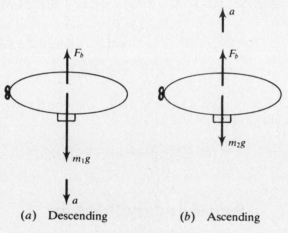

(a) Descending   (b) Ascending

Fig. 4-7

4.10. Objects A and B, each of mass m, are connected by a light inextensible cord. They are constrained to move on a frictionless ring in a vertical plane, as shown in Fig. 4-8. The objects are released from rest at the positions shown. Find the tension in the cord just after release.

At the moment of release A is constrained to move horizontally and B vertically, so that the two initial accelerations are tangential as shown. Furthermore, the two accelerations have the same magnitude, a, since otherwise the cord would have to stretch. Thus, the horizontal force equation for A and the vertical force equation for B, at the indicated positions, are

$$T \sin 45° = ma \qquad mg - T \sin 45° = ma$$

Eliminating a,

$$T = \frac{mg}{2 \sin 45°} = \frac{mg}{\sqrt{2}}$$

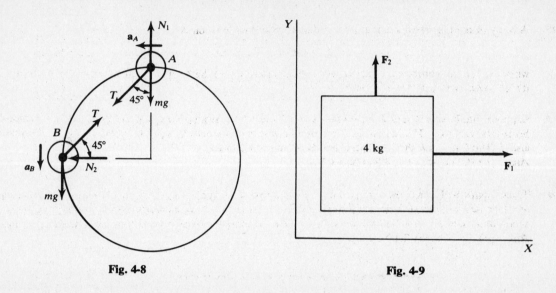

Fig. 4-8   Fig. 4-9

4.11. A body of mass m moves along X such that at time t its position is

$$x(t) = \alpha t^4 - \beta t^3 + \gamma t$$

where α, β, γ are constants. (*a*) Calculate the acceleration of the body.   (*b*) What is the force acting on it?

(*a*) $\dot{x} = 4\alpha t^3 - 3\beta t^2 + \gamma$ and $\ddot{x} = 12\alpha t^2 - 6\beta t$

(*b*) $F_x = m\ddot{x} = 12m\alpha t^2 - 6m\beta t$

4.12. Figure 4-9 shows a block of mass 4 kg acted on by the two forces, $\mathbf{F}_1 = 4\mathbf{i}$ N and $\mathbf{F}_2 = 2\mathbf{j}$ N. Calculate the acceleration of the block.

$$\mathbf{a} = \frac{1}{m}\sum \mathbf{F} = \frac{1}{4}(4\mathbf{i} + 2\mathbf{j}) = \mathbf{i} + \frac{1}{2}\mathbf{j} \quad \text{m/s}^2$$

Supplementary Problems

4.13. Shortly after leaping from an airplane a 91.8 kg man has an upward force of 225 N exerted on him by the air.   Find the resultant force on the man.   *Ans.*   674.64 N, downward

4.14. To measure the mass of a box, we push it along a smooth surface, exerting a net horizontal force of 150 N. The acceleration is observed to be 3.0 m/s². What is the mass of the box?   *Ans.*   50 kg

4.15. A 40 kg trunk sliding across a floor slows down from 5.0 m/s to 2.0 m/s in 6.0 s.   Assuming that the force acting on the trunk is constant, find its magnitude and its direction relative to the velocity vector of the trunk.   *Ans.*   20 N, opposite to the velocity

4.16. A resultant force of 20 N gives a body of mass m an acceleration of 8.0 m/s², and a body of mass m' an acceleration of 24 m/s².   What acceleration will this force cause the two masses to acquire if fastened together?   *Ans.*   6 m/s²

4.17. The 4.0 kg head of a sledge hammer is moving at 6.0 m/s when it strikes a spike, driving it into a log; the duration of the impact (or the time for the sledge hammer to stop after contact) is 0.0020 s.   Find (*a*) the time average of the impact force, (*b*) the distance the spike penetrates the log.
Ans.   (*a*) 12 kN;   (*b*) 6 mm

4.18. A body of mass m moves along Y such that at time t its position is

$$y(t) = at^{3/2} - bt + c$$

where a, b, c are constants. (a) Calculate the acceleration of the body. (b) What is the force acting on it? *Ans.* (a) $\frac{3}{4}at^{-1/2}$; (b) $\frac{3}{4}mat^{-1/2}$

4.19. Suppose that blocks A and B have masses of 2 kg and 6 kg, respectively, and are in contact on a smooth horizontal surface. If a horizontal force of 6 N pushes them, calculate (a) the acceleration of the system and (b) the force that the 2 kg block exerts on the other block.
Ans. (a) 0.75 m/s^2; (b) 4.5 N or -1.5 N

4.20. Three identical blocks, each of mass 0.6 kg, are connected by light strings as shown in Fig. 4-10. Assume that they lie on a smooth, horizontal surface and are observed to have an acceleration of 4.0 m/s^2 under the action of a force F. (a) Find the value of F. (b) Calculate the tension in each of the connecting strings.
Ans. (a) 7.2 N; (b) 2.4 N, 4.8 N

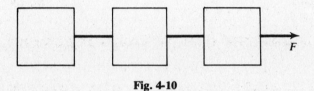

Fig. 4-10

4.21. What force in addition to $\mathbf{F}_1$ and $\mathbf{F}_2$ must be applied to the body in Fig. 4-9 so that: (a) it does not accelerate? (b) it has an acceleration of 4 m/s^2 along $-X$? Assume that $F_1 = 4$ N and $F_2 = 2$ N.
Ans. (a) $-4\mathbf{i} - 2\mathbf{j}$ N; (b) $-16\mathbf{i} - 2\mathbf{j}$ N

Chapter 5

Newton's Laws of Motion: More Advanced Problems

In this chapter, Newton's second law of motion for a particle,

$$\mathbf{F} = m\mathbf{a} = m\,\frac{d\mathbf{v}}{dt}$$

is studied for various physical systems, each with its own force function $\mathbf{F}$.

5.1 CENTER OF MASS

The *center of mass* of a system of particles of masses $m_1, m_2, \ldots, m_n$ has a position vector $\mathbf{r}_{cm}$ given by

$$M\mathbf{r}_{cm} = m_1\mathbf{r}_1 + m_2\mathbf{r}_2 + \cdots + m_n\mathbf{r}_n \qquad \text{or} \qquad \mathbf{r}_{cm} = \frac{1}{M}\sum m_i\mathbf{r}_i \qquad (5.1)$$

where $M = m_1 + m_2 + \cdots + m_n$. The center of mass of a body is that point at which the body's entire mass can be regarded as being concentrated, in light of (5.2) below.

The *center of gravity* of a body is that point at which the body's entire weight can be regarded as being concentrated. A rigid body can be suspended in any orientation from its center of gravity without tending to rotate. The center of gravity and center of mass of a body coincide in a uniform gravitational field, and the two points may be treated as identical in nearly all problems. The weight of a body can be considered as a downward force acting from its center of gravity.

5.2 SYSTEMS OF INTERACTING PARTICLES

For a system of interacting particles,

$$\sum \mathbf{F}_{ext} = M\mathbf{a}_{cm} \qquad (5.2)$$

where $\sum \mathbf{F}_{ext}$ is the vector sum of the external forces only, M is the system's total mass, and $\mathbf{a}_{cm}$ is the acceleration of the center of mass. In applying (5.2) to a complex system, we must make clear exactly which particles are to be included in the system and which are to be excluded.

5.3 FRICTION FORCES

When two objects are in contact, the force exerted on one object by the other has a component $\mathbf{N}$ normal to the surfaces in contact and a component $\mathbf{f}$, parallel to the surfaces in contact. The latter is called the *force of friction*; it is directed so as to oppose the tendency to slip or the slipping of one surface relative to the other. The following approximate empirical laws relate f to N:

$$f = \mu_k N \text{ (sliding)}$$

$$f \le \mu_s N \text{ (no sliding)} \qquad (5.3)$$

where μ_k and μ_s are the *coefficients of kinetic and static friction*, respectively.

Static friction occurs between surfaces at rest relative to each other.   When an increasing force is applied to an object at rest the force of static friction initially increases to prevent motion.   Finally a certain limiting force will be reached which static friction cannot exceed, and the object will begin to move.   Once the object begins moving, the force of *kinetic friction* remains almost constant at a value that is usually smaller than the maximum value of static friction.

There should be no rolling friction, since there should be no relative motion between the surfaces of contact of one object rolling on another.   Actually, however, a wheel or similar body is slightly flattened when it rests on a surface, which itself becomes slightly deformed.   A resistive force, *rolling friction*, then arises when the wheel rolls, because both it and the surface on which it moves must be continually deformed and because there is some relative motion between the surfaces of contact due to the deformation.   Coefficients of rolling friction are much smaller than coefficients of sliding friction.

5.4 UNIFORM CIRCULAR MOTION

Applied to a particle moving in a *circle* with a *constant speed*, Newton's second law becomes

$$\sum F = \frac{mv^2}{r} = m\omega^2 r \qquad (5.4)$$

where $\sum F$ is the resultant *radial force* toward the center of the circle and v is the speed of the particle. ω is the rotational speed of the particle and is measured in radians per second (rad/s).   This *inwardly* directed (radial) force that must be applied to keep a body moving in a circle is called a *centripetal force*.   See Fig. 5-1.

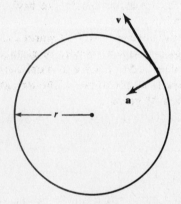

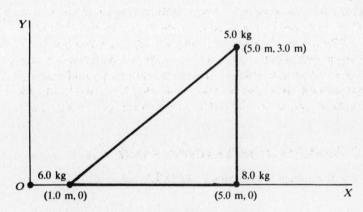

Fig. 5-1. Directions of Velocity and Acceleration in Uniform Circular Motion

Fig. 5-2

Solved Problems

5.1.   Locate the center of mass of the three-particle system shown in Fig. 5-2.

From Fig. 5-2, $m_1 = 6.0$ kg, $r_1 = 1.0i$; $m_2 = 8.0$ kg, $r_2 = 5.0i$; $m_3 = 5.0$ kg, $r_3 = 5.0i + 3.0j$.   Then, from (5.1),

$$(6.0 + 8.0 + 5.0)r_{cm} = (6.0)(1.0i) + (8.0)(5.0i) + (5.0)(5.0i + 3.0j)$$

$$19.0 r_{cm} = 71.0i + 15.0j$$

$$r_{cm} = \frac{71.0}{19.0}i + \frac{15.0}{19.0}j = 3.7i + 0.8j$$

or $x_{cm} = 3.7$ m, $y_{cm} = 0.8$ m.

5.2. A 10.0 kg block and a 20.0 kg block are placed on a smooth table and connected by a spring. The 20 kg block is then pushed east by a horizontal force of 120 N. (a) Find the acceleration of the center of mass of the two blocks. (b) The velocity of the center of mass after 2.0 s has elapsed is 8.0 m/s, east. At this time the 20 kg block has a velocity of 6.0 m/s, east. What is the velocity of the 10 kg block?

(a) Choose the $+X$ direction to be east. Then

$$\mathbf{a}_{cm} = \frac{\Sigma\,\mathbf{F}_{ext}}{M} = \frac{120\mathbf{i}}{10.0 + 20.0} = 4.0\mathbf{i}\ \text{m/s}^2$$

(b) Differentiation of (5.1) with respect to time gives

$$M\mathbf{V}_{cm} = m_1\mathbf{v}_1 + m_2\mathbf{v}_2 + \cdots + m_n\mathbf{v}_n \qquad\qquad (5.5)$$

Hence, after 2.0 s,

$$(10.0 + 20.0)(8.0\mathbf{i}) = (10.0)\mathbf{v}_1 + (20.0)(6.0\mathbf{i})$$

$$(10.0)\mathbf{v}_1 = 240.0\mathbf{i} - 120.0\mathbf{i}$$

$$\mathbf{v}_1 = 12.0\mathbf{i}\ \text{m/s}$$

5.3. Two bodies with masses 10 and 12 kg are connected by a light, inextensible string passing over a smooth fixed pulley, Fig. 5-3(a). Find (a) the velocities at the end of 3 s and (b) the distance moved in 3 s. (c) If at the end of 3 s the string is cut, find the distances moved by the bodies in the next 6 s. See Fig. 5-3(b).

(a) The force equations for each body are

$$\left(\sum F\right)_1 = m_1 g - T = m_1 a \qquad \text{and} \qquad \left(\sum F\right)_2 = T - m_2 g = m_2 a$$

The two accelerations are equal in magnitude but opposite in direction. In each force equation the direction of motion is taken as positive. Add these equations to get

$$a = \frac{(m_1 - m_2)g}{m_1 + m_2} = \frac{12 - 10}{12 + 10}(9.8) = 0.89\ \text{m/s}^2$$

Since the acceleration is constant, the common speed at the end of 3 s is

$$v = v_0 + at = 0 + (0.89)(3) = 2.67\ \text{m/s}$$

Body 1 moves down and body 2 moves up.

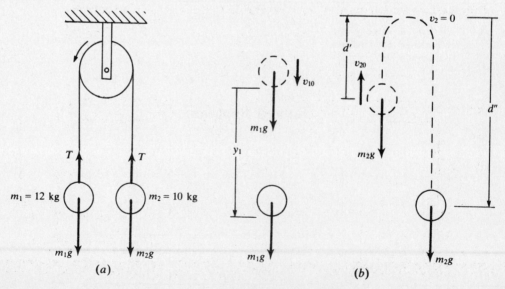

(a)   (b)

Fig. 5-3

(b)   The distance moved by each body in 3 s is

$$s = v_0 t + \frac{1}{2} a t^2 = (0)(3) + \frac{1}{2}(0.89)(3)^2 = 4 \text{ m}$$

(c)   If the string is cut, the bodies fall freely with initial velocities $v_{10} = 2.67$ m/s and $v_{20} = 2.67$ m/s, in each case taking the initial direction of motion as positive.   For body 1, the distance in 6 s is then

$$y_1 = v_{10}t + \frac{1}{2} g t^2 = (2.67)(6) + \frac{1}{2}(9.8)(6)^2 = 182.4 \text{ m}$$

Body 2 travels upward a distance

$$d' = \frac{v_{20}^2}{2g} = \frac{(2.67)^2}{2(9.8)} = 0.4 \text{ m}$$

before coming to a stop and then falling downward.   The time of travel upward before coming to a stop for body 2 is

$$t_{\text{up}} = \frac{v_{20}}{g} = \frac{2.67}{9.8} = 0.27 \text{ s}$$

Body 2 then travels downward 5.73 s for a distance

$$d'' = \left| \frac{1}{2}(-g)t^2 \right| = \frac{1}{2}(9.8)(5.73)^2 = 159.2 \text{ m}$$

The total distance traveled by body 2 is then

$$d = d' + d'' = 0.4 + 159.2 = 159.6 \text{ m}$$

5.4.   A body of mass m_3 is in motion on a smooth horizontal table.   On each side is attached a light, inelastic string passing over a smooth pulley at an edge of the table.   The other ends of the strings are attached to bodies of masses m_1 and m_2, which move vertically.   See Fig. 5-4.   Find the accelerations of bodies 1, 2, and 3, and the tensions in the strings.

Since the lengths of the strings are fixed, $x + y_2 = \text{constant}$ and $y_1 + y_2 = \text{constant}$.   Thus,

$$\ddot{x} + \ddot{y}_2 = 0 \qquad \text{or} \qquad a_3 + a_2 = 0$$

and $\qquad\qquad\qquad \ddot{y}_1 + \ddot{y}_2 = 0 \qquad \text{or} \qquad a_1 + a_2 = 0$

For each body write Newton's second law, arbitrarily choosing a positive direction:

$$m_1 g - T_1 = m_1 a_1 \qquad m_2 g - T_2 = m_2 a_2 \qquad T_1 - T_2 = m_3 a_3$$

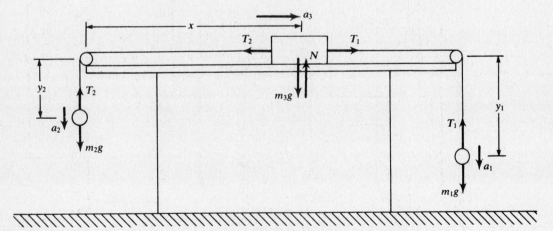

Fig. 5-4

We now have five equations in the five unknowns a_1, a_2, a_3, T_1, T_2. Solving,

$$a_1 = -a_2 = a_3 = \frac{(m_1 - m_2)g}{m_1 + m_2 + m_3}$$

$$T_1 = \frac{m_1(m_3 + 2m_2)g}{m_1 + m_2 + m_3} \qquad T_2 = \frac{m_2(m_3 + 2m_1)g}{m_1 + m_2 + m_3}$$

From the expression for the accelerations it is seen that if $m_1 > m_2$, the direction of $\mathbf{a}_2$ in Fig. 5-4 should be reversed; whereas if $m_2 > m_1$, $\mathbf{a}_1$ should be reversed. In the latter case, $\mathbf{a}_3$ should also be reversed.

5.5.   In the pulley system shown in Fig. 5-5, the movable pulleys A, B, C are of mass 1 kg each. D and E are fixed pulleys. The strings are vertical and inextensible. Find the tension in the string and the accelerations of the pulleys.

Write y_A, y_B, y_C for the positions of the centers of the pulleys A, B, C at time t; a_A, a_B, a_C are the accelerations at time t.

Following the string from the end at the center of A to the end at the center of B, we get

$$(y_B - y_A) + y_B + 2y_A + y_C + (y_C - y_B) = \text{constant} \qquad \text{or} \qquad y_A + y_B + 2y_C = \text{constant}$$

Take the second time-derivative of this equation to get

$$a_A + a_B + 2a_C = 0$$

There is just one string and, thus, one tension T. The force equations are

$$T + mg - 2T = ma_A \qquad T + mg - 2T = ma_B \qquad mg - 2T = ma_C$$

Substituting $m = 1$ kg and solving the four equations for the four unknowns a_A, a_B, a_C, T, we obtain

$$a_A = a_B = -a_C = \frac{g}{3} = 3.3 \text{ m/s} \qquad T = 6.5 \text{ N}$$

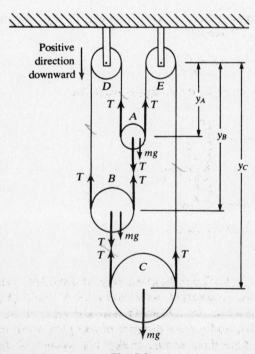

Fig. 5-5

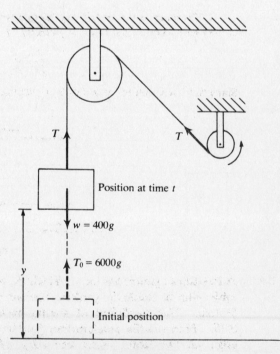

Fig. 5-6

5.6.   A body of mass 400 kg is suspended at the lower end of a light vertical chain and is being pulled up vertically (see Fig. 5-6).   Initially the body is at rest and the pull on the chain is $6000g$ (N).   The pull gets smaller uniformly at the rate of $360g$ (N) per each meter through which the body is raised.   What is the velocity of the body when it has been raised 10 m?

   At time t, let y be the height (in meters) of the body above its initial position.   The pull in the chain is then

$$T = (6000 - 360y)g$$

and Newton's second law gives

$$T - 400g = 400\ddot{y} \qquad \text{or} \qquad (5600 - 360y)g = 400\ddot{y}$$

This equation may be changed into one for $\dot{y} = v$ (the velocity of the body) by use of the identity $2\ddot{y} = d(v^2)/dy$.   Thus

$$200\frac{d(v^2)}{dy} = (5600 - 360y)g \qquad \text{or} \qquad d(v^2) = g(28 - 1.8y)\, dy$$

Let V be the velocity at height 10 m.   Then, on integrating

$$\int_0^{v^2} d(v^2) = g \int_0^{10} (28 - 1.8y)\, dy$$

$$V^2 = g[28y - 0.9y^2]_0^{10} = g[28(10) - 0.9(100)] = 190g$$

$$V = +\sqrt{190g} = +43.2 \text{ m/s}$$

   The choice of the + sign for V (upward motion) should be checked.   For $0 \le y \le 10$, the net force, $(5600 - 360y)g$, is positive, and so the acceleration is positive.   Then, since the body started from rest, V must be positive.   Had the force changed sign in the interval of integration, the direction of motion might have reversed, and the final velocity might have been negative.

5.7.   In Fig. 5-7, find the acceleration of the cart that is required to prevent block B from falling.   The coefficient of static friction between the block and the cart is μ_s.

   If the block is not to fall, the friction force, f, must balance the block's weight: $f = mg$.   But the horizontal motion of the block is given by $N = ma$.   Therefore,

$$\frac{f}{N} = \frac{g}{a} \qquad \text{or} \qquad a = \frac{g}{f/N}$$

Since the maximum value of f/N is μ_s, we must have $a \ge g/\mu_s$ if the block is not to fall.

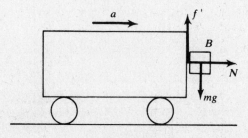

Fig. 5-7

5.8.   A flat dinner plate rests on a tablecloth, with its center 0.3 m from the edge of the table.   The tablecloth is suddenly yanked horizontally with a constant acceleration of 9.2 m/s² [Fig. 5-8(a)].   The coefficient of sliding friction between the tablecloth and the plate is $\mu_k = 0.75$.   Find (a) the acceleration, (b) the velocity, and (c) the distance of the *plate* from the edge of the table, when the edge of the tablecloth passes under the center of the plate.   Assume that the tablecloth just fits the tabletop.

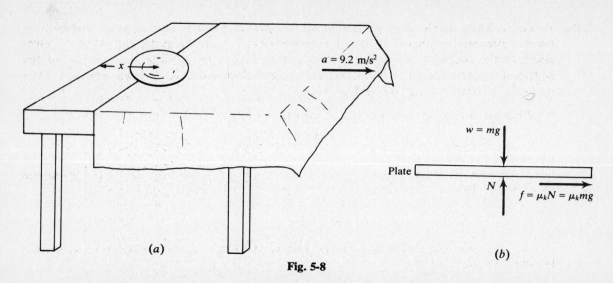

Fig. 5-8

(a)   From Fig. 5-8(b), the force equation for the plate is

$$\mu mg = m\ddot{x}_p \qquad \text{or} \qquad \ddot{x}_p = \mu g = (0.75)(9.8) = 7.35 \text{ m/s}^2$$

The plate slips, since $\ddot{x}_p$ is less than 9.2 m/s².

(b)   At the time the edge of the tablecloth is at the center of the plate, the cloth and the plate are at the same distance from the edge of the table:

$$x_p = x_c$$

$$0.3 + \frac{1}{2}(7.4)t^2 = 0 + \frac{1}{2}(9.2)t^2$$

Solving, $t = 0.57$ s, and $v_p = 0 + (7.35)(0.57) = 4.19$ m/s.

(c) $$x_p = 0.3 + 0(0.57) + \frac{1}{2}(7.35)(0.57)^2 = 1.49 \text{ m}$$

5.9.   If the system shown in Fig. 5-9(a) is given an acceleration, find the forces on the sphere, assuming no friction.

From Fig. 5-9(b), $\Sigma F_{\text{ver}} = R_1 \cos 30° - w = ma_{\text{ver}} = 0$ and $\Sigma F_{\text{hor}} = R_2 - R_1 \sin 30° = ma$.   Thus, the acting forces are

$$R_1 = \frac{w}{\cos 30°} = 1.15w \qquad R_2 = R_1 \sin 30° + \frac{w}{g}a = (1.15\,W)(0.5) + \frac{w}{g}a = w\left(0.58 + \frac{a}{g}\right)$$

and the weight, w.

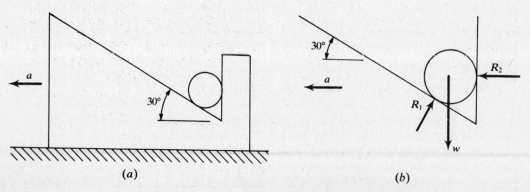

Fig. 5-9

5.10. In Fig. 5-10(a), block 1 is one-fourth the length of block 2 and weighs one-fourth as much. Assume that there is no friction between block 2 and the surface on which it moves and that the coefficient of sliding friction between blocks 1 and 2 is $\mu_k = 0.2$. After the system is released, find the distance block 2 has moved when only one-fourth of block 1 is still on block 2. Block 1 and block 3 have the same mass.

From Fig. 5-10(b), the equations of motion are:

$$\sum F_1 = T - \mu_k w_1 = ma_1 \qquad \sum F_2 = \mu_k w_1 = 4ma_2 \qquad \sum F_3 = w_1 - T = ma_1$$

Solve the first and third equations simultaneously to get $a_1 = (g/2)(1 - \mu_k)$; from the second equation, $a_2 = (g/4)\mu_k$. Then the displacements of blocks 1 and 2 are given by $x = \frac{1}{2}at^2$, i.e.

$$x_1 = \frac{g}{4}(1 - \mu_k)t^2 \qquad x_2 = \frac{g}{8}\mu_k t^2$$

At the instant that one-fourth of block 1 remains on block 2, $x_2 + \ell = x_1 + (\ell/16)$, where ℓ is the length of block 2. Therefore,

$$\frac{g}{8}\mu_k t^2 + \ell = \frac{g}{4}(1 - \mu_k)t^2 + \frac{\ell}{16} \qquad \text{or} \qquad t^2 = \frac{15\ell}{2g(2 - 3\mu_k)}$$

and

$$x_2 = \left(\frac{g}{8}\mu_k\right)\frac{15\ell}{2g(2 - 3\mu_k)} = \frac{15\mu}{16(2 - 3\mu_k)}\ell = \frac{\ell}{7.47}$$

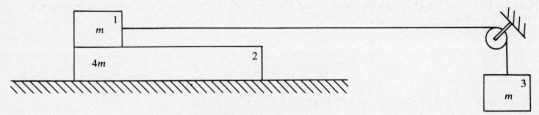

(a) Configuration at $t = 0$

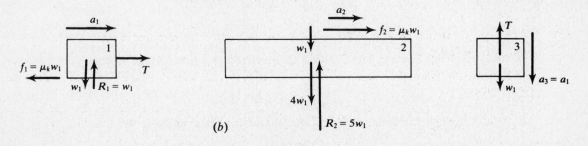

(b)

Fig. 5-10

5.11. Two bodies, of masses m_1 and m_2, are released from the position shown in Fig. 5-11(a). If the mass of the smooth-topped table is m_3, find the reaction of the floor on the table. Assume that the table does not move.

From Fig. 5-11(b), the force equations for the bodies are:

$$\text{body 1:} \quad \sum F_{\text{ver}} = w_1 - T = m_1 a$$

$$\text{body 2:} \quad \sum F_{\text{hor}} = T = m_2 a$$

$$\text{table:} \quad \sum F_{\text{ver}} = N - T - w_2 - w_3 = 0$$

$$\sum F_{\text{hor}} = T - f = 0$$

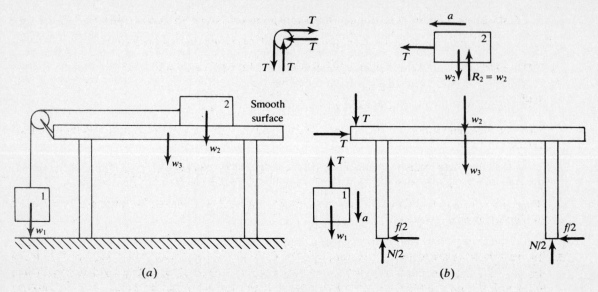

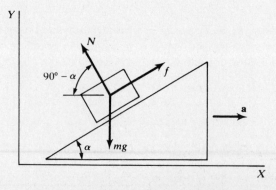

Fig. 5-11

where N and f are the vertical and horizontal (frictional) components of the force exerted by the floor on the table.

From the first two equations,

$$a = \frac{w_1}{m_1 + m_2} = \frac{m_1 g}{m_1 + m_2}$$

Then,

$$f = T = m_2 a = \frac{m_1 m_2 g}{m_1 + m_2}$$

and, finally,

$$N = T + m_2 g + m_3 g = \left(\frac{m_1 m_2}{m_1 + m_2} + m_2 + m_3\right) g$$

5.12. The inclined plane shown in Fig. 5-12 has an acceleration **a** to the right. Show that the block will slide on the plane if

$$a > g \tan (\theta - \alpha)$$

where $\mu_s = \tan \theta$ is the coefficient of static friction for the contacting surfaces.

Fig. 5-12

If the block is not to slide, it must have the same acceleration as the plane.   Hence

$$f \cos \alpha - N \sin \alpha = ma \qquad f \sin \alpha + N \cos \alpha - mg = 0$$

From these, $$f = m(a \cos \alpha + g \sin \alpha) \qquad N = m(g \cos \alpha - a \sin \alpha)$$

and $$\frac{f}{N} = \frac{a \cos \alpha + g \sin \alpha}{g \cos \alpha - a \sin \alpha} = \frac{a + g \tan \alpha}{g - a \tan \alpha}$$

Now the maximum value of f/N in the absence of slipping is $\mu_s = \tan \theta$.   Thus the acceleration a must satisfy

$$\frac{a + g \tan \alpha}{g - a \tan \alpha} \le \tan \theta \qquad \text{or} \qquad a \le g \frac{\tan \theta - \tan \alpha}{1 + \tan \theta \tan \alpha} = g \tan(\theta - \alpha)$$

If $a > g \tan(\theta - \alpha)$, the block will slide.   (Notice that for $a = 0$, the no-slip condition becomes $\alpha \le \theta$, which properly defines the angle θ.)

5.13.   In the turntable arrangement shown in Fig. 5-13, block A has a mass of 0.9 kg, block B has a mass of 1.7 kg, and the blocks are 13 cm from the axis of rotation.   The coefficient of static friction between the blocks, and between the blocks and the turntable, is $\mu_s = 0.1$.   Consider the friction and the mass of the pulley in Fig. 5-13(a) as negligible.   Find the angular speed of rotation of the turntable for which the blocks just begin to slide.

In this problem, everything depends on correctly predicting the direction of the frictional force between A and B.   Since B is more massive than A, we extrapolate to the case where A is very light: B would tend to move radially outward, pulling A radially inward.   The friction force f between the two surfaces would act to oppose their relative motion; it would act radially inward on B and radially outward on A, as shown in Fig. 5-13(c).

The force equations for no slipping are then:

$$\sum F_B = T + f + f' = m_B r \omega^2 \qquad \sum F_A = T - f = m_A r \omega^2$$

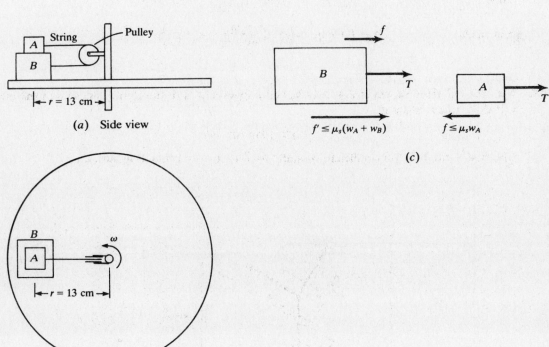

(a)   Side view

(c)

(b)   Top view

Fig. 5-13

By subtraction,

$$2f + f' = (m_B - m_A)r\omega^2$$

It is seen that ω can increase until both f and f' attain their maximum values. Thus

$$2\mu_s m_A g + \mu_s(m_A + m_B)g = (m_B - m_A)r\omega_{max}^2$$

or
$$\omega_{max} = \left[\frac{\mu_s g(3m_A + m_B)}{r(m_B - m_A)}\right]^{1/2} = \left[\frac{(0.1)(9.8)(2.7 + 1.7)}{(0.13)(1.7 - 0.9)}\right]^{1/2} = 6.4 \text{ rad/s}$$

5.14. A smooth horizontal tube of length ℓ rotates about a vertical axis as shown in Fig. 5-14(a). A particle placed at the extreme end of the tube is projected toward O with a velocity $\ell\omega$, while at the same time the tube rotates about the axis with constant angular speed ω. Show that the particle will have moved half the length of the tube during a time $(1/\omega)(\ln 2)$ and that it will not reach O in a finite time.

Since the tube is smooth, there is *no radial force* on the particle; the force, and hence the acceleration, is purely in the circumferential direction. This suggests viewing the motion in the noninertial frame that rotates with the tube, thereby "getting rid of" the circumferential force.

When the particle is at a distance r from O in the noninertial frame [Fig. 5-14(b)], the only force on it is the inertial force ("centrifugal force") $mr\omega^2$, directed as shown. Equation (*4.1*) becomes

$$m\ddot{r} = 0 + mr\omega^2 \qquad \text{or} \qquad \ddot{r} = \omega^2 r$$

Multiplying by $\dot{r}\,dt = dr$ and integrating,

$$\frac{1}{2}\int d(\dot{r}^2) = \omega^2 \int r\,dr$$

$$\frac{1}{2}\dot{r}^2 = \frac{1}{2}\omega^2 r^2 + C$$

When $r = \ell$, $\dot{r} = -\ell\omega$, which gives $C = 0$ and

$$\dot{r} = -\omega r$$

the minus sign being taken because r is decreasing. Finally,

$$\int \frac{dr}{r} = -\omega \int dt$$

$$\ln r = -\omega t + c'$$

When $r = \ell$, $t = 0$, whence $c' = \ln \ell$ and

$$t = \frac{1}{\omega}\ln\frac{\ell}{r}$$

When $r = \ell/2$, $t = (\ln 2)/\omega$. As $r \to 0$, $t \to \infty$, and the particle will not reach O in a finite time.

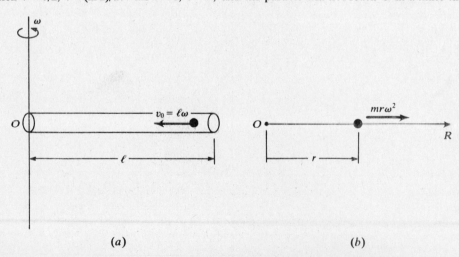

(a)   (b)

Fig. 5-14

Supplementary Problems

5.15. Show that the acceleration of the center of mass in Problem 4.9 does not change when the ballast is ejected.   Use this fact to confirm the value of m found in Problem 4.9.

5.16. Three blocks, of masses 2.0, 4.0, and 6.0 kg, arranged in the order lower, middle, and upper, respectively, are connected by strings on a frictionless inclined plane of 60°.   A force of 120 N is applied upward along the incline to the uppermost block, causing an upward movement of the blocks.   The connecting cords are light.   What is the tension in the cord between (a) the upper and middle blocks, and (b) the lower and middle blocks?   (c) What is the acceleration of the blocks?
Ans.   (a) 60 N; (b) 20 N; (c) 1.513 m/s^2

5.17. A skier goes down a hillside, which makes an angle θ with respect to the horizontal. If μ_k is the coefficient of sliding friction between skis and slope, show that the acceleration of the skier is $a = g(\sin\theta - \mu_k \cos\theta)$.

5.18. The breaking strength of a steel cable is 2.0×10^4 N.   If one pulls horizontally with this cable, what is the maximum horizontal acceleration which can be given to a 8000 kg body resting on a rough horizontal surface if the coefficient of kinetic friction is 0.15?   *Ans.*   1.03 m/s^2

5.19. Assuming that the earth ($m = 6.0 \times 10^{24}$ kg) orbits with uniform speed about the sun in a circular path of radius $R = 1.5 \times 10^8$ km, calculate the force required to sustain this motion.   *Ans.* 3.6×10^{22} N

5.20. A small block is a distance r from the center of a turntable.   The coefficient of static friction between the block and the turntable surface is μ_s.   What is the maximum angular velocity of the turntable if the block is not to slide?   *Ans.* $(\mu_s g/r)^{1/2}$

5.21. A small object of mass 0.1 kg is suspended by a light string of length 0.20 m and moving with uniform speed in a horizontal circular orbit of radius 0.10 m.   Calculate (a) the tension in the string, (b) the speed of the object.   *Ans.*   (a) 1.132 N; (b) 0.752 m/s

5.22. Consider a bead of mass m that is free to move on a thin fixed wire bent into a circle of radius R.   If initially the bead is given a push so that it starts to travel at a speed v_0, and if μ_k is the coefficient of sliding friction, calculate its speed at any subsequent time t.   Neglect gravity.   (*Hint*:   Newton's second law, written for the radial and tangential directions, gives

$$N = mR\dot\theta^2 \qquad -\mu_k N = mR\ddot\theta$$

where $R\theta$ is the arc length described in time t.)

Ans. $v = \dfrac{v_0}{1 + (\mu_k v_0/R)t}$

Chapter 6

Momentum, Impulse, and Relative Motion

6.1 LINEAR MOMENTUM

The *momentum* of a particle of mass m moving with velocity $\mathbf{v}$ is

$$\mathbf{p} = m\mathbf{v}$$

The units of momentum are $\text{kg} \cdot \text{m/s}$. The direction of the momentum of a particle is the same as the direction in which it is moving.

Newton's second law may be restated as: *The time rate of change of momentum of a particle is equal to the resultant force acting on the particle,*

$$\sum \mathbf{F} = \frac{d\mathbf{p}}{dt}$$

The total momentum $\mathbf{P}$ of a system of particles is the vector sum of the momentum vectors of the individual particles:

$$\mathbf{P} = \mathbf{p}_1 + \mathbf{p}_2 + \cdots + \mathbf{p}_n = M\mathbf{V}_{cm}$$

where the second equality follows from (*5.5*). Then (*5.2*) becomes

$$\sum \mathbf{F}_{ext} = \frac{d\mathbf{P}}{dt}$$

where $\sum \mathbf{F}_{ext}$ is the vector sum of the external forces acting on the particles of the system.

6.2 IMPULSE

The *impulse* of a force $\mathbf{F}$ in a time interval $\Delta t = t - t_0$ is the *vector* defined by

$$\mathbf{I} = \int_{t_0}^{t} \mathbf{F}(t')\, dt' = \bar{\mathbf{F}}\, \Delta t$$

where $\bar{\mathbf{F}}$ is the average value of $\mathbf{F}$ during the time interval Δt. The *impulse-momentum theorem* states that the impulse of the resultant force acting on a particle is equal to the change in the particle's momentum:

$$\int_{t_i}^{t_f} \mathbf{F}\, dt = \mathbf{p}_f - \mathbf{p}_i$$

The left-hand side is the impulse of the force $\mathbf{F}$ and the right-hand side is the resultant change in linear momentum. The units of impulse are the $\text{N} \cdot \text{s}$ ($1\,\text{N} \cdot \text{s} = 1\,\text{kg} \cdot \text{m/s}$).

6.3 CONSERVATION OF LINEAR MOMENTUM

The law of conservation of linear momentum states that:

$$\text{if} \quad \sum \mathbf{F}_{ext} = 0, \quad \text{then} \quad \mathbf{P} = \text{constant vector}$$

This implies that, for any process occurring within the system, $\mathbf{P}_{initial} = \mathbf{P}_{final}$.

6.4 RELATIVE MOTION

If a particle has velocity $\mathbf{v}$ relative to frame S and velocity $\mathbf{v}'$ relative to frame S', then

$$\mathbf{v} = \mathbf{V}_{S'} + \mathbf{v}'$$

where $\mathbf{V}_{S'}$ is the velocity of frame S' relative to frame S.

If S is an inertial frame and if $\mathbf{V}_{S'}$ is constant, then S' is also an inertial frame, and a particle has the same acceleration relative to both frames. (Compare Example 4.1.).

Solved Problems

6.1. A missile launcher, of mass 4400 kg, fires horizontally a rocket of mass 110 kg and recoils up a smooth inclined plane, rising to a height of 4 m (see Fig. 6-1). Find the initial speed of the rocket.

Since gravity was the only force acting on the launcher opposing its motion, the speed at which it starts up the inclined plane can be found from

$$0^2 = V^2 + 2(-g \sin \theta)s = V^2 - 2gh$$

or $\qquad\qquad V = \sqrt{2gh} = \sqrt{2(9.8)(4)} = 8.85 \text{ m/s}$

All forces involved in the launching are internal forces. For the horizontal momentum: $p_{\text{initial}} = p_{\text{final}}$. Choosing the rocket's direction as positive,

$$0 = m_r v_0 - m_l V$$

$$v_0 = \frac{m_l}{m_r} V = \left(\frac{4400}{110}\right)(8.85) = 354 \text{ m/s}$$

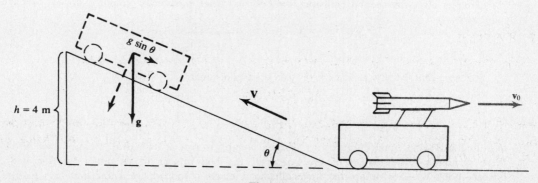

Fig. 6-1

6.2. Suppose that a boy stands at one end of a boxcar sitting on a railroad track. Let the mass of the boy and the boxcar be M. He throws a ball of mass m with velocity $\mathbf{v}_0$ toward the other end, where it bounces off the wall and travels back down the length (L) of the car, striking the opposite side and coming to rest. If there is no friction in the wheels of the boxcar, describe the motion of the boxcar.

All forces are internal (Fig. 6-2). Therefore, if $\mathbf{V}$ and $\mathbf{v}$ are the velocities of the boxcar-plus-boy and the ball, conservation of momentum gives

$$M\mathbf{V} + m\mathbf{v} = 0 \qquad \text{or} \qquad \mathbf{V} = -\frac{m}{M}\mathbf{v}$$

at all times.

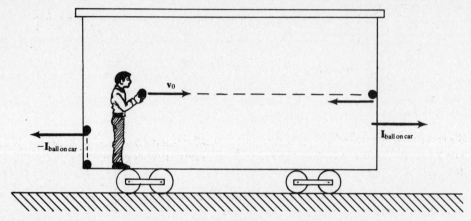

Fig. 6-2

Before the first collision, $\mathbf{v} = \mathbf{v}_0$, and so

$$\mathbf{V} = -\frac{m}{M}\,\mathbf{v}_0 \qquad 0 < t < \frac{L}{\left(1 + \dfrac{m}{M}\right)v_0}$$

where we have on the right the time for the ball to reach the boxcar wall, traveling at speed

$$v_0 + V = v_0 + \frac{m}{M}\,v_0$$

with respect to the floor of the boxcar.

Assuming that the first collision is perfectly elastic, the effect of the first collision will be simply to reverse both velocity vectors. Thus,

$$\mathbf{V} = +\frac{m}{M}\,\mathbf{v}_0 \qquad \frac{L}{\left(1 + \dfrac{m}{M}\right)v_0} < t < \frac{2L}{\left(1 + \dfrac{m}{M}\right)v_0}$$

Finally, after the second collision, the ball and boxcar have a common velocity. Hence, $\mathbf{V} = -(m/M)\mathbf{V}$, or

$$\mathbf{V} = 0 \qquad t > \frac{2L}{\left(1 + \dfrac{m}{M}\right)v_0}$$

It is seen that the boxcar first moves to the left a distance

$$\left(\frac{m}{M}\,v_0\right)\frac{L}{\left(1 + \dfrac{m}{M}\right)v_0} = \left(\frac{m}{M+m}\right)L$$

and then moves an equal distance to the right, coming to rest at its starting point. This result—that there is no net displacement of the boxcar if the ball returns to its initial location within the boxcar—holds whether or not the first collision is elastic. Indeed, we know that because the total momentum of the system is zero, the center of mass must remain at rest ($\mathbf{P} = M\mathbf{V}_{\text{cm}}$).

6.3. Suppose that two putty balls move along a frictionless floor, as shown in Fig. 6-3. They stick together after collision. If ball A is moving to the left at 15 m/s and ball B to the right at 25 m/s, and if their masses are equal, find their common velocity after collision.

The two balls constitute the system. No external horizontal forces act on the system, so that horizontal momentum is conserved: $\mathbf{p}_{\text{initial}} = \mathbf{p}_{\text{final}}$.

$$-m_A v_A + m_B v_B = (m_A + m_B)v \qquad \text{or} \qquad v_B - v_A = 2v$$

since $m_A = m_B$. Thus,

$$v = \frac{v_B - v_A}{2} = \frac{25 - 15}{2} = 5 \text{ m/s}$$

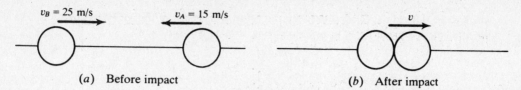

(a)   Before impact   (b)   After impact

Fig. 6-3

6.4.   Suppose that the two putty balls of Problem 6.3 collide obliquely, as shown in Fig. 6-4(a), and stick together after collision.   Find their velocity after impact.   Take $v_A = v_B = 45$ m/s and $\theta = 45°$.

The vector conservation of momentum is shown in Fig. 6-4(b).   Since $m_A = m_B$ and $v_A = v_B$, we have at once that

$$\phi = \frac{\theta}{2} = 22.5°$$

Then,

$$m_A v_A \cos \frac{\theta}{2} + m_B v_B \cos \frac{\theta}{2} = (m_A + m_B)v$$

or

$$v = v_A \cos \frac{\theta}{2} = 45(0.924) = 41.58 \text{ m/s}$$

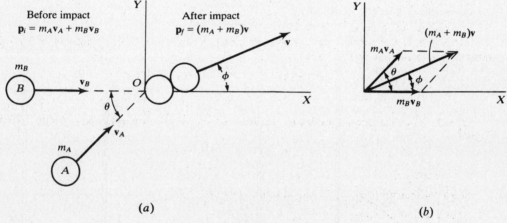

(a)   (b)

Fig. 6-4

6.5.   If a rocket moves in interstellar space, far from attracting bodies, there are no external forces acting on it.   If M is the instantaneous mass of the rocket and unused fuel, m is the mass of gas discharged by the rocket per second, and V is the speed of the discharged gas relative to the rocket, find the velocity v of the rocket at any time.

Let us view the system (gas cloud + rocket) in an inertial frame, with the direction of the rocket chosen as positive.   The total momentum of the system is constant.   Now, in a small time interval dt, a mass $m\, dt$ of gas, traveling in the negative direction with speed $V - v$ (relative to our reference frame), joins the cloud.   The cloud's change in momentum is therefore $m(v - V)\, dt$.   By conservation, this change must exactly cancel the change in the rocket's momentum, $d(Mv)$; that is,

$$m(v - V)\, dt + d(Mv) = 0$$

But $dM = -m\, dt$, and so

$$(V - v)\, dM + d(Mv) = 0 \qquad \text{or} \qquad V\, dM - v\, dM + M\, dv + v\, dM = 0$$

or

$$dv = -V \frac{dM}{M}$$

which is the differential relation between the velocity and mass of the rocket. Letting v_0 and M_0 denote the velocity and mass at $t = 0$, and supposing V constant, we have

$$\int_{v_0}^{v} dv = -V \int_{m_0}^{M} \frac{dM}{M} \qquad \text{or} \qquad v = v_0 + V \ln \frac{M_0}{M}$$

6.6. A uniform rope, of mass m per unit length, hangs vertically from a support so that the lower end just touches the tabletop shown in Fig. 6-5(a). If it is released, show that at the time a length y of the rope has fallen, the force on the table is equivalent to the weight of a length $3y$ of the rope.

The descending part of the rope is in free fall; it has speed $v = \sqrt{2gy}$ at the instant all of its points have descended a distance y. The length of the rope which lands on the table during an interval dt following this instant is $v\,dt$. The increment of momentum imparted to the table by this length in coming to rest is $m(v\,dt)v$. Thus, the rate at which momentum is transferred to the table is

$$\frac{dp}{dt} = mv^2 = (2my)g$$

and this is the force arising from stopping the downward fall of the rope. Since a length of rope y, of weight $(my)g$, already lies on the tabletop, the total force on the tabletop is

$$(2my)g + (my)g = (3my)g$$

or the weight of a length $3y$ of rope.

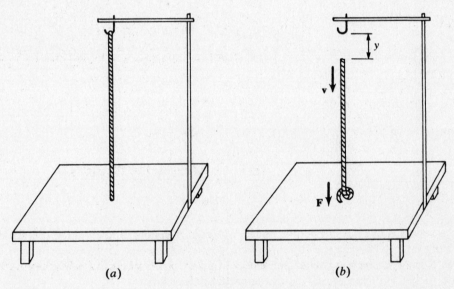

(a)   (b)

Fig. 6-5

6.7. A missile of mass M, moving with velocity **v** ($v = 200$ m/s), explodes in mid-air, breaking into two parts [Fig. 6-6(a)], of masses $M/4$ and $3M/4$. If the smaller piece flies off at an angle of 60° with respect to the original direction of motion with a speed of 400 m/s, find the initial velocity of the other piece.

Over a very small time interval surrounding the moment of explosion, the effect of gravity (an external force) can be neglected. Then all forces are internal and momentum is conserved.

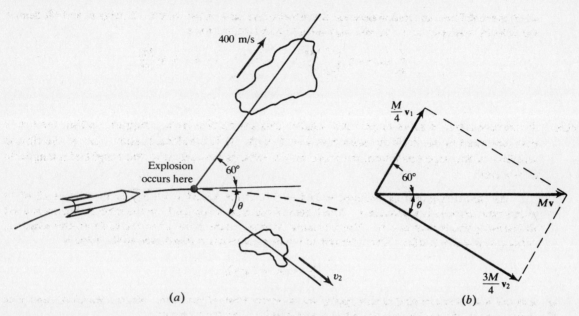

Fig. 6-6

The vector diagram for momentum conservation is shown in Fig. 6-6(b). We have:

$$\frac{3M}{4}\mathbf{v}_2 = M\mathbf{v} - \frac{M}{4}\mathbf{v}_1 \qquad \text{or} \qquad \mathbf{v}_2 = \frac{4}{3}\mathbf{v} - \frac{1}{3}\mathbf{v}_1$$

Then,

$$v_2^2 = \mathbf{v}_2 \cdot \mathbf{v}_2 = \frac{16}{9}(\mathbf{v} \cdot \mathbf{v}) - \frac{8}{9}(\mathbf{v} \cdot \mathbf{v}_1) + \frac{1}{9}(\mathbf{v}_1 \cdot \mathbf{v}_1)$$

$$= \frac{16}{9}(200)^2 - \frac{8}{9}(200)(400)(\cos 60°) + \frac{1}{9}(400)^2 = \frac{48 \times 10^4}{9}$$

$$v_2 = \frac{400}{\sqrt{3}} = 231 \text{ m/s}$$

and

$$\mathbf{v} \cdot \mathbf{v}_2 = \frac{4}{3}\mathbf{v} \cdot \mathbf{v} - \frac{1}{3}\mathbf{v} \cdot \mathbf{v}_1$$

$$(200)\left(\frac{400}{\sqrt{3}}\right)(\cos \theta) = \frac{4}{3}(200)^2 - \frac{1}{3}(200)(400)(\cos 60°)$$

$$\cos \theta = \frac{\sqrt{3}}{2}$$

$$\theta = 30°$$

One might also have found v_2 and θ by applying the law of cosines and the law of sines.

6.8. A 0.11 kg baseball is thrown with a speed of 17 m/s toward a batter. After the ball is struck by the bat, it has a speed of 34 m/s in the direction shown in Fig. 6-7(a). If ball and bat are in contact for 0.025 s, find the magnitude of the average force exerted on the ball by the bat.

The impulse is $\mathbf{I} = \bar{\mathbf{F}} \Delta t$. The impulse-momentum relation is shown in Fig. 6-7(b). From the law of cosines:

$$I^2 = p_i^2 + p_f^2 - 2p_i p_f \cos 120° = [(0.11)(17)]^2 + [(0.11)(34)]^2 - 2[(0.11)(17)][(0.11)(34)](-0.5)$$

$$I = (0.11)(17)(\sqrt{7}) = 4.947 \text{ N} \cdot \text{s}$$

and

$$\bar{F} = \frac{I}{\Delta t} = \frac{4.947}{0.025} = 197.90 \text{ N}$$

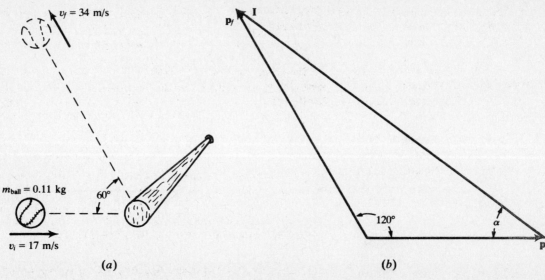

Fig. 6-7

6.9. A bucket filled to overflowing with water sits on one platform of a balance, as shown in Fig. 6-8.   A constant stream of water is poured from a height of 10 m into the bucket and overflows on one side of the balance.   The water is poured in at 0.5 kg/s.   If the platforms are in balance without the flow of water, how much more does the bucket "weigh" when the water is flowing?

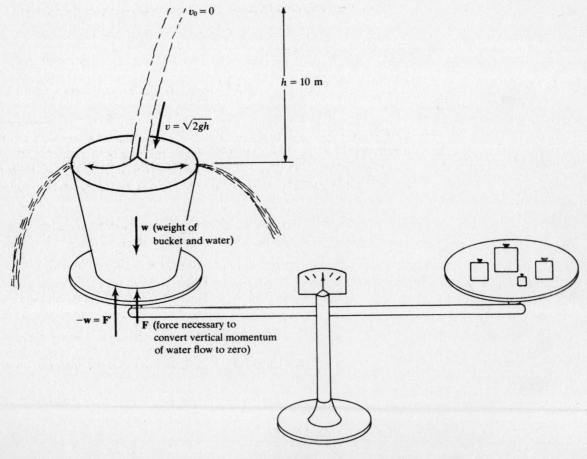

Fig. 6-8

The balance supports the weight of the filled bucket and supplies the impulse to stop the downward flow of water. In time Δt the vertical momentum of a mass $(0.5 \text{ kg/s}) \Delta t$ of water, falling at speed $v = \sqrt{2gh}$, is changed to zero; thus,

$$I = F \Delta t = (0.5 \Delta t)v \qquad \text{or} \qquad F = 0.5v = 0.5\sqrt{2(9.8)(10)} = 7 \text{ N}$$

Now, a mass of 1 kg weighs 9.8 N; so the bucket appears to have a weight corresponding to a mass of about 0.7 kg more than it actually possesses.

6.10. A helicopter is trying to land on a submarine deck which is moving south at 17 m/s. A 12 m/s wind is blowing into the west. If to the submarine crew the helicopter is descending vertically at 5 m/s, find its speed (*a*) relative to the water, and (*b*) relative to the air. See Fig. 6-9.

(*a*)
$$\mathbf{v}_{\text{hel/water}} = \mathbf{v}_{\text{sub/water}} + \mathbf{v}_{\text{hel/sub}} = 17\mathbf{j} + (-5)\mathbf{k} = 17\mathbf{j} - 5\mathbf{k} \quad \text{m/s}$$

(*b*)
$$\mathbf{v}_{\text{hel/air}} = \mathbf{v}_{\text{hel/water}} + \mathbf{v}_{\text{water/air}} = \mathbf{v}_{\text{hel/water}} - \mathbf{v}_{\text{air/water}}$$
$$= (17\mathbf{j} - 5\mathbf{k}) - 12\mathbf{i} = -12\mathbf{i} + 17\mathbf{j} - 5\mathbf{k} \quad \text{m/s}$$

6.11. Rain, pouring down at an angle α with the vertical, has a constant speed of 10 m/s. A woman runs against the rain with a speed of 8 m/s and sees the rain make an angle β with the vertical. Find the relation between α and β.

From the vector diagram, Fig. 6-10,

$$\tan \beta = \frac{v_{\text{woman}} + v_{\text{rain}} \sin \alpha}{v_{\text{rain}} \cos \alpha} = \frac{8 + 10 \sin \alpha}{10 \cos \alpha}$$

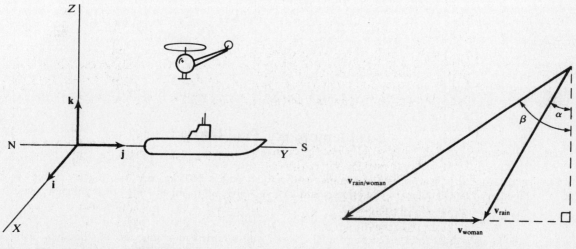

Fig. 6-9   Fig. 6-10

6.12. An elevator is moving upward at a constant speed of 4 m/s. A light bulb falls out of a socket in the ceiling of the elevator. A man in the building watching the cage sees the bulb rise for $(4/9.8)$ s and then fall for $(4/9.8)$ s; at $t = (4/9.8)$ s the bulb appears to be at rest to the man. Compute the velocity of the bulb at $t = (4/9.8)$ s from the view of an observer in the elevator.

Taking *up* as positive,

$$v_{\text{bulb/elev}} = v_{\text{bulb/bldg}} - v_{\text{elev/bldg}} = 0 - 4 = -4 \text{ m/s}$$

Alternatively, in the inertial frame of the elevator,

$$v = v_0 + at = 0 + (-9.8)\left(\frac{4}{9.8}\right) = -4 \text{ m/s}$$

6.13. An armored car 2 m long and 3 m wide is moving at 13 m/s when a bullet hits it in a direction making an angle arctan (3/4) with the car (Fig. 6-11). The bullet enters one edge of the car at the corner and passes out at the diagonally opposite corner. Neglecting any interaction between bullet and car, find the time for the bullet to cross the car.

Call the speed of the bullet V. Because of the motion of the car, the velocity of the bullet relative to the car in the direction of the length of the car is $V \cos \theta - 13$, and the velocity in the direction of the width of the car is $V \sin \theta$. Then, from $s = vt$,

$$2 = (V \cos \theta - 13)t \qquad 3 = (V \sin \theta)t$$

Eliminate V to find

$$t = \frac{1}{13}\left(\frac{3}{\tan \theta} - 2\right) = \frac{2}{13} = 0.1538 \text{ s}$$

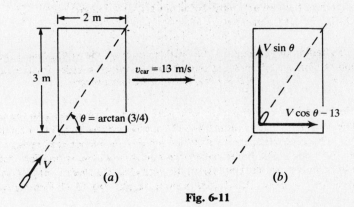

(a)   (b)

Fig. 6-11

Supplementary Problems

6.14. A 1200 kg car is moving east at 30.0 m/s and collides with a 3600 kg truck moving at 20.0 m/s in a direction 60° north of east. The vehicles interlock and move off together. Find their common velocity. *Ans.* 19.84 m/s at 40.9° north of east

6.15. A 60 kg man dives from the stern of a 90 kg boat with a horizontal component of velocity of 3.0 m/s north. Initially the boat was at rest. Find the magnitude and direction of the velocity acquired by the boat. *Ans.* 2.0 m/s south

6.16. A boy of mass m is standing on a toboggan of mass M which is moving with constant velocity V_i over the surface of a frozen lake. The boy then runs along the toboggan in the direction opposite to V_i and acquires a speed v *relative to the toboggan* as he jumps off the rear of the toboggan. What is the final speed of the toboggan relative to the ice?

Ans. $V_f = V_i + \dfrac{mv}{M + m}$

6.17. A 1.0 kg object, A, with a velocity of 4.0 m/s to the right, strikes a second object, B, of 3.0 kg, originally at rest. In the collision, A is deflected from its original direction through an angle of 50°; its speed after the collision is 2.0 m/s. Find (a) the angle between B's velocity after the collision and the original direction of A and (b) the speed of B after the collision. *Ans.* (a) 29.4°; (b) 1.04 m/s

6.18. The uranium-238 nucleus is unstable and decays into a thorium-234 nucleus and an α-particle. The α-particle is emitted with a speed of 1.4×10^6 m/s. What is the recoil speed of the thorium-234 nucleus, assuming the uranium-238 atom to be at rest at the time of decay? The thorium-234 and α-particle masses are in the ratio 234 to 4. *Ans.* 2.393×10^4 m/s

6.19. An object at rest explodes into three pieces of equal mass. One moves east at 20 m/s; a second moves southeast at 30 m/s. What is the velocity of the third piece? *Ans.* 46 m/s at 27° north of west

6.20. A 1.0 kg steel ball 4.0 m above the floor is released, falls, strikes the floor, and rises to a maximum height of 2.5 m. Find the momentum transferred from the ball to the floor in the collision.
Ans. 15.9 kg · m/s downward

6.21. A girl standing on a train moving at 40 m/s throws a ball straight up at 10 m/s relative to the train, just as the train passes a crossing. (*a*) How far from the crossing is the girl when she catches the ball? (*b*) What is the path of the ball as viewed by an observer on the ground?
Ans. (*a*) 81.63 m; (*b*) a parabola

6.22. Hailstones are falling at an angle of 60° with respect to the vertical at a speed of 40 m/s. In what direction and at what speed must a ground observer travel in order for the hailstones to appear to fall "straight down"? *Ans.* 34.6 m/s "into" the rain of hailstones

6.23. A 3.0 kg block slides on a frictionless horizontal surface, first moving to the left at 50 m/s. It collides with a spring as it moves left, compresses the spring, and is brought to rest momentarily. The body continues to be accelerated to the right by the force of the compressed spring. Finally, the body moves to the right at 40 m/s. The block remains in contact with the spring for 0.020 s. (*a*) What were the magnitude and direction of the impulse of the spring on the block? (*b*) What was the spring's average force on the block? *Ans.* (*a*) 270 N · s to the right; (*b*) 13.5 kN to the right

<u>Chapter 7</u>

Motion in a Plane
Along a Curved Path

7.1 CONSTANT ANGULAR SPEED

A particle moving with constant angular speed ω in a circle of radius r, Fig. 7-1, has *period*

$$T = \frac{1}{f} = \frac{2\pi}{\omega}$$

where f is the *frequency*; its *coordinates* are

$$x = r \cos(\omega t + \theta_0) \qquad y = r \sin(\omega t + \theta_0)$$

where θ_0 represents the initial angular position; its *acceleration* is $\mathbf{a} = -\omega^2\mathbf{r}$, with components

$$a_x = -\omega^2 x \qquad a_y = -\omega^2 y$$

The period is measured in seconds (s) and the frequency in revolutions per second (rev/s). The SI unit of ω is the radian per second (rad/s).

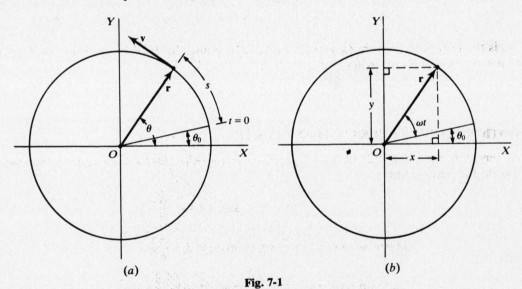

(a) (b)

Fig. 7-1

7.2 ANGULAR MOTION WITH VARYING SPEED

Angular motion is described by

$$angular\ displacement \equiv \theta - \theta_0$$

$$angular\ speed \equiv \omega = \frac{d\theta}{dt}$$

$$angular\ acceleration \equiv \alpha = \frac{d\omega}{dt} = \frac{d^2\theta}{dt^2}$$

In a motion parallel to the XY plane, the rotation of an entire rigid body is characterized by these angular quantities. The angular displacement is measured in radians and the angular acceleration is in rad/s^2.

Constant Angular Acceleration

The formulas relating the angular displacement, angular velocity, and angular acceleration of a rotating body for the case $\alpha =$ constant are:

$$\theta = \theta_0 + \omega_0 t + \frac{1}{2}\alpha t^2$$

$$\omega = \omega_0 + \alpha t$$

$$\omega^2 = \omega_0^2 + 2\alpha(\theta - \theta_0)$$

$$\theta - \theta_0 = \frac{1}{2}(\omega_0 + \omega)t$$

where θ_0 and ω_0 are the initial angular position and initial angular speed.

Relations Between Linear and Angular Quantities in Circular Motion

For a particle moving at a constant distance r from a fixed axis of rotation,

$$s = (\theta - \theta_0)r$$

$$v = \omega r$$

$$a_s = \alpha r$$

where a_s is the linear acceleration tangent to the circular path. In the first equation it is supposed that the initial value of the arc length s is zero [see Fig. 7-1(a)].

7.3 MOTION ALONG A GENERAL PLANE CURVE

In a general motion (Fig. 7-2), described in terms of a particle's distance $s = s(t)$ traveled along a curved path, the particle has

$$speed \equiv v = \frac{ds}{dt}$$

$$tangential \ component \ of \ acceleration \equiv a_s = \frac{d^2 s}{dt^2}$$

$$normal \ component \ of \ acceleration \equiv a_n = \frac{v^2}{\rho} = \rho\omega^2$$

where ρ is the radius of curvature of the path and where $\omega = v/\rho$ is defined as the particle's angular speed of rotation about an axis through the instantaneous center of curvature. Newton's second law gives

$$F_n = ma_n \qquad \text{and} \qquad F_s = ma_s$$

where the resultant force acting on the particle has a normal component F_n and a tangential component F_s. Note that a positive F_n produces acceleration *toward* the center of curvature.

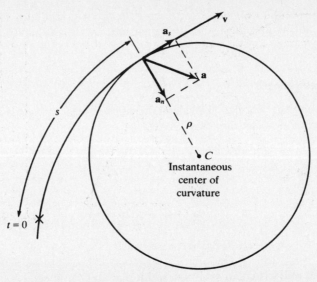

Fig. 7-2

Solved Problems

7.1. A watch has a second hand which is 2.0 cm long.   (a) What is the frequency of revolution of the hand?   (b) What is the speed of the tip of the second hand relative to the watch?

(a) If T is the period in seconds,

$$f = \frac{1}{T} = \frac{1}{60} = 0.017 \, \text{rev/s}$$

(b)
$$v = r\omega = r2\pi f = (2.0 \times 10^{-2})(2\pi)\left(\frac{1}{60}\right) = 2.1 \times 10^{-3} \, \text{m/s}$$

7.2. An automobile moves around a curve of radius 300 m at a constant speed of 60 m/s [Fig. 7-3(a)].   (a) Calculate the resultant change in velocity (magnitude and direction) when the car goes around the arc of 60°.   (b) Compare the magnitude of the instantaneous acceleration of the car to the magnitude of the average acceleration over the 60° arc.

(a) From Fig. 7-3(b), $\Delta v = 60$ m/s and Δv makes a 120° angle with v_A.

(b) The instantaneous acceleration has magnitude

$$a = \frac{v^2}{r} = \frac{(60)^2}{300} = 12 \, \text{m/s}^2$$

The time average acceleration (which, at constant speed, is the same as the average with respect to arc length) is $\bar{a} = \Delta v/\Delta t$.   Since

$$\Delta t = \frac{\Delta s}{v} = \frac{r \, \Delta \theta}{v} = \frac{300(\pi/3)}{60} = \frac{5\pi}{3} \, \text{s}$$

we have:
$$\bar{a} = \frac{\Delta v}{\Delta t} = \frac{60}{5\pi/3} = 11.5 \, \text{m/s}^2$$

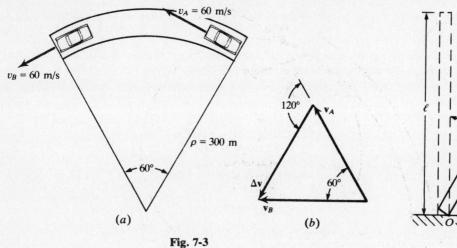

Fig. 7-3

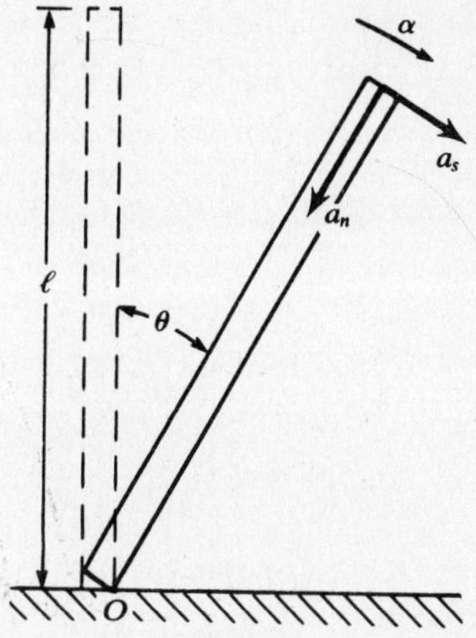

Fig. 7-4

7.3. The angular acceleration of the toppling pole shown in Fig. 7-4 is given by $\alpha = k \sin \theta$, where θ is the angle between the axis of the pole and the vertical, and k is a constant. The pole starts from rest. Find (*a*) the tangential and (*b*) the centripetal acceleration of the upper end of the pole in terms of k, θ and ℓ (the length of the pole).

(*a*)
$$a_s = \frac{dv}{dt} = \frac{d}{dt}(\ell\omega) = \ell\alpha = \ell k \sin \theta$$

(*b*) From $d\omega/dt = \alpha$,

$$d\omega = k \sin \theta \, dt = k \sin \theta \, \frac{dt}{d\theta} \, d\theta = \frac{k}{\omega} \sin \theta \, d\theta$$

Then
$$\int_0^\omega \omega \, d\omega = k \int_0^\theta \sin \theta \, d\theta \qquad \text{or} \qquad \omega^2 = 2k(1 - \cos \theta)$$

and $a_n = \ell\omega^2 = 2k\ell(1 - \cos \theta).$

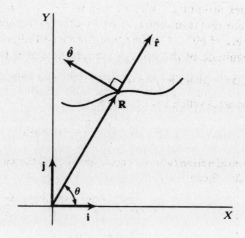

Fig. 7-5

7.4. Find (*a*) the velocity and (*b*) the acceleration in polar coordinates for an object moving in a curved path in a plane.

(*a*) Consider the motion of a particle along the curve $\mathbf{R} = \mathbf{R}(t)$ shown in Fig. 7-5. At a point of the curve, the unit vectors $\hat{\mathbf{r}}$, $\hat{\boldsymbol{\theta}}$ are given in terms of the unit vectors $\mathbf{i}$, $\mathbf{j}$ by

$$\hat{\mathbf{r}} = \mathbf{i} \cos \theta + \mathbf{j} \sin \theta \qquad \hat{\boldsymbol{\theta}} = -\mathbf{i} \sin \theta + \mathbf{j} \cos \theta$$

The velocity is given by

$$\mathbf{v} = \frac{d\mathbf{R}}{dt} = \frac{d(R\hat{\mathbf{r}})}{dt} = \frac{dR}{dt} \hat{\mathbf{r}} + R \frac{d\hat{\mathbf{r}}}{dt}$$

But

$$\frac{d\hat{\mathbf{r}}}{dt} = -\mathbf{i}(\sin \theta) \frac{d\theta}{dt} + \mathbf{j}(\cos \theta) \frac{d\theta}{dt} = \omega \hat{\boldsymbol{\theta}}$$

and so

$$\mathbf{v} = \dot{R}\hat{\mathbf{r}} + R\omega \hat{\boldsymbol{\theta}}$$

It is seen that the velocity has a radial component $\dot{R}$ and an angular component $R\omega$.

(*b*)

$$\mathbf{a} = \frac{d\mathbf{v}}{dt} = \frac{d}{dt} \left(\frac{dR}{dt} \hat{\mathbf{r}} + R \frac{d\hat{\mathbf{r}}}{dt} \right) = \frac{d^2R}{dt^2} \hat{\mathbf{r}} + 2 \frac{dR}{dt} \frac{d\hat{\mathbf{r}}}{dt} + R \frac{d^2\hat{\mathbf{r}}}{dt^2}$$

Substituting $d\hat{\mathbf{r}}/dt = \omega \hat{\boldsymbol{\theta}}$ and

$$\frac{d^2\hat{\mathbf{r}}}{dt^2} = \frac{d(\omega \hat{\boldsymbol{\theta}})}{dt} = \alpha \hat{\boldsymbol{\theta}} + \omega \frac{d\hat{\boldsymbol{\theta}}}{dt}$$

$$= \alpha \hat{\boldsymbol{\theta}} + \omega[-\mathbf{i}(\cos \theta)\omega - \mathbf{j}(\sin \theta)\omega] = \alpha \hat{\boldsymbol{\theta}} - \omega^2 \hat{\mathbf{r}}$$

we obtain

$$\mathbf{a} = (\ddot{R} - R\omega^2)\hat{\mathbf{r}} + (R\alpha + 2\dot{R}\omega)\hat{\boldsymbol{\theta}}$$

The radial component of acceleration consists of two parts: $\ddot{R}$, the linear acceleration in the direction of increasing R due to change in radial speed, and $-R\omega^2$, the familiar centripetal acceleration due to the change in direction of the velocity vector. Likewise, the angular component of acceleration consists of two parts: $R\alpha$, the linear acceleration due to change in angular speed, and $2\dot{R}\omega$, an acceleration due to the *joint* change in radius and angle. This term is called the *coriolis acceleration*.

Observe that the angular component can be written as the single term

$$\frac{1}{R} \frac{d}{dt} (R^2\omega)$$

It is important to realize that, in general, $\hat{\mathbf{r}}$ and $\hat{\boldsymbol{\theta}}$ are not normal and tangential to the path. Hence, the above acceleration components are different from those given in Section 7.3. (Of course, the *resultant* accelerations are necessarily the same.)

7.5. A bead slides on a long bar with constant speed v_0 relative to the bar. From Fig. 7-6, $v_0 = \dot{r}$, where r is the distance of the bead from the axle through the end of the bar. At the same time, the bar rotates about the axle with constant angular speed ω_0. Find (*a*) the velocity, (*b*) the acceleration, and (*c*) the path of the bead.

Use the results of Problem 7.4 ($\ddot{R} = \alpha = 0$).

(*a*)

$$\mathbf{v} = v_0\hat{\mathbf{r}} + r\omega_0\hat{\boldsymbol{\theta}}$$

(*b*)

$$\mathbf{a} = -r\omega_0^2\hat{\mathbf{r}} + 2v_0\omega_0\hat{\boldsymbol{\theta}}$$

(*c*) Integrating $\dot{r} = v_0$, $\dot{\theta} = \omega_0$,

$$r = r_0 + v_0 t \qquad \theta = \theta_0 + \omega_0 t$$

Elimination of t gives the equation of the path:

$$r - r_0 = \frac{v_0}{\omega_0} (\theta - \theta_0)$$

which is a spiral.

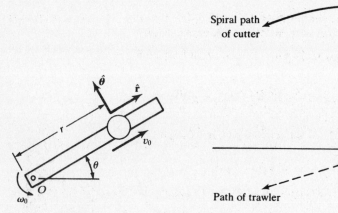

Fig. 7-6

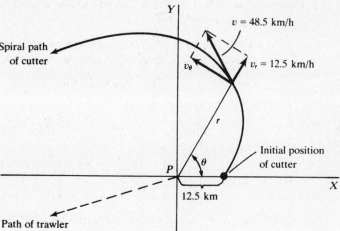

Fig. 7-7

7.6. A Coast Guard cutter in a fog at sea is notified by radio that an illegal trawler is at a particular position P, 12.5 km due west of the cutter. The trawler also hears the message and heads off immediately at 12.5 km/h. The captain of the cutter anticipates this speed, but does not know the direction the trawler takes. He waits for 1 h and then begins to spiral around P at 48.5 km/h, with a component of velocity directed away from P equal to 12.5 km/h. What is the maximum time that it takes, after the message is received, to catch the trawler?

From Fig. 7-7,

$$v_r = \frac{dr}{dt} = 12.5 \text{ km/h} \qquad v_\theta = r\frac{d\theta}{dt} = \sqrt{(48.5)^2 - (12.5)^2} = 46.85 \text{ km/h}$$

From the first equation, $r = 12.5t$, where we have used the initial condition: $r = 12.5$ km at $t = 1$ h. Substituting for r in the second equation and integrating,

$$12.5t\frac{d\theta}{dt} = 46.85$$

$$\int_0^\theta d\theta = 3.75 \int_1^t \frac{dt}{t}$$

$$\theta = 3.75 \ln t$$

The spiral path of the cutter must cross the radial path of the trawler at some moment, $t = \tau$, during the first revolution. At that moment, both ships will be at the same distance from P, so that the cutter will have indeed caught the trawler. Since $\theta \le 2\pi$ for $t = \tau$,

$$3.75 \ln \tau \le 2\pi \qquad \text{or} \qquad \tau \le e^{2\pi/3.75} = 5.34 \text{ h}$$

7.7. A wet, open umbrella is held upright as shown in Fig. 7-8(a) and is twirled about the handle at a uniform rate of 21 revolutions in 44 s. If the rim of the umbrella is a circle 1 m in diameter, and the height of the rim above the floor is 1.5 m, find where the drops of water spun off the rim hit the floor.

The angular speed of the umbrella is

$$\omega = \frac{\Delta\theta}{\Delta t} = \frac{21(2\pi)}{44} = 3 \text{ rad/s}$$

Then the tangential speed of the water drops on leaving the rim of the umbrella is $v_0 = r\omega = (0.5)(3) = 1.5$ m/s.

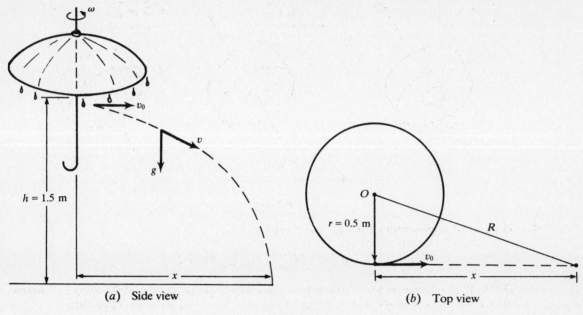

(a) Side view   (b) Top view

Fig. 7-8

To calculate the time for a drop to reach the floor use $h = \frac{1}{2}gt^2$:

$$t = \sqrt{\frac{2h}{g}} = \sqrt{\frac{2(1.5)}{9.8}} = 0.553 \text{ s}$$

The horizontal range of the drop is then $x = v_0t = (1.5)(0.55) = 0.83$ m; and the locus of the drops is a circle of radius

$$R = \sqrt{(0.5)^2 + (0.83)^2} = 0.97 \text{ m}$$

7.8. A wheel turning with angular speed of 30 rev/s is brought to rest with a constant acceleration. It turns 60 rev before it stops. (a) What is its angular acceleration? (b) What time elapses before it stops?

(a) The angular acceleration may be found from $\omega^2 = \omega_0^2 + 2\alpha(\theta - \theta_0)$:

$$\alpha = \frac{\omega^2 - \omega_0^2}{2(\theta - \theta_0)} = \frac{0^2 - [(30)(2\pi)]^2}{2(60)(2\pi)} = -47 \text{ rad/s}^2$$

(b) The time is found from $\omega = \omega_0 + \alpha t$:

$$t = \frac{0 - (30)(2\pi)}{-47} = 4 \text{ s}$$

7.9. A spinning wheel initially has an angular velocity of 50 rad/s east; 20 s later its angular velocity is 50 rad/s west. If the angular acceleration is constant, what are (a) the magnitude and direction of the angular acceleration, (b) the angular displacement over 20 s, and (c) the angular speed at 30 s?

(a) The direction of the angular acceleration is west, as shown in Fig. 7-9(a), since

$$\boldsymbol{\alpha} \, \Delta t = \boldsymbol{\omega}_f - \boldsymbol{\omega}_i = \boldsymbol{\omega}_f + (-\boldsymbol{\omega}_i)$$

and both vectors on the right are to the west. The magnitude of the angular acceleration is

$$\alpha = \frac{\omega_f - \omega_i}{\Delta t} = \frac{50 - (-50)}{20} = 5 \text{ rad/s}$$

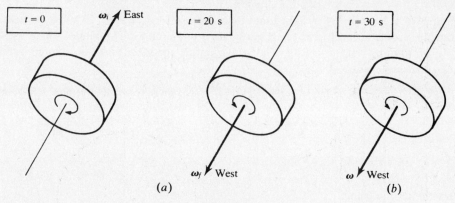

Fig. 7-9

(b)   The angular displacement, from $\omega_f^2 = \omega_i^2 + 2\alpha(\theta - \theta_0)$, is

$$\theta - \theta_0 = \frac{\omega_f^2 - \omega_i^2}{2\alpha} = \frac{(50)^2 - (50)^2}{2(5)} = 0$$

This result also follows from the fact that the average angular velocity, $\frac{1}{2}(\omega_i + \omega_f)$, is zero. The sole effect of the angular acceleration over the 20 s interval is to reverse the axis of rotation.

(c)   From Fig. 7-9(b), the angular speed at the end of 30 s is

$$\omega = \omega_0 + \alpha t = 50 + 5(30 - 20) = 100 \text{ rad/s}$$

Once $\boldsymbol{\alpha}$ and $\boldsymbol{\omega}$ are parallel, the angular speed, but not the direction of the rotation axis, changes.

7.10. In Fig. 7-10, as the block descends, the rigid rotor winds up on its rope and, thus, ascends. Find the relations between the linear and angular accelerations and between the linear and angular speeds.

If θ is the angle through which the rotor has turned from its initial position,

$$y_1 = y_{10} - r\theta \qquad y_2 = y_{20} + R\theta - r\theta$$

since the length of rope that is wound on the smaller cylinder is $r\theta$ and the length of rope that is unwound from the larger cylinder is $R\theta$. Take first and second time-derivatives of y_1 and y_2:

$$v_1 = \dot{y}_1 = -r\dot{\theta} = -r\omega \qquad v_2 = \dot{y}_2 = (R - r)\dot{\theta} = (R - r)\omega$$

$$a_1 = \ddot{y}_1 = -r\ddot{\theta} = -r\alpha \qquad a_2 = \ddot{y}_2 = (R - r)\ddot{\theta} = (R - r)\alpha$$

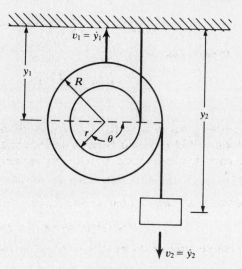

Fig. 7-10

7.11. Figure 7-11 shows a ray of light that passes from air into water. The ray is bent upon passing into the water, according to Snell's law ($\sin \theta = n \sin \psi$). The angle θ increases at a constant rate of 10 rad/s, and $n = 1.3$. Find the angular speed ω and the angular acceleration α of the refracted ray for $\theta = 30°$.

Take first and second time-derivatives of $\sin \theta = n \sin \psi$ to get $\omega = \dot{\psi}$ and $\alpha = \ddot{\psi}$; recall that $\ddot{\theta} = 0$.

$$\dot{\theta} \cos \theta = n\dot{\psi} \cos \psi \qquad \text{or} \qquad \dot{\psi} = \frac{\dot{\theta} \cos \theta}{n \cos \psi} = \frac{\dot{\theta} \cos \theta}{\sqrt{n^2 - \sin^2 \theta}}$$

$$-\dot{\theta}^2 \sin \theta = n\ddot{\psi} \cos \psi - n\dot{\psi}^2 \sin \psi \qquad \text{or} \qquad \ddot{\psi} = \frac{n\dot{\psi}^2 \sin \psi - \dot{\theta}^2 \sin \theta}{n \cos \psi} = \frac{(\dot{\psi}^2 - \dot{\theta}^2) \sin \theta}{\sqrt{n^2 - \sin^2 \theta}}$$

Substituting the data,

$$\dot{\psi} = \frac{10(\sqrt{3}/2)}{\sqrt{(1.3)^2 - (1/2)^2}} = 7.22 \text{ rad/s}$$

$$\ddot{\psi} = \frac{[(7.22)^2 - (10)^2](1/2)}{\sqrt{(1.3)^2 - (1/2)^2}} = -20.0 \text{ rad/s}^2$$

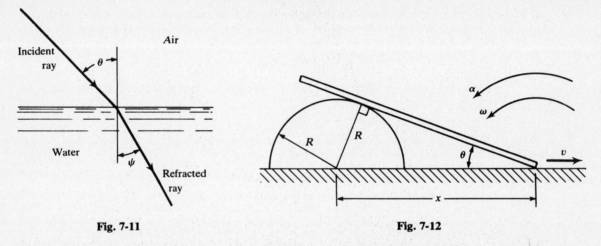

Fig. 7-11   **Fig. 7-12**

7.12. A rod leans against a cylindrical body as shown in Fig. 7-12, and its right end slides to the right on the floor with a constant speed v. Find (a) the angular speed ω and (b) the angular acceleration α, in terms of v, x, and R.

(a) From the geometry, $x = R/\sin \theta$. Also, $\omega = -\dot{\theta}$. Therefore,

$$v = \dot{x} = \frac{d}{dt}\left(\frac{R}{\sin \theta}\right) = \frac{-R\dot{\theta} \cos \theta}{\sin^2 \theta} = \frac{\omega R \cos \theta}{\sin^2 \theta}$$

$$\omega = \frac{v \sin^2 \theta}{R \cos \theta} = \frac{Rv}{x\sqrt{x^2 - R^2}}$$

(b)
$$\dot{\alpha} = \omega = \frac{d}{dt}\left(\frac{Rv}{x\sqrt{x^2 - R^2}}\right) = \frac{-Rv^2(2x^2 - R^2)}{x^2(x^2 - R^2)^{3/2}}$$

7.13. A particle of mass m moves without friction along a semicubical parabolic curve, $y^2 = ax^3$, with constant speed v. Find the reaction force of the curve on the particle.

The local radius of curvature of the curve is

$$\rho = \frac{[1 + (dy/dx)^2]^{3/2}}{d^2y/dx^2} = \frac{[1 + (9/4)ax]^{3/2}}{(3/4)a^{1/2}x^{-1/2}}$$

In this motion of the particle, the curve exerts a normal or centripetal force causing the particle momentarily to move in an arc of a circle of radius ρ (see Fig. 7-2).   Thus,

$$F = \frac{mv^2}{\rho} = \frac{3}{4} a^{1/2} x^{-1/2} \left(1 + \frac{9}{4} ax \right)^{-3/2} mv^2$$

7.14.   A particle whose mass is 2 kg moves with a speed of 44 m/s on a curved path.   The resultant force acting on the particle at a particular point of the curve is 30 N at 60° to the tangent to the curve, as shown in Fig. 7-13.   At that point, find (a) the radius of curvature of the curve and (b) the tangential acceleration of the particle.

(a)
$$\rho = \frac{mv^2}{F_n} = \frac{2(44)^2}{30 \sin 60°} = 149 \text{ m}$$

(b)
$$a_s = \frac{F_s}{m} = \frac{30 \cos 60°}{2} = 7.5 \text{ m/s}^2$$

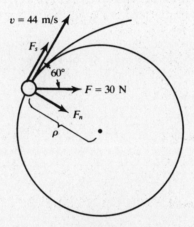

Fig. 7-13

7.15.   A bug is crawling with constant speed v along the spoke of a bicycle wheel, of radius a, while the bicycle moves down the road with constant speed V.   Find the accelerations of the bug, as observed by a man standing beside the road, along and perpendicular to the spoke of the wheel.

Choose a coordinate system that travels with the center of the wheel; accelerations in this coordinate system are the same as in the ground system, because the two systems have a constant relative velocity.   Applying the results of Problem 7.4, with $\omega = V/a$, $\dot{R} = v$, we find

$$a_r = -R \frac{V^2}{a^2} \qquad a_\theta = 2v \frac{V}{a}$$

Supplementary Problems

7.16.   Find the time it takes a body to travel around a circular path of radius r if (a) the speed is constant at v_0, (b) the tangential component a_s of the acceleration is constant and the body starts from rest.
Ans.   (a) $2\pi r/v_0$;   (b) $\sqrt{4\pi r/a_s}$

7.17.   The shaft of an engine, which is rotating at 40 rev/s, is given an angular acceleration of 1.0 rad/s² for a 10 s interval.   (a) What will be the angular displacement of the shaft during this interval?   (b) What will be the final angular speed?   Ans.   (a) 2563.3 rad;   (b) 261.3 rad/s

7.18. A wheel rotates with a constant angular acceleration of 8 rad/s²; it undergoes an angular displacement of 140 rad in a certain time interval of 5.0 s. If the wheel had started from rest, how long did it rotate before the 5.0 s interval began?   *Ans.*   1.0 s

7.19. The orbit of the earth about the sun is approximately circular, of radius 1.5×10^8 km. Treat the earth as a particle and calculate both its angular speed about the sun and its speed in orbit around the sun.   *Ans.* 2.0×10^{-7} rad/s; 30 km/s

7.20. An auto moves along a road which is curved into a circular arc of radius r. The road is banked at an angle θ to the horizontal. Find the speed of the auto for which there is no frictional force exerted on the auto by the road.   *Ans.* $v = \sqrt{rg \tan \theta}$

7.21. A particle of mass m, initially at rest, moves in a circular path of radius r. The resultant force acting on the particle has a tangential component given by $F_s = Kt$. Express the time required for the particle to return to its starting point in terms of r, K, and m.   *Ans.* $(12\pi rm/K)^{1/3}$

7.22. Because of the earth's rotation about its axis, an object located on the earth at a latitude λ moves on the circumference of a circle of radius $R \cos \lambda$, where $R \approx 6400$ km is the radius of the earth. Find the centripetal acceleration of the object in terms of λ and the period of rotation of the earth, P.   *Ans.* $(4\pi^2 R \cos \lambda)/P^2$

7.23. Refer to Problem 5.22. Calculate as a function of time the magnitude of the bead's total acceleration.

Ans. $a = \dfrac{v_0^2 \sqrt{1 + \mu_k^2}}{R[1 + (\mu_k v_0/R)t]^2}$

Chapter 8

Work, Kinetic Energy, and Power

8.1 WORK

The *work* done by a force $\mathbf{F}$ on a particle which moves from A to B along a specified path is

$$W_{AB} = \int_A^B \mathbf{F} \cdot d\mathbf{s} = \int_A^B F_s\, ds$$

where $F_s = F \cos \theta$ is the component of $\mathbf{F}$ in the direction of motion (Fig. 8-1). Since $d\mathbf{s} = dx\,\mathbf{i} + dy\,\mathbf{j} + dz\,\mathbf{k}$, we may also write

$$dW = \mathbf{F} \cdot d\mathbf{s} = F_x\,dx + F_y\,dy + F_z\,dz$$

The unit of work is the *joule* (J), where $1\,\text{J} = 1\,\text{kg} \cdot \text{m}^2/\text{s}^2 = 1\,\text{N} \cdot \text{m}$.

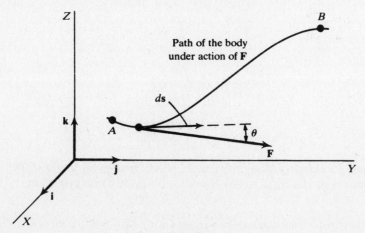

Fig. 8-1

8.2 ENERGY

Energy is that property which enables an object to do work. The energy a body has by virtue of its motion is called *kinetic energy*. A particle of mass m traveling at speed v has a kinetic energy given by

$$K = \frac{1}{2} m v^2$$

The unit of energy is the same as that of work, the joule.

8.3 WORK-ENERGY PRINCIPLE

The work-energy principle for a particle states that the work W_{AB} done by the *resultant* force acting on the particle is equal to the change in the particle's kinetic energy:

$$W_{AB} = K_B - K_A$$

71

EXAMPLE 8.1. For a particle moving at constant acceleration (under a constant force) along a straight line, AB,

$$v_B^2 = v_A^2 + 2as$$

$$\frac{1}{2}mv_B^2 = \frac{1}{2}mv_A^2 + Fs$$

$$K_B = K_A + W_{AB}$$

8.4 POWER

Power is the time rate of doing work:

$$P = \frac{dW}{dt} = \mathbf{F} \cdot \mathbf{v} = Fv\cos\theta = F_x\frac{dx}{dt} + F_y\frac{dy}{dt} + F_z\frac{dz}{dt}$$

where $\mathbf{F}$ and $\mathbf{v}$ are the instantaneous force and velocity, respectively, and θ is the angle between $\mathbf{F}$ and $\mathbf{v}$. If the power does not vary with time, $P = W/t$.

The units for power are joules per second. This combination of units is termed the *watt*, abbreviated W. Therefore,

$$1\text{ W} = 1\text{ J/s} = 1\text{ kg} \cdot \text{m}^2/\text{s}^3$$

Solved Problems

8.1. A box is dragged across a floor by a rope which makes an angle of 60° with the horizontal. The tension in the rope is 100 N while the box is dragged 15 m. How much work is done?

Only the horizontal component of the tension, $T_x = 100\cos 60°$, does work. Thus,

$$W = T_x x = (100\cos 60°)(15) = 750\text{ J}$$

8.2. A pistol fires a 3 gram bullet with a speed of 400 m/s. The pistol barrel is 13 cm long. (*a*) How much energy is given to the bullet? (*b*) What average force acted on the bullet while it was moving down the barrel? (*c*) Was this force equal in magnitude to the force of the expanding gases on the bullet?

(*a*) The kinetic energy of the bullet on leaving the barrel is

$$K_f = \frac{1}{2}mv^2 = \frac{1}{2}(0.003)(400)^2 = 240\text{ J}$$

(*b*) The work done on the bullet is equal to the change in its kinetic energy,

$$W = \bar{F}x = K_f - K_i$$

where $\bar{F}$ is the average (with respect to distance x) force exerted on the bullet. Thus,

$$\bar{F} = \frac{K_f - K_i}{x} = \frac{240 - 0}{0.13} = 1846\text{ N}$$

since the bullet was at rest initially.

(*c*) No, since there are frictional forces in play as the bullet moves down the barrel.

8.3.   A 4 kg rifle fires a 6 gram bullet with a speed of 500 m/s.   What kinetic energy is acquired (*a*) by the bullet?   (*b*) by the rifle?   (*c*) Find the ratio of the distance the rifle moves backward while the bullet is in the barrel to the distance the bullet moves forward.   See Fig. 8-2.

(*a*)   The kinetic energy of the bullet is

$$K_b = \frac{1}{2} m_b v_b^2 = \frac{1}{2} (0.006)(500)^2 = 750 \text{ J}$$

(*b*)   By conservation of momentum, since no external forces act on the system (the bullet and rifle),

$$\mathbf{P}_i = \mathbf{P}_f$$

$$0 = m_r v_r + m_b v_b$$

$$v_r = -\frac{m_b}{m_r} v_b$$

The kinetic energy of the rifle then is

$$K_r = \frac{1}{2} m_r \frac{m_b^2}{m_r^2} v_b^2 = \frac{1}{2} \frac{(0.006)^2}{4} (500)^2 = 1.125 \text{ J}$$

(*c*)   From (*b*),

$$0 = m_r v_r + m_b v_b = m_r \frac{\Delta x_r}{\Delta t} + m_b \frac{\Delta x_b}{\Delta t} \qquad \text{or} \qquad 0 = m_r \Delta x_r + m_b \Delta x_b$$

which expresses that the center of mass of the system remains at rest.   Solving,

$$\frac{|\Delta x_r|}{|\Delta x_b|} = \frac{m_b}{m_r} = \frac{6}{4000} = \frac{1}{667}$$

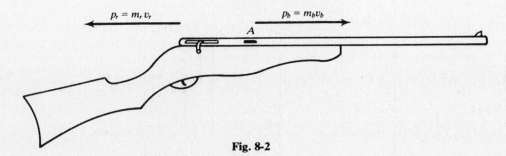

$$p_r = m_r v_r \qquad\qquad p_b = m_b v_b$$
$$A$$

Fig. 8-2

8.4.   A bullet having a speed of 153 m/s crashes through a plank of wood.   After passing through the plank, its speed is 130 m/s.   Another bullet, of the same mass and size but traveling at 92 m/s, is fired at the plank.   What will be this second bullet's speed after tunneling through?   Assume that the resistance of the plank is independent of the speed of the bullet.

The plank does the same amount of work on the two bullets, and therefore decreases their kinetic energies equally.

$$\frac{1}{2} m (153)^2 - \frac{1}{2} m (130)^2 = \frac{1}{2} m (92)^2 - \frac{1}{2} mv^2$$

$$v^2 = 1955$$

$$v = 44.2 \text{ m/s}$$

8.5.   A carnival hand lets fall a heavy mallet of mass M from a height y upon the top of a tent stake of mass m, and drives it into the ground a distance d.   Find (*a*) the resistance of the ground, assuming it to be constant and the stake and mallet to stay together on impact; (*b*) the time the stake is in motion; and (*c*) the kinetic energy lost at impact.   See Fig. 8-3.

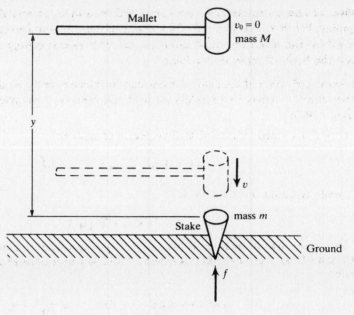

Fig. 8-3

(a) The speed of the mallet on just striking the stake is $v = \sqrt{2gy}$. Momentum is conserved at the instant of collision, so that

$$Mv = (M+m)v'$$

where v' is the speed of the stake-plus-mallet just after impact.

The resultant (upward) force on the stake-plus-mallet is $\Sigma F = f - (M+m)g$, where f is the resistive force of the ground. Then the work-energy principle gives

$$\Delta K = \left(\sum F\right)(-d)$$

$$0 - \frac{1}{2}(M+m)v'^2 = [f - (M+m)g](-d)$$

$$f = (M+m)g + (M+m)\frac{v'^2}{2d}$$

Substituting $v' = Mv/(M+m)$ and $v^2 = 2gy$,

$$f = (M+m)g + (M+m)\frac{M^2v^2}{2(M+m)^2d} = (M+m)g + \frac{M^2}{M+m}\frac{gy}{d}$$

(b) Now, $\Sigma F = \Delta p / \Delta t$, where Δt is the time interval from just after (or just before, since momentum is conserved) the impact to cessation of motion of the stake-plus-mallet. Then,

$$\Delta t = \frac{\Delta p}{\sum F} = \frac{0 - [-(M+m)v']}{f - (M+m)g} = \frac{Mv}{\left(\dfrac{M^2}{M+m}\right)\dfrac{gy}{d}} = \frac{M+m}{M}d\sqrt{\frac{2}{gy}}$$

(c) Just before impact the kinetic energy of the system was $\frac{1}{2}Mv^2$; just afterwards it was

$$\frac{1}{2}(M+m)v'^2 = \frac{1}{2}\frac{M^2}{M+m}v^2$$

So the amount lost by the mallet was

$$\frac{1}{2}Mv^2 - \frac{1}{2}\frac{M^2}{M+m}v^2 = \frac{m}{M+m}\left(\frac{1}{2}Mv^2\right)$$

or the fraction lost was $m/(M+m)$.

8.6. Two bodies, of masses m and $2m$, are connected by a light, inextensible cord passing over a smooth pulley, Fig. 8-4. At the end of 4 s a body of mass m is suddenly joined to the ascending body. Find (a) the resulting speed and (b) how much kinetic energy is lost by the descending body when the body of mass m is added.

Because the only effect of the pulley is to change the direction of the tension in the cord, the system may be conveniently analyzed as a single body having mass $m_A + m_B$ acted on by the single force $w_A - w_B$ [Fig. 8-4(c)].

(a) For $t < 4$ s, $m_A = 2m_B = 2m$, and so the equation of motion is

$$mg = 3ma \qquad \text{or} \qquad a = \frac{g}{3}$$

The speed just before $t = 4$ s is then

$$v = 0 + at = \frac{4g}{3} \text{ (m/s)}$$

We assume that the addition of mass at $t = 4$ s is equivalent to a sticky collision between the system and a body of mass m which is *at rest*. Then, by conservation of momentum, the new speed is given by

$$3mv + 0 = 4mv' \qquad \text{or} \qquad v' = \frac{3}{4}v = g$$

(b) The loss in kinetic energy of A is

$$\frac{1}{2}(2m)v^2 - \frac{1}{2}(2m)v'^2 = \frac{7}{9}mg^2 \text{ (J)}$$

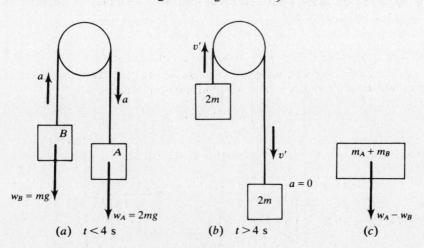

(a) $t < 4$ s   (b) $t > 4$ s   (c)

Fig. 8-4

8.7. In Fig. 8-5, evaluate the work done by the weight mg acting on a particle of mass m, as the particle is moved (by the application of other forces) from: (a) A to B; (b) B to A; (c) A to B to C; (d) A to C directly; (e) A to B to C to A.

(a) For the path AB, $m\mathbf{g}$ is in the opposite direction to the direction of motion. Thus, $W_{AB} = -mgy$.

(b) $$W_{BA} = -W_{AB} = mgy$$

(c) $$W_{ABC} = W_{AB} + W_{BC} = -mgy + 0 = -mgy$$

(d) The component of force in the direction of motion is $-mg \cos \phi$ and $\overline{AC} = \Delta s = y/(\cos \phi)$.

$$W_{AC} = (-mg \cos \phi)\left(\frac{y}{\cos \phi}\right) = -mgy$$

(e) $$W_{ABCA} = W_{AB} + W_{BC} + W_{CA} = -mgy + 0 + mgy = 0$$

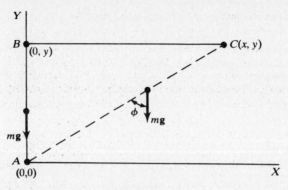

Fig. 8-5

8.8. Find the work done by a force given in SI units by $F_x = 5.0x - 4.0$, when this force acts on a particle that moves from $x = 1.0$ m to $x = 3.0$ m.

$$W = \int F_x \, dx = \int_{1.0}^{3.0} (5.0x - 4.0) \, dx = [2.5x^2 - 4.0x]_{1.0}^{3.0} = 12 \text{ J}$$

8.9. A body of mass m, after falling from rest through a distance y, begins to raise a body of mass M ($M > m$) that is connected to it by means of a light inextensible string passing over a fixed smooth pulley. (a) Find the time it will take for the body of mass M to return to its original position. (b) Find the fraction of kinetic energy lost when the body of mass M is jerked into motion. See Fig. 8-6.

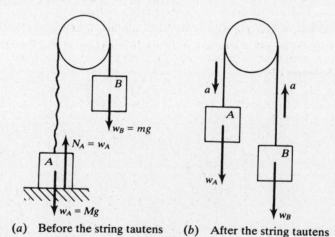

(a) Before the string tautens   (b) After the string tautens

Fig. 8-6

This problem is basically the same as Problem 8.6: only *internal* forces are exerted as the mass of the system is increased, so that momentum is conserved for the instant.

(a) The speed of body B just before the string becomes taut is $v = \sqrt{2gy}$, and its momentum, which is the momentum of the system, is mv. Immediately after the string becomes taut, the speed of the system (the common speed of the two bodies) is v'; by momentum conservation,

$$mv = (M + m)v' \qquad \text{or} \qquad v' = \frac{m}{M + m} v$$

Moreover, the acceleration of the system is given by

$$\sum F = mg - Mg = (M + m)a \qquad \text{or} \qquad a = -\frac{M-m}{M+m}g$$

where the positive direction is that used above for the momentum.

Applying the constant-acceleration formula $s = v_0t + \frac{1}{2}at^2$, we find that the system returns to its original position when

$$0 = v't + \frac{1}{2}at^2 \qquad \text{or} \qquad t = -\frac{2v'}{a} = \frac{2m}{M-m}\sqrt{\frac{2y}{g}}$$

(b) The fractional loss of kinetic energy is

$$\frac{\frac{1}{2}mv^2 - \frac{1}{2}(M+m)v'^2}{\frac{1}{2}mv^2} = \frac{M}{M+m}$$

8.10. A smooth track in the form of a quarter-circle of radius 6 m lies in the vertical plane (see Fig. 8-7). A particle of weight 4 N moves from P_1 to P_2 under the action of forces $\mathbf{F}_1$, $\mathbf{F}_2$, and $\mathbf{F}_3$. Force $\mathbf{F}_1$ is always toward P_2 and is always 20 N in magnitude; force $\mathbf{F}_2$ always acts horizontally and is always 30 N in magnitude; force $\mathbf{F}_3$ always acts tangentially to the track and is of magnitude $(15 - 10s)$ N when s is in m.   If the particle has speed 4 m/s at P_1, what will its speed be at P_2?

The work done by $\mathbf{F}_1$ is:

$$W_1 = \int_{P_1}^{P_2} F_1 \cos \theta \, ds$$

From Fig. 8-7, $ds = (6 \text{ m})\,d(-2\theta) = -12\,d\theta$, and $F_1 = 20$.   Hence,

$$W_1 = -240 \int_{\pi/4}^{0} \cos \theta \, d\theta = 240 \sin \frac{\pi}{4} = 120\sqrt{2} \text{ J}$$

The reader may notice that $W_1 = (20 \text{ N})(6\sqrt{2} \text{ m})$, just as if the chord P_1P_2, and not the circular arc, were the path of integration.   The reason for this is that $\mathbf{F}_1$ is a *conservative* force (see Section 9.1).   For such a force, the work is the same along all paths joining two given points.

The work done by $\mathbf{F}_3$ is:

$$W_3 = \int F_3 \, ds = \int_0^{6(\pi/2)} (15 - 10s) \, ds = [15s - 5s^2]_0^{3\pi} = -302.8 \text{ J}$$

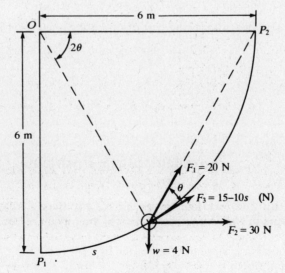

Fig. 8-7

To calculate the work done by $\mathbf{F}_2$ and by $\mathbf{w}$, it is convenient to take the projection of the path in the direction of the force, instead of vice versa. Thus,

$$W_2 = F_2(\overline{OP_2}) = 30(6) = 180 \text{ J}$$

and

$$W = (-w)(\overline{P_1O}) = (-4)(6) = -24 \text{ J}$$

The total work done is

$$W_1 + W_3 + W_2 + W = 23 \text{ J}$$

Then, by the work-energy principle,

$$K_{P_2} - K_{P_1} = 23 \text{ J}$$

$$\frac{1}{2}\left(\frac{4}{9.8}\right)v_2^2 - \frac{1}{2}\left(\frac{4}{9.8}\right)(4)^2 = 23$$

$$v_2 = 11.3 \text{ m/s}$$

8.11. An engine pumps water continuously through a hose. If the speed with which the water passes through the hose nozzle is v, and if k is the mass per unit length of the water jet as it leaves the nozzle, find the rate at which kinetic energy is being imparted to the water.

During a small elapsed time, Δt, the mass of water ejected is $k(v\,\Delta t)$. The kinetic energy of this water is $\frac{1}{2}(kv\,\Delta t)v^2$. The rate at which kinetic energy is imparted is then

$$\lim_{\Delta t \to 0} \frac{\frac{1}{2}kv^3\,\Delta t}{\Delta t} = \frac{1}{2}kv^3$$

8.12. The 4 metric ton (4000 kg) hammer of a pile driver is lifted 1.0 m in 2.0 s. What power does the engine furnish to the hammer? Assume there is no acceleration of the hammer while it is being lifted.

$$P = \frac{W}{t} = \frac{Fs}{t} = \frac{mgs}{t} = \frac{(4.0 \times 10^3)(9.8)(1.0)}{2.0} = 19.6 \text{ kW}$$

8.13. While a boat is being towed at a speed of 20 m/s, the tension in the towline is 6 kN. What is the power supplied to the boat by the towline?

$$P = Fv = (6 \times 10^3)(20) = 120 \text{ kW}$$

8.14. At 8¢ per kilowatt hour, what is the cost of operating a 5.0 hp motor for 2.0 h? (1 hp = 746 W.)

$$\text{cost} = (0.08)(5.0)(746 \times 10^{-3})(2.0) = \$0.60$$

Supplementary Problems

8.15. A girl exerts a 20.0 N horizontal force as she pushes a box across a horizontal tabletop through a distance of 1.80 m. How much work is done on the box by this force?   *Ans.*   36.0 J

8.16. The force acting on a suspended particle is given by $F_y = -ky + mg$. Find the work done by this force as the particle moves from y_1 to y_2.

Ans. $-\dfrac{k}{2}(y_2^2 - y_1^2) + mg(y_2 - y_1)$

8.17. A railroad car is accelerated from rest until it has a kinetic energy of 0.5 MJ. (*a*) What work is done by the force exerted on the car? (*b*) If the force has a constant value of 1 kN and is in the direction of motion, find the distance traveled by the car while attaining this kinetic energy.
   Ans. (*a*) 0.5 MJ; (*b*) 500 m

8.18. The kinetic energy of a particle is given by $K = \frac{1}{2}mv^2$ and the momentum is given by $p = mv$. By using differential calculus show that:

$$(a) \quad p = \frac{dK}{dv} \qquad (b) \quad Fv = \frac{dK}{dt} \qquad (c) \quad Fv = \frac{1}{2m}\frac{d(p^2)}{dt}$$

(F = force, t = time).

8.19. A block slides on a rough plane along a path of length Δs. Find the work done by (*a*) the normal component **N** of the force exerted on the block by the plane, (*b*) the force of kinetic friction **f** exerted on the block by the plane.   *Ans.* (*a*) 0; (*b*) $-f\,\Delta s$

8.20. A constant horizontal force of 900 N pushes on a box as it slides up a plane inclined at an angle of 60° to the horizontal. What is the work done by this force during a 3.0 m displacement of the box along the inclined plane?   *Ans.* 1.35 kJ

8.21. What power must a motor supply to lift a 70 kg man at a constant speed of 1.0 m/s?   *Ans.* 686 W

8.22. In Problem 8.10, does the particle move *solely* under the action of the four given forces? Explain.
   Ans. No; additional radial forces must operate to provide the necessary centripetal component. For instance, at P_1 the necessary centripetal force is

$$\frac{mv^2}{r} = \frac{(4/9.8)(4)^2}{6} = 1.09 \text{ N}$$

whereas $F_1 \cos 45° - w = 10.14$ N.

Chapter 9

Potential Energy and Energy Conservation

9.1 CONSERVATIVE FORCES

A force **F** that can be derived from a function of position U by

$$F_s = -\frac{dU}{ds} \qquad (9.1)$$

is called a *conservative* force. Here, F_s is the component of the force in the direction of $d\mathbf{s}$, an arbitrary small displacement away from the point of observation. In vector form,

$$dU = -\mathbf{F} \cdot d\mathbf{s} \qquad (9.2)$$

The work done by a conservative force is the negative of the change in the associated function U:

$$W_{AB} = \int_A^B F_s \, ds = \int_A^B -\frac{dU}{ds} \, ds = -(U_B - U_A) \qquad (9.3)$$

This work is independent of the path joining the endpoints A and B; for a closed path, this work is zero. The work done by a *nonconservative* force (such as the force of kinetic friction) depends on the path taken between the endpoints.

9.2 POTENTIAL ENERGY

A body that is acted upon by a conservative force has energy by virtue of its position. This energy, called *potential energy*, is measured by the function U associated with the conservative force.

EXAMPLE 9.1. The Hooke's law force, $F_x = -kx$, is conservative. Thus a Hookean spring stretched a distance x has potential energy $U = \frac{1}{2}kx^2$. Another conservative force is a uniform gravitational field g directed vertically downward; the gravitational potential energy of a particle of mass m at altitude y is $U = mgy$.

9.3 CONSERVATION OF ENERGY

The *law of conservation of energy* for a particle is: For any two points A and B on the particle's path,

$$K_A + U_A = K_B + U_B \qquad \text{or} \qquad \Delta K + \Delta U = 0 \qquad (9.4)$$

provided no work is done on the particle by nonconservative forces.

When a force of kinetic friction does work of magnitude $|W_f|$ as a particle moves from A to B, we have

$$K_A + U_A = K_B + U_B + |W_F|$$

and we say that energy has been dissipated by friction. More generally, if the total work done on the particle from A to B by forces not accounted for in U is W', then

$$K_A + U_A = K_B + U_B - W' \qquad \text{or} \qquad \Delta K + \Delta U = W' \qquad (9.5)$$

Solved Problems

9.1. A box weighing 200 N is dragged up an incline 10 m long and 3 m high. The average force (parallel to the plane) is 120 N. (*a*) How much work is done? (*b*) What is the change in the potential energy of the box? in its kinetic energy? (*c*) What is the frictional force on the box?

(*a*) The work done by the dragging force is

$$W_{i \to f} = \bar{F}s = (120)(10) = 1200 \text{ J}$$

(*b*) The change in potential energy is

$$\Delta U = U_f - U_i = wh - 0 = (200)(3) = 600 \text{ J}$$

(*c*) Because the box starts and finishes at rest, $\Delta K = 0$. The total work done by nonconservative forces is $W_{i \to f} + W$, where W is the work done on the box by friction. Then

$$\Delta K + \Delta U = W_{i \to f} + W$$

$$0 + 600 = 1200 + W$$

$$W = -600 \text{ J}$$

But $W = -fs$ (the friction force f opposes the motion of the box), and so

$$f = \frac{-600}{-10} = 60 \text{ N}$$

9.2. A rock weighing 20 N falls from a height of 16 m and sinks 0.6 m into the ground. From energy considerations, find the average force f between the rock and the ground as the rock sinks. See Fig. 9-1.

Between A and C, nonconservative work $W' = -fh'$ is done on the rock.

$$\Delta K + \Delta U = W'$$

$$0 + [mg(-h') - mgh] = -fh'$$

$$f = \frac{mg(h + h')}{h'} = \frac{20(16.6)}{0.6} = 553.33 \text{ N}$$

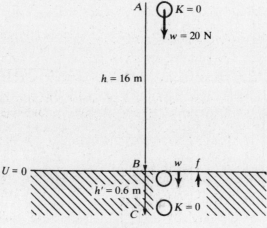

Fig. 9-1

9.3. A load W is suspended from a construction supply car by a cable of length d. The car and load are moving at a constant speed v_0. The car is stopped by a bumper and the load on the cable swings out, as shown in Fig. 9-2(b). (a) What is the angle through which the load swings? (b) If the angle is 60° and $d = 5$ m, what was the initial speed of the car?

(a) The cable does no work on the load, so the load's energy is conserved.

$$K_i + U_i = K_f + U_f$$

$$\frac{1}{2}\frac{W}{g}v_0^2 + 0 = 0 + W(d - d\cos\theta)$$

$$v_0^2 = 2gd(1 - \cos\theta) = 4gd\sin^2\frac{\theta}{2}$$

$$\theta = 2\arcsin\left(\frac{v_0}{2\sqrt{gd}}\right)$$

(b)
$$v_0 = 2\sqrt{gd}\sin\frac{\theta}{2} = 2\sqrt{(9.8)(5)}\left(\frac{1}{2}\right) = 7 \text{ m/s}$$

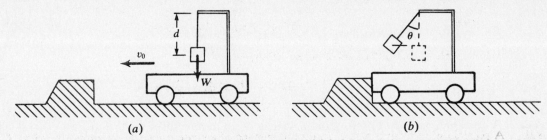

(a)   (b)

Fig. 9-2

9.4. A force $\mathbf{F} = x^2y^2\mathbf{i} + x^2y^2\mathbf{j}$ (N) acts on a particle which moves in the XY plane. (a) Determine if $\mathbf{F}$ is conservative and (b) find the work done by $\mathbf{F}$ as it moves the particle from A to C (Fig. 9-3) along each of the paths ABC, ADC, and AC.

(a) If $\mathbf{F}$ is conservative, then

$$F_x = -\frac{\partial U}{\partial x} \qquad F_y = -\frac{\partial U}{\partial y}$$

and so
$$\frac{\partial F_x}{\partial y} = -\frac{\partial^2 U}{\partial y\,\partial x} = -\frac{\partial^2 U}{\partial x\,\partial y} = \frac{\partial F_y}{\partial x}$$

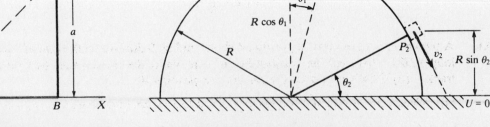

Fig. 9-3   **Fig. 9-4**

But, for the given force,

$$\frac{\partial F_x}{\partial y} = 2x^2 y \qquad \frac{\partial F_y}{\partial x} = 2xy^2$$

Hence the given force is not conservative.

(b) The work done by **F** is given by

$$W = \int \mathbf{F} \cdot d\mathbf{s} = \int (x^2 y^2 \mathbf{i} + x^2 y^2 \mathbf{j}) \cdot (dx\,\mathbf{i} + dy\,\mathbf{j}) = \int x^2 y^2\,dx + \int x^2 y^2\,dy$$

Along AB, $y = 0$ and so $W_{AB} = 0$. Along BC, $dx = 0$ and

$$W_{BC} = \int_0^a a^2 y^2\,dy = \frac{a^5}{3}$$

Thus
$$W_{ABC} = W_{AB} + W_{BC} = \frac{a^5}{3}\ (\text{J})$$

Along AD, $x = 0$ and so $W_{AD} = 0$. Along DC, $dy = 0$ and

$$W_{DC} = \int_0^a x^2 a^2\,dx = \frac{a^5}{3}$$

Thus
$$W_{ADC} = W_{AD} + W_{DC} = \frac{a^5}{3}\ (\text{J})$$

Along AC, $x = y$ and $dx = dy$. Thus,

$$W_{AC} = 2\int_0^a x^4\,dx = \frac{2a^5}{5}\ (\text{J})$$

9.5. A particle moves from rest at P_1 on the surface of a smooth circular cylinder of radius R (Fig. 9-4). At P_2 the particle leaves the cylinder. Find the equation relating θ_1 and θ_2 as shown.

As the normal force does no work on the particle, its energy is conserved.

$$(K + U)_{P_1} = (K + U)_{P_2}$$

$$0 + mgR\cos\theta_1 = \frac{1}{2}mv_2^2 + mgR\sin\theta_2$$

$$v_2^2 = 2gR(\cos\theta_1 - \sin\theta_2)$$

At P_2, the normal force exerted by the surface vanishes, leaving the radial component of the particle's weight, $mg\sin\theta_2$, as the instantaneous centripetal force. Then, by Newton's second law,

$$mg\sin\theta_2 = \frac{mv_2^2}{R}$$

$$mg\sin\theta_2 = \frac{m}{R}[2gR(\cos\theta_1 - \sin\theta_2)]$$

$$\sin\theta_2 = 2\cos\theta_1 - 2\sin\theta_2$$

$$\sin\theta_2 = \frac{2}{3}\cos\theta_1$$

9.6. A toy car starts from rest at position *1* shown in Fig. 9-5(a) and rolls without friction along the loop *12324*. (a) Find the smallest height h at which the car can start without falling off the track. (b) What speed does the car have at position *4*?

(a) When h has its critical value, the car will just lose contact with the track at position *3*. Then, at *3*,

$$mg = \frac{mv_3^2}{r} \qquad \text{or} \qquad v_3^2 = gr$$

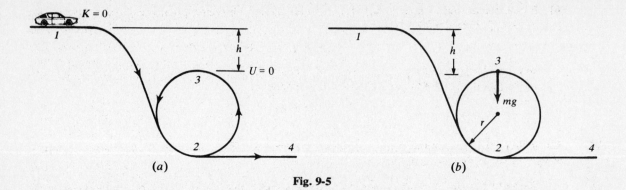

Fig. 9-5

Further, the potential energy of the car at position *1* is the same as the kinetic energy of the car at position *3*:

$$mgh = \frac{1}{2} mv_3^2$$

$$h = \frac{v_3^2}{2g} = \frac{r}{2}$$

(b) Applying conservation of energy between positions *1* and *4*,

$$0 + mgh = \frac{1}{2} mv_4^2 + mg(-2r)$$

$$v_4^2 = 2g(h + 2r) = 5gr$$

$$v_4 = \sqrt{5gr}$$

9.7. An auto, starting from rest, reaches a kinetic energy K by accelerating without skidding along a horizontal road. (*a*) Forgetting about air resistance, find the work done by the external forces which accelerate the car. (*b*) Is the result in (*a*) consistent with the conservation of energy?

(*a*) The external forces that act on the auto are shown in Fig. 9-6. The net external force is $\mathbf{f}_1 + \mathbf{f}_2$ (the sum of the forces of static friction exerted on the tires by the road). This net force accelerates the auto, i.e.

$$\mathbf{f}_1 + \mathbf{f}_2 = M\mathbf{a}_{cm}$$

where $\mathbf{a}_{cm}$ is the acceleration of the center of mass of the auto. Since the portions of the tires instantaneously in contact with the roadway are at rest relative to the road (no skidding), the forces $\mathbf{f}_1$ and $\mathbf{f}_2$ acting on these parts do no work. Thus, $W_{ext} = 0$.

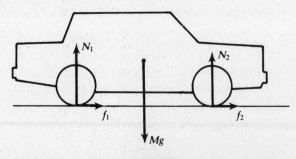

Fig. 9-6

(b) If no external work is done on the system, where does its kinetic energy K come from? We might speak of "internal work" W', such that

$$\Delta K = K - 0 = W'$$

More likely, we would identify W' with the decrease in an "internal energy" Φ (the energy content of the gasoline) and write the conservation of energy as

$$\Delta K + \Delta \Phi = 0$$

9.8. An artillery piece, of mass m_1, fires a shell, of mass m_2. Find the ratio of the kinetic energies of the artillery piece and of the shell.

The system made up of the artillery piece and the shell is initially at rest, so that by conservation of momentum

$$\mathbf{p}_1 + \mathbf{p}_2 = 0 \qquad \text{or} \qquad \mathbf{p}_1 = -\mathbf{p}_2$$

Then

$$\frac{K_1}{K_2} = \frac{p_1^2/2m_1}{p_2^2/2m_2} = \frac{m_2}{m_1}$$

9.9. Calculate the forces $F(y)$ associated with the following one-dimensional potential energies: (a) $U = -\omega y$, (b) $U = ay^3 - by^2$, (c) $U = U_0 \sin \beta y$.

(a)
$$F = -\frac{dU}{dy} = \omega$$

(b)
$$F = -\frac{dU}{dy} = -3ay^2 + 2by$$

(c)
$$F = -\frac{dU}{dy} = -\beta U_0 \cos \beta y$$

9.10. How much work is done in moving a body of mass $1.0\,\text{kg}$ from an elevation of $2\,\text{m}$ to an elevation of $20\,\text{m}$, (a) by the gravitational field of the earth? (b) by an external agent?

(a)
$$W = -\Delta U = -[(1.0)(9.8)(20) - (1.0)(9.8)(2)] = -176.4 \text{ J}$$

The work is negative because the force opposes the motion.

(b)
$$W' = \Delta K + \Delta U = \Delta K + 176.4 \text{ J}$$

Unlike the gravitational work, the external work depends on the speed of the body. If the body is unaccelerated ($\Delta K = 0$), then $W' = 176.4$ J, the negative of the gravitational work.

9.11. Consider, in Fig. 9-7, a pendulum of length ℓ suspended at a distance $\ell - \ell_1$ vertically above a small peg C. Suppose that the bob is initially displaced by an angle β_0 and then released from rest. Find the speed v of the bob at the instant shown in the figure, when it is moving in a circular path of radius ℓ_1 and has an angular displacement θ with respect to the vertical.

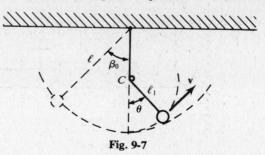

Fig. 9-7

The tension that the string exerts on the bob carries out no work on the bob throughout its motion. The weight of the bob is the only force doing work. Take the lowest point in the path as the reference point of gravitational potential energy. The potential energy of the bob at the instant shown is $mg\ell_1(1 - \cos \theta)$. Since it starts out at rest and its initial potential energy is $mg\ell(1 - \cos \beta_0)$, the conservation of energy expression is

$$0 + mg\ell(1 - \cos \beta_0) = \frac{1}{2} mv^2 + mg\ell_1(1 - \cos \theta)$$

$$v = \{2g[\ell(1 - \cos \beta_0) - \ell_1(1 - \cos \theta)]\}^{1/2}$$

9.12. A 20 gram bullet is fired horizontally with a speed of 600 m/s into a 7 kg block sitting on a tabletop; the bullet (b) lodges in the block (B). If the coefficient of kinetic friction between the block and the tabletop is 0.4, find the distance the block will slide.

By conservation of momentum, the momentum of the system block-plus-bullet just after the interaction is $p = m_b v_{0b}$; hence the kinetic energy of the system is

$$K = \frac{p^2}{2(m_B + m_b)} = \frac{m_b^2 v_{0b}^2}{2(m_B + m_b)}$$

The friction force does work $W_f = -fs = -\mu_k(m_B + m_b)gs$ in stopping the block. Hence

$$\Delta K = W_f$$

$$0 - \frac{m_b^2 v_{0b}^2}{2(m_B + m_b)} = -\mu_k(m_B + m_b)gs$$

$$s = \frac{1}{2\mu_k g}\left(\frac{m_b v_{0b}}{m_B + m_b}\right)^2 = \frac{1}{2(0.4)(9.8)}\left[\frac{(0.020)(600)}{7.020}\right]^2 = 0.372 \text{ m}$$

Supplementary Problems

9.13. Consider an elastic (energy is conserved), one-dimensional collision between a particle of mass m_1 and an initially stationary particle of mass m_2. What is the fractional decrease in the kinetic energy of the first particle?

Ans. $\dfrac{K_{1i} - K_{1f}}{K_{1i}} = \dfrac{4m_1 m_2}{(m_1 + m_2)^2}$

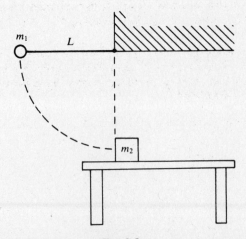

Fig. 9-8

9.14. A ball of mass m_1 is fastened to a cord of length L (see Fig. 9-8) and released when the cord is horizontal. When the cord is vertical the ball makes an elastic (energy conserved) collision with a block of mass m_2, which is resting on a smooth table. Find the speeds of the ball and the block just after the collision.

Ans. $\sqrt{2gL}\ \dfrac{|m_1 - m_2|}{m_1 + m_2}$; $\sqrt{2gL}\ \dfrac{2m_1}{m_1 + m_2}$

9.15. In Problem 9.6, what should be the starting height so that the car pushes against the top of the track with a force equal to its weight?   Ans. $h = r/2$

9.16. A small block of mass m slides down a quarter-circular path of radius R cut into a large block of mass M (Fig. 9-9). The large block is at rest on a table initially, and both objects move without friction. Assume that both blocks are sitting still initially and that the block of mass m starts from the top of the path. If $m = M$, find the speed of the small block when it leaves the larger block.
Ans. $v = \sqrt{gR}$

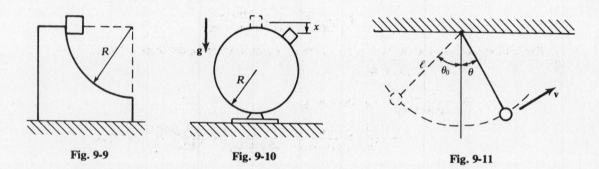

Fig. 9-9   Fig. 9-10   Fig. 9-11

9.17. A block slides from rest from the top of a frictionless sphere of radius R (Fig. 9-10). How far below the top of the sphere does the block lose contact with the sphere?   Ans. $x = R/3$

9.18. Derive the answer to Problem 9.17 from the result of Problem 9.5. Why is the difference between cylinder and sphere irrelevant in this case?

9.19. A pendulum consists of a light rigid rod of length ℓ, pivoted at one end and with a mass m attached at the other end. The pendulum is released from rest at angle θ_0, as shown in Fig. 9-11. What is the speed of mass m when the rod is at angle θ?   Ans. $v = [2g\ell(\cos\theta - \cos\theta_0)]^{1/2}$

9.20. A block of mass m slides down a plane that makes an angle θ with the horizontal. Find the speed of the block after it has descended through height h, assuming that it starts from rest and that the coefficient of sliding friction is μ.   Ans. $v = [2gh(1 - \mu\cot\theta)]^{1/2}$

9.21. A Hookean spring, of mass M and stiffness constant k, is compressed vertically against the floor to a length $L/2$, where L is its natural length. (a) When the spring is released, how much work is done on it by the floor? (b) How high does the center of mass of the spring rise above its original position?
Ans.   (a) zero;   (b) $h = kL^2/8Mg$

Chapter 10

Statics of Rigid Bodies

A *rigid body* is one in which each pair of points maintains a constant separation. The center of mass, as defined in (*5.1*), is, in the case of a rigid body, fixed with respect to the body. (This is true whether or not the center of mass is itself a point of the body.)

In actual calculations of the center of mass, it is often simplest to find the components x_{cm}, y_{cm}, z_{cm} of $\mathbf{r}_{\text{cm}}$:

$$x_{\text{cm}} = \frac{\sum m_i x_i}{\sum m_i} \qquad y_{\text{cm}} = \frac{\sum m_i y_i}{\sum m_i} \qquad z_{\text{cm}} = \frac{\sum m_i z_i}{\sum m_i}$$

For a continuous body, the sums are replaced by integrals:

$$x_{\text{cm}} = \frac{\int x\, dm}{\int dm} \qquad y_{\text{cm}} = \frac{\int y\, dm}{\int dm} \qquad z_{\text{cm}} = \frac{\int z\, dm}{\int dm}$$

10.1 TORQUE

Consider a force **F** lying in a plane and an axis perpendicular to the plane through the point O of the plane (Fig. 10-1). The perpendicular distance from the line of action of **F** to the axis is called the *moment arm, D*. The magnitude τ of the *torque* of **F** about the axis is defined to be the *product of the magnitude of the force and the moment arm*; that is,

$$\tau = FD$$

A torque on any point of a rigid body is experienced by the body as a whole; the torque measures the effectiveness of the given force in producing rotation of the body about the given axis. The unit for torque is the newton-meter (N · m).

Torque can have either a clockwise or a counterclockwise sense. To distinguish between the two possible senses of rotation we adopt the convention that a *counterclockwise torque is positive* and that a *clockwise torque is negative*.

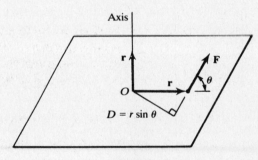

Fig. 10-1

A more compact expression for the torque of a force is

$$\boldsymbol{\tau} = \mathbf{r} \times \mathbf{F}$$

where the magnitude of the torque is $\tau = rF \sin \theta = FD$ and where the axis of induced rotation (the direction of $\boldsymbol{\tau}$) is perpendicular to the plane determined by $\mathbf{r}$ and $\mathbf{F}$.

EXAMPLE 10.1. Let us compute the torque about some fixed point O that a body of mass M experiences in a uniform gravitational field $\mathbf{g}$.

An element dm of the body has weight $dm\, \mathbf{g}$; so the differential torque is $d\boldsymbol{\tau} = \mathbf{r} \times dm\, \mathbf{g}$, where $\mathbf{r}$ is the radius vector from O to the element dm. Integrating over the entire body,

$$\boldsymbol{\tau} = \int \mathbf{r} \times dm\, \mathbf{g} = \left(\int \mathbf{r} dm \right) \times \mathbf{g} = (M\mathbf{r}_{cm}) \times \mathbf{g} = \mathbf{r}_{cm} \times M\mathbf{g}$$

Thus, the gravitational torque may be obtained by considering the entire weight, $M\mathbf{g}$, of the body to act at the center of mass.

10.2 CONDITIONS FOR EQUILIBRIUM

A rigid body is *in equilibrium* if its center of mass is at rest or moving with constant velocity ($\mathbf{v}_{cm} = $ constant, $\mathbf{a}_{cm} = 0$) and there is no rotation of the body about any axis. For equilibrium, two conditions must be satisfied. The first condition is that $\Sigma \mathbf{F}_{ext} = 0$, or

> algebraic sum of X components of external forces = 0
> algebraic sum of Y components of external forces = 0
> algebraic sum of Z components of external forces = 0

The second condition of equilibrium is $\Sigma \boldsymbol{\tau}_{ext} = 0$, or

> algebraic sum of X components of external torques = 0
> algebraic sum of Y components of external torques = 0
> algebraic sum of Z components of external torques = 0

In Problem 10.6 it is shown that if the first condition is met, then the second condition may be applied with the torques computed about any chosen point. The point is usually taken to lie on the line of action of an unknown force, thereby eliminating the torque due to that force.

Many problems involve external forces that are all coplanar. With that plane taken as the XY plane, the first condition reduces to two equations and the second condition to a single equation (all torques are parallel to the Z axis).

Solved Problems

10.1. Four particles, of masses 1 kg, 2 kg, 3 kg, and 4 kg, are at the vertices of a rectangle of sides a and b (see Fig. 10-2). If $a = 1$ m and $b = 2$ m, find the location of the center of mass.

Set up a Cartesian coordinate system in the plane of the rectangle, with the origin at the 1 kg particle. The coordinates of the four particles are, in increasing order of their masses, $(0,0)$, $(a,0)$, (a,b), and $(0,b)$. The total mass M is

$$M = m_1 + m_2 + m_3 + m_4 = 10 \text{ kg}$$

Substituting into the center-of-mass equations, we find

$$x_{cm} = \frac{\sum m_i x_i}{\sum m_i} = \frac{1}{10}(0 + 2a + 3a + 0) = 0.5 \text{ m} \qquad y_{cm} = \frac{\sum m_i y_i}{\sum m_i} = \frac{1}{10}(0 + 0 + 3b + 4b) = 1.4 \text{ m}$$

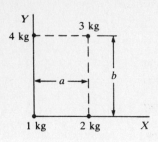

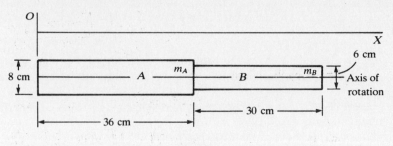

Fig. 10-2   Fig. 10-3

10.2. Figure 10-3 is the side view of a small machine shaft made of homogeneous material.   Find its center of mass.

In calculating the center of mass of a body, any part of the body may be treated as if all its mass were concentrated at its own center of mass.   Thus, the given shaft is equivalent to a point mass m_A at $x_A = 36/2 = 18$ cm and a point mass m_B at $x_B = 36 + (30/2) = 51$ cm.   Then,

$$x_{cm} = \frac{m_A x_A + m_B x_B}{m_A + m_B} = \frac{m_A(18) + m_B(51)}{m_A + m_B} = \frac{18(m_A/m_B) + 51}{(m_A/m_B) + 1}$$

$$y_{cm} = 0$$

But, since the material is homogeneous,

$$\frac{m_A}{m_B} = \frac{V_A}{V_B} = \frac{36(8)^2}{30(6)^2} = \frac{32}{15}$$

and $$x_{cm} = \frac{18(32/15) + 51}{(32/15) + 1} = 28.53 \text{ cm} \qquad y_{cm} = 0$$

In writing $y_{cm} = 0$, we made use of the fact that if a body possesses an axis of mass-symmetry, the center of mass necessarily lies on that axis.

10.3. Find the center of mass of a quadrant of a thin elliptical section made of material of mass per unit area σ.   See Fig. 10-4.

The quadrant is bounded by the ellipse

$$\frac{x^2}{a^2} + \frac{y^2}{b^2} = 1$$

and the coordinate axes; its area is $A = \pi ab/4$.

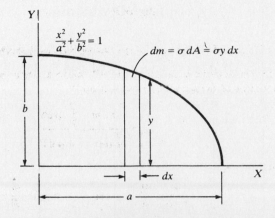

Fig. 10-4

$$x_{cm} = \frac{\int x\, dm}{\int dm} = \frac{\int_0^a \sigma xy\, dx}{\sigma A} = \frac{1}{A}\int_0^a xy\, dx$$

But, along the ellipse,

$$\frac{2x\, dx}{a^2} + \frac{2y\, dy}{b^2} = 0 \qquad \text{or} \qquad x\, dx = -\frac{a^2}{b^2} y\, dy$$

and so

$$x_{cm} = \frac{1}{A}\left(-\frac{a^2}{b^2}\right)\int_b^0 y^2\, dy = \frac{1}{A}\left(-\frac{a^2}{b^2}\right)\left(-\frac{b^3}{3}\right) = \frac{4a}{3\pi}$$

By symmetry, $y_{cm} = 4b/3\pi$.

10.4. Locate the center of mass of a thin hemispherical shell of radius R (Fig. 10-5).

The coordinates of the center of mass are $x_{cm} = y_{cm} = 0$ and

$$z_{cm} = \frac{\int z\, dm}{\int dm} = \frac{\int z\sigma\, dA}{\sigma A}$$

where σ is the mass per unit area of the thin shell. Since

$$dA = 2\pi R^2 \sin\theta\, d\theta = A \sin\theta\, d\theta$$

and $z = R\cos\theta$, we have

$$z_{cm} = \int_0^{\pi/2} R\cos\theta \sin\theta\, d\theta = -\frac{R}{2}\left[\cos^2\theta\right]_0^{\pi/2} = \frac{R}{2}$$

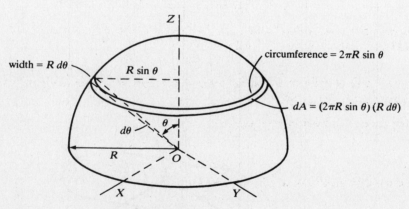

Fig. 10-5

10.5. Find the center of mass of a right circular cone of height h, radius R, and constant density ρ.

In Fig. 10-6, the base of the cone lies in the XY plane and the axis of symmetry is along Z. The coordinates of the center of mass are $x_{cm} = y_{cm} = 0$ and

$$z_{cm} = \frac{\int z\, dm}{\int dm} = \frac{\int z\rho\, dV}{\int \rho\, dV}$$

By similar triangles,

$$\frac{r}{h-z} = \frac{R}{h} \qquad \text{or} \qquad z = h - \frac{h}{R}r$$

so that $dV = -(\pi h/R)r^2\, dr$ and

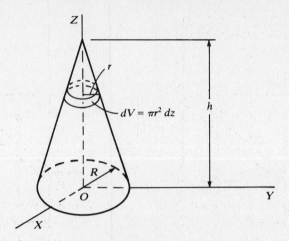

Fig. 10-6

$$z_{cm} = \frac{\int_R^0 \left(h - \frac{h}{R}r\right)r^2 \, dr}{\int_R^0 i^2 \, dr} = \frac{-h\frac{R^3}{3} + h\frac{R^3}{4}}{-\frac{R^3}{3}} = \frac{h}{4}$$

10.6. Show that if the resultant external force on a rigid body is zero, the resultant torque has a fixed value, independent of the point about which the torque is computed.

Refer to Fig. 10-7, which shows two arbitrary points, P and Q, which are fixed in space but do not necessarily belong to the rigid body. The torque about P is given by

$$\tau_P = \int \mathbf{r}_P \times d\mathbf{F} = \int (\mathbf{D} + \mathbf{r}_Q) \times d\mathbf{F}$$

$$= \mathbf{D} \times \int d\mathbf{F} + \int \mathbf{r}_Q \times d\mathbf{F}$$

$$= 0 + \tau_Q = \tau_Q$$

since, by hypothesis, $\int d\mathbf{F} = 0$.

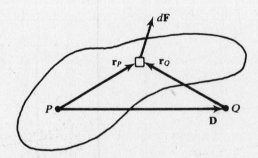

Fig. 10-7

10.7. A light rod [Fig. 10-8(a)], which is suspended by a rope, itself suspends a 3114 N load at point B. The ends of the rod are in contact with smooth vertical walls. (a) If $\ell = 0.25$ m, find the tension in the rope CD and the reaction forces at A and E. (b) Determine the maximum value of ℓ if the largest reaction force at E is 2224 N.

(a) The system is in static equilibrium. The reaction forces shown in Fig. 10-8(b) are in the X direction because the walls are assumed frictionless. The force condition in the Y direction gives

$$\sum F_y = T - 3114 = 0 \qquad \text{or} \qquad T = 3144 \text{ N}$$

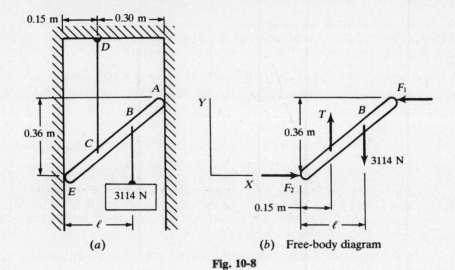

(a)   (b)   Free-body diagram

Fig. 10-8

Also, upon taking the sum of the torques about E (counterclockwise is positive), we get

$$\sum \tau_E = F_1(0.36) - 3114(0.25) + 3114(0.15) = 0 \qquad \text{or} \qquad F_1 = 865 \text{ N}$$

The force condition in the X direction then gives

$$\sum F_x = F_2 - F_1 = 0 \qquad \text{or} \qquad F_2 = 865 \text{ N}$$

(b)   With $F_2 = F_1 = 2224$ N, the torque condition gives

$$\sum \tau_E = (2224)(0.36) - (3114)\ell + (3114)(0.15) = 0 \qquad \text{or} \qquad \ell = 0.41 \text{ m}$$

10.8. Two wheels, of weights W and $3W$, are connected by a bar of negligible weight and are free to roll on the 45° inclines shown in Fig. 10-9(a).   Find the angle ϕ that the bar makes with the horizontal when the system is in static equilibrium.

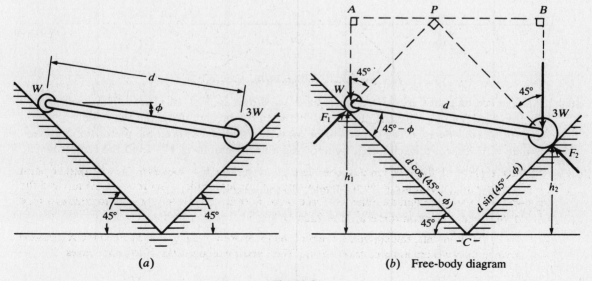

(a)   (b)   Free-body diagram

Fig. 10-9

Assuming that the inclines are smooth, the reaction forces F_1 and F_2 indicated in Fig. 10-9(b) will be normal; the wheels will slip, rather than roll, along the inclines. Take the sum of the torques about the point P shown. Forces W and $3W$ have respective moment arms

$$\overline{AP} = [d \sin(45° - \phi)] \sin 45° \qquad \overline{BP} = [d \cos(45° - \phi)] \sin 45°$$

Thus (counterclockwise torques are positive):

$$Wd \sin(45° - \phi) \sin 45° - 3Wd \cos(45° - \phi) \sin 45° = 0$$
$$\tan(45° - \phi) = 3$$
$$45° - \phi = 71.6°$$
$$\phi = -26.6°$$

A negative value for ϕ means that, at equilibrium, the heavier wheel lies higher than the lighter wheel. Such a result, which is contrary to the way we have drawn Fig. 10-9, and contrary to our physical intuition, ought to be looked into. Let us calculate the gravitational potential energy of the system, relative to point C in Fig. 10-9(b).

$$U = Wh_1 + 3Wh_2 = Wd \cos(45° - \phi) \sin 45° + 3Wd \sin(45° - \phi) \sin 45°$$

From the graph of this function, Fig. 10-10, it is seen that at $\phi = -26.6°$ the potential energy is a *maximum*. This means that the system is in equilibrium, but the equilibrium is *unstable*. At the slightest application of negative torque, the system would leave the position $\phi = -26.6°$ and go to the position of *stable* equilibrium, $\phi = 45°$ (where the potential energy is a minimum). Notice that when the heavier wheel is at point C it is in contact with both inclines, so that the net reaction force on it has an unknown direction. That is the reason why the above torque equation, involving the point P, failed to give the value $\phi = 45°$.

To complete the solution of this problem, the reader should examine the possibility $\phi = -45°$.

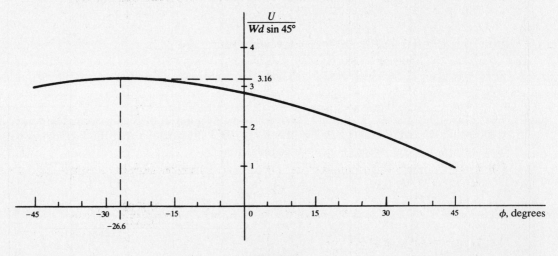

Fig. 10-10

10.9. A large set of plates is stacked on a table as shown in Fig. 10-11(a). The coefficients of friction between two adjacent plates, and between the bottom plate and the tabletop, are 0.25 and 0.15, respectively. The stack is to be moved to the left, without the plates tipping over or sliding with respect to each other, by applying a horizontal force **F**. Find the largest height h at which **F** may be applied.

In order to move the stack as a rigid body the following must be true:

(1) The force **F** must just overcome the friction force between the bottom plate and the tabletop. See Fig. 10-11(b).

$$\sum F_x = F - f = 0 \qquad \text{or} \qquad F = f = \mu_1 N = \mu_1 W = 0.15 W$$

where $N = W$ because $\sum F_y = N - W = 0$.

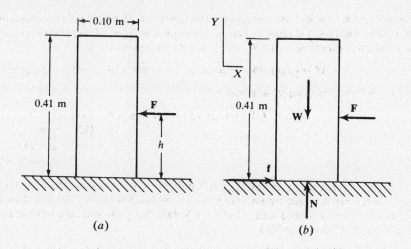

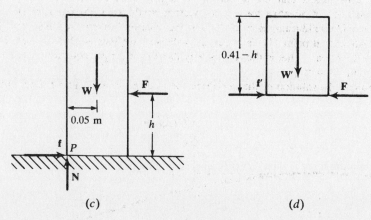

Fig. 10-11

(2) h must be limited so that the stack will not tip over.   When the stack is just on the verge of rotating about P, Fig. 10-11(c),

$$\sum \tau_P = Fh - W(0.05) = 0 \qquad \text{or} \qquad h = \frac{W(0.05)}{0.15\,W} = 0.33 \text{ m}$$

The stack will tip if $h > 0.33$ m.

(3) h must also be limited so that there will be no sliding of plates relative to each other.   The plate to which **F** is applied and the plates above it may be considered a free body [Fig. 10-11(d)], of weight

$$W' = W\frac{0.41 - h}{0.41}$$

Then, from $\sum F_x = F - f' = 0$,

$$F = f'$$

$$\mu_1 W = \mu_2 W'$$

$$0.15\,W = 0.25\,W\frac{0.41 - h}{0.41}$$

$$h = 0.164 \text{ m}$$

If $h > 0.164$ m is applied, the plates above and including the plate where the $0.15\,W$ force will slide and the lower plates will not move.

Therefore, for all plates to move together, $h_{\max} = 0.16$ m.

10.10. A rigid T-bar, 10 cm on each arm, rests between two vertical walls, as shown in Fig. 10-12. The left wall is smooth; the coefficients of static friction between the bar and floor, and between the bar and right wall, are 0.35 and 0.50, respectively. The bar is subject to a vertical load of 1 N, as shown. What is the smallest value of the vertical force F for which the bar will be in static equilibrium in the position shown?

The solution of this problem is greatly simplified by an intuitive consideration of the situation when F is very small. The 1 N force then sets up a counterclockwise torque that, because of the low frictional resistance offered by the left wall and the floor, immediately causes the bar to lose contact with the right wall. Therefore, if a value of F can be found that puts the bar in equilibrium with its end just touching the right wall (i.e. $N_3 = f_3 = 0$), this value of F may be presumed to be the desired minimum.

With $N_3 = f_3 = 0$, the force conditions are

$$N_2 - 1 - F = 0 \qquad\qquad (1)$$

$$N_1 - f_2 = 0 \qquad\qquad (2)$$

and the torque condition (about the contact point with the floor) is

$$-N_1(0.08) + 1(0.03) - F(0.01) = 0 \qquad\qquad (3)$$

Elimination of F and N_1 between these three equations gives

$$f_2 = 0.5 - 0.125 N_2$$

But the largest possible value of f_2 is $\mu_2 N_2 = 0.35 N_2$. Hence,

$$0.5 - 0.125 N_2 \le 0.35 N_2 \qquad \text{or} \qquad N_2 \ge \frac{0.5}{0.475} = \frac{20}{19} \text{ N}$$

and, from (1),

$$F = N_2 - 1 \ge \frac{1}{19} \text{ N}$$

The minimum force is thus $(1/19)$ N, corresponding to which

$$N_1 = \frac{7}{19} \text{ N} = f_2 \qquad N_2 = \frac{20}{19} \text{ N}$$

The force may be increased above this value, still keeping $N_3 = f_3 = 0$, up to $F = 3$ N, at which point N_1 and f_2 vanish. Thus there is a whole range of solutions such that the right wall might just as well not be there.

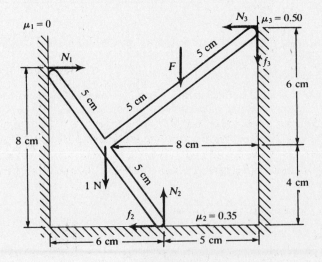

Fig. 10-12

10.11. A uniform beam, which weighs 400 N and is 5.0 m long, is hinged to a wall at its lower end by a frictionless hinge. A horizontal rope 3.0 m long is fastened between the upper end of the beam and the wall. Find the force T [see Fig. 10-13(b)] exerted on the beam by the rope, and also find the horizontal and vertical components of the force exerted on the beam by the hinge.

Let a frame of reference be set up with the X axis directed horizontally to the right and the Y axis directed vertically upward.

Replace the force exerted on the beam by the hinge by its horizontal component F_x and its vertical component F_y. The weight acts at the center of mass of the beam. The rope pulls on the beam and therefore exerts a force T on the beam which acts to the left.

Consider torques about a horizontal axis perpendicular to the beam and passing through the hinge. The moment arms of F_x and F_y are zero and these forces have zero torque about this axis. The 400 N weight has a moment arm of 1.5 m. The moment arm of the force T is the distance along the wall from the hinge to the rope,

$$\sqrt{(5.0)^2 - (3.0)^2} = 4.0 \text{ m}$$

The torque condition is then

$$\sum \tau = T(4) - 400(1.5) = 0 \qquad \text{or} \qquad T = 150 \text{ N}$$

and the force conditions give

$$\sum F_x = F_x - T = 0 \qquad \text{or} \qquad F_x = T = 150 \text{ N}$$
$$\sum F_y = F_y - 400 = 0 \qquad \text{or} \qquad F_y = 400 \text{ N}$$

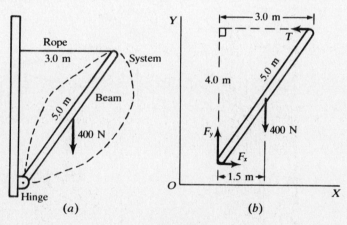

Fig. 10-13

10.12. A 244 N warehouse door closes itself by the attachment to it of a $112\sqrt{2}$ N weight [see Fig. 10-14(a)]. The door is held open by a force $\mathbf{F}$ applied at the knob along a normal to the door. Find the magnitude of $\mathbf{F}$ and the components of the reaction forces at the hinges C and D when the door is opened such that $\theta = 90°$. Assume that there is no axial thrust at the hinge C.

Since the door is in static equilibrium:

$$0 = \sum \tau_D = \sum \mathbf{r} \times \mathbf{F}$$

$$= (4\mathbf{j} + 6\mathbf{k}) \times 112(\mathbf{i} - \mathbf{j}) + (5\mathbf{k}) \times (C_x\mathbf{i} + C_y\mathbf{j}) + (2\mathbf{j} + 2.5\mathbf{k}) \times (-224\mathbf{k}) + (3.5\mathbf{j} + 2\mathbf{k}) \times (-F\mathbf{i})$$

$$= 672\mathbf{i} + 672\mathbf{j} - 448\mathbf{k} - 5C_y\mathbf{i} + 5C_x\mathbf{j} - 448\mathbf{i} - 2F\mathbf{j} + 3.5F\mathbf{k}$$

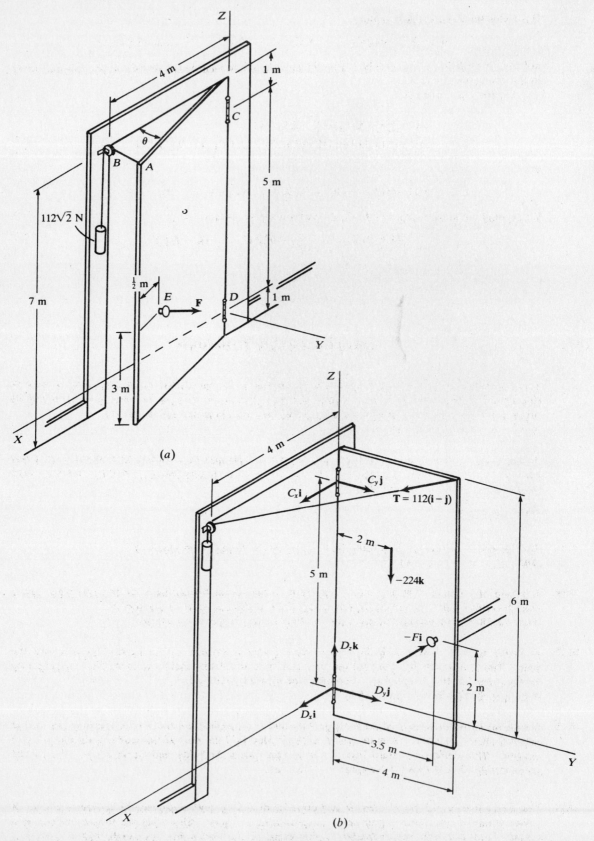

(a)

(b)

Fig. 10-14

This yields the three scalar equations

$$672 - 5C_y - 448 = 0 \qquad 672 + 5C_x - 2F = 0 \qquad -448 + 3.5F = 0$$

Solving, $F = 128.0$ N, $C_y = 44.8$ N, $C_x = -83.2$ N (the minus sign means that $\mathbf{C}_x$ in the figure is drawn in the wrong direction).

The force condition is

$$0 = \sum \mathbf{F} = \mathbf{C} + \mathbf{D} + \mathbf{T} - 224\mathbf{k} - F\mathbf{i}$$
$$= C_x\mathbf{i} + D_x\mathbf{i} + 112\mathbf{i} - F\mathbf{i} + D_y\mathbf{j} - 112\mathbf{j} + C_y\mathbf{j} + D_z\mathbf{k} - 224\mathbf{k}$$

or

$$C_x + D_x + 112 - F = 0 \qquad D_y - 112 + C_y = 0 \qquad D_z - 224 = 0$$

Substituting the known values of F, C_x, C_y and solving, we obtain

$$D_x = 99.2 \text{ N} \qquad D_y = 67.2 \text{ N} \qquad D_z = 224 \text{ N}$$

Supplementary Problems

10.13. Three particles, of masses 2 kg, 4 kg, and 6 kg, are located at the vertices of an equilateral triangle of side 0.5 m. Find the center of mass of this collection, giving its coordinates in terms of a system with its origin at the 2 kg particle and with the 4 kg particle located along the positive X axis.
Ans. $x_{cm} = 0.29$ m, $y_{cm} = 0.22$ m

10.14. A thin bar of length L has a mass per unit length, λ, that increases linearly with distance from one end. If its total mass is M and its mass per unit length at the lighter end is λ_0, find the distance of the center of mass from the lighter end.
Ans. $\dfrac{2L}{3} - \dfrac{\lambda_0 L^2}{6M}$

10.15. Find the center of mass of a uniform solid hemisphere of radius R and mass M.
Ans. $x_{cm} = y_{cm} = 0$, $z_{cm} = \frac{3}{8}R$

10.16. A sphere of radius 0.10 m and mass 10 kg rests in the corner formed by a 30° inclined plane and a smooth vertical wall. Calculate the forces that the two surfaces exert on the sphere.
Ans. 56.58 N perpendicular to the wall; 113.16 N perpendicular to the plane

10.17. A 13.0 m uniform ladder weighing 300 N rests against a smooth wall at height 12.0 m above the floor. The floor is rough. Find (a) the frictional force and the normal force exerted on the ladder by the floor, and (b) the normal force exerted on the ladder by the wall.
Ans. (a) 62.5 N, 300 N; (b) 62.5 N

10.18. A uniform ladder is supported on a rough floor and leans against a smooth wall, touching the wall at height h above the floor. A man climbs up the ladder until the base of the ladder is on the verge of slipping. The coefficient of static friction between the foot of the ladder and the floor is μ. What is the *horizontal* distance moved by the man? *Ans.* μh

10.19. Two forces have equal magnitudes, F, and opposite directions, with a distance D between the lines of action of the forces. Such a pair of forces constitutes a *couple*. Show that the torque exerted by a couple about any axis perpendicular to the plane of the forces has magnitude FD, and thus is independent of the position of the axis.

Chapter 11

Motion of a Rigid Body

11.1 MOMENT OF INERTIA

The rotational analog of mass is a quantity called the *moment of inertia*. The greater the moment of inertia of a body, the greater its resistance to change in its angular speed. The moment of inertia of a system of particles *about a given axis* is

$$I = m_1 r_1^2 + m_2 r_2^2 + \cdots + m_n r_n^2$$

where the particles of the system have masses $m_1, m_2, \ldots, m_n$ and are located at the respective distances $r_1, r_2, \ldots, r_n$ from the axis. The moment of inertia of a body with a continuous distribution of mass is

$$I = \int r^2 \, dm$$

It is sometimes convenient to express the moment of inertia I of a body of mass M in terms of a length k, called the *radius of gyration*, which is defined by

$$I = Mk^2$$

This implies that a single particle of mass M at distance k from the axis has a moment of inertia equal to that of the body in question.

The units for moment of inertia are $kg \cdot m^2$.

11.2 THEOREMS ON MOMENTS OF INERTIA

Decomposition theorem. If a system is composed of two parts, with moments of inertia I_1 and I_2, the moment of inertia of the complete system is

$$I = I_1 + I_2$$

Parallel-axis theorem. A system's moment of inertia I_{cm} about an axis through its center of mass, and its moment of inertia I about a parallel axis, are related by

$$I = I_{cm} + MD^2$$

where M is the total mass and D is the distance between the two parallel axes.

Perpendicular-axis theorem. If a flat plate in the XY plane has moments of inertia I_{OX}, I_{OY}, and I_{OZ} about the mutually perpendicular axes OX, OY, and OZ, then

$$I_{OZ} = I_{OX} + I_{OY}$$

11.3 KINETIC ENERGY OF A MOVING RIGID BODY

The kinetic energy of a rigid body *rotating* about a fixed axis is

$$K = \frac{1}{2} I \omega^2$$

where I is the moment of inertia of the body about the fixed axis and ω is the angular speed of the body, also calculated about the fixed axis.

In *any motion* of a rigid body, the total kinetic energy is the sum of the kinetic energy associated with the translation of the center of mass and the kinetic energy associated with the rotation of the body about the center of mass.   In particular, if the motion is parallel to the XY plane,

$$K = \frac{1}{2} M V_{cm}^2 + \frac{1}{2} I_{cm}\omega^2$$

where I_{cm} and ω are calculated about an axis through the center of mass that moves parallel to the Z axis.

11.4 TORQUE AND ANGULAR ACCELERATION

When a rigid body rotates *about an axis fixed in an inertial frame or about an axis through the center of mass and fixed in direction,*

$$\tau_{ext} = I\alpha$$

where α is the angular acceleration of the body, I is its moment of inertia, and τ_{ext} is the algebraic sum of the torques of the *external* forces, all three quantities being calculated about the given axis.   This relation is the rotational counterpart to Newton's second law.

EXAMPLE 11.1.   Like Newton's second law, $\tau_{ext} = I\alpha$ does not generally hold in a noninertial frame.   Consider a pencil which is held horizontal and then released.   The pencil stays horizontal as it falls.   Thus, relative to an end of the pencil, there is no rotation ($\alpha = 0$), even though there is a gravitational torque acting ($\tau_{ext} \neq 0$).

Note, however, that if the axis is chosen *through the center of mass* of the pencil, both α and τ_{ext} are zero, and so $\tau_{ext} = I_{cm}\alpha$.

The rate at which work is done on a rigid body by an external torque is given by

$$\text{instantaneous power} = \tau_{ext}\omega$$

provided that the external forces actually move through an angle about the rotation axis.

EXAMPLE 11.2.   It was seen in Problem 9.7 that the frictional forces, and hence the frictional torques, exerted on the car by the road did no work on the car.   The forces had fixed points of application relative to the car's axles.

Solved Problems

11.1. Find the moment of inertia of the system shown in Fig. 11-1, about (*a*) an axis OZ perpendicular to the plane of the figure and passing through the origin, (*b*) an axis CZ' passing through the center of mass and parallel to the axis OZ.

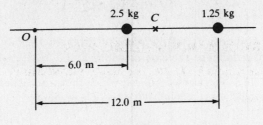

Fig. 11-1

(a) The moment of inertia about the axis OZ is

$$I = (2.5)(6.0)^2 + (1.25)(12.0)^2 = 270 \text{ kg} \cdot \text{m}^2$$

(b) The center of mass C is between the two particles and is 2.0 m from the 2.5 kg particle. The moment of inertia of the system about the axis CZ' is

$$I_{cm} = (2.5)(2.0)^2 + (1.25)(4.0)^2 = 30 \text{ kg} \cdot \text{m}^2$$

11.2. Four particles, of masses 2 kg, 4 kg, 6 kg, and 8 kg, are at the vertices of a square of side $s = 1.0$ m. Assuming that the mass of the connecting rods can be ignored, find the moment of inertia of the structure about an axis perpendicular to the plane of the square and going through the intersection of its diagonals.

Since each diagonal of the square is of length $s\sqrt{2}$, the perpendicular distance of each of the four particles from the axis is $\frac{1}{2}(1)\sqrt{2}$ m. Thus

$$I = \sum m_i r_i^2 = (2 + 4 + 6 + 8)(\tfrac{1}{2}) = 10 \text{ kg} \cdot \text{m}^2$$

11.3. Find the moment of inertia of a uniform rectangular flat plate about an axis along the edge of width a (Fig. 11-2). The plate has a mass M and a length L.

The important step in calculating a moment of inertia is the choosing of a typical element, dm. All parts of dm must be at essentially the *same distance* r from the axis. Since the plate is uniform, the mass of any part is proportional to its area. For the element shown in Fig. 11-2,

$$\frac{dm}{M} = \frac{a\, dr}{aL} \qquad \text{or} \qquad dm = \frac{M}{L}\, dr$$

The moment of inertia of this element about the axis at a distance r is

$$r^2\, dm = \frac{M}{L} r^2\, dr$$

The sum of the contributions from all the elements that comprise the plate is

$$I = \int_0^L \frac{M}{L} r^2\, dr = \frac{1}{3} ML^2$$

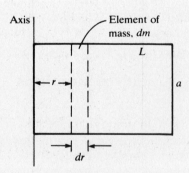

Fig. 11-2

11.4. Find the moment of inertia I_{cm} of the plate of Problem 11.3 about an axis passing through its center of mass and parallel to the side of width a.

From Problem 11.3, the moment of inertia of the plate about an axis along the edge is $I = \frac{1}{3}ML^2$. Using the parallel-axis theorem with $D = L/2$, we have

$$\frac{1}{3} ML^2 = I_{cm} + M\left(\frac{L}{2}\right)^2 \qquad \text{or} \qquad I_{cm} = \frac{1}{12} ML^2$$

11.5. Calculate the moment of inertia of a very thin, cylindrical shell of radius R and mass m, about the symmetry axis of the cylinder.

Since every element of mass of the shell is at the same perpendicular distance R from the axis,

$$I = mR^2$$

11.6. Calculate the moment of inertia of a solid homogeneous cylinder of mass M and radius R, about the axis of symmetry of the cylinder.

Set up a coordinate system with the Z axis along the axis of symmetry (see Fig. 11-3). Because the cylinder is homogeneous, the moment of inertia of a thin cylindrical shell of radius r and thickness dr is

$$dI = r^2\,dm = r^2\left(\frac{2\pi r\,dr}{\pi R^2}M\right) = \frac{2M}{R^2}r^3\,dr$$

and the moment of inertia of the entire cylinder is

$$I = \int dI = \frac{2M}{R^2}\int_0^R r^3\,dr = \frac{MR^2}{2}$$

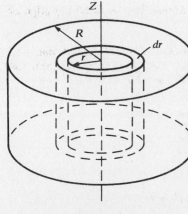

Fig. 11-3

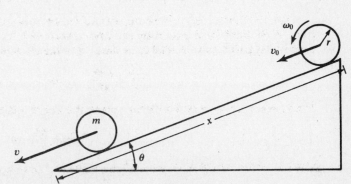

Fig. 11-4

11.7. A solid homogeneous cylinder rolls without slipping down an inclined plane, starting with angular speed ω_0 and linear speed v_0 (see Fig. 11-4). (a) Find the linear speed v of the cylinder after rolling a distance x. (b) What friction force acts on the cylinder?

(a) Because there is no slipping, the friction force does no work on the cylinder. Then, by conservation of energy,

$$(K + U)_{\text{top}} = (K + U)_{\text{bottom}}$$

$$\frac{1}{2}mv_0^2 + \frac{1}{2}I\omega_0^2 + mgx\sin\theta = \frac{1}{2}mv^2 + \frac{1}{2}I\omega^2 + 0$$

$$\frac{1}{2}mv_0^2 + \frac{1}{2}\left(\frac{mr^2}{2}\right)\frac{v_0^2}{r^2} + mgx\sin\theta = \frac{1}{2}mv^2 + \frac{1}{2}\left(\frac{mr^2}{2}\right)\frac{v^2}{r^2}$$

since $\omega = v/r$ (no slipping) and $I = mr^2/2$ for a solid cylinder about its central axis (Problem 11.6). Thus

$$v^2 = v_0^2 + \frac{4}{3}gx\sin\theta$$

(b) A comparison between the result of (a) and the constant-acceleration formula $v^2 = v_0^2 + 2ax$ indicates that the center of mass of the cylinder moves down the plane at constant acceleration

$$a = \frac{2}{3}g\sin\theta$$

But the net force down the plane is $(mg \sin \theta) - f$. Thus,

$$(mg \sin \theta) - f = m\left(\frac{2}{3} g \sin \theta\right) \qquad \text{or} \qquad f = \frac{1}{3} mg \sin \theta$$

11.8. A light thread is unwound from a spool by a constant force of 4.4 N (Fig. 11-5). The spool weighs 1.1 N and its radius of gyration with respect to its axis is 0.01 m. Friction keeps it from slipping. Find the speed of its center after moving 1.8 m.

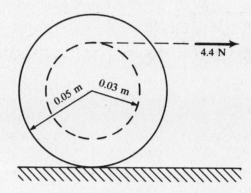

Fig. 11-5

The angle the spool has turned through while the spool has moved 1.8 m is

$$\theta = \frac{s}{r} = \frac{1.8}{0.05} = 36 \text{ rad}$$

While the center of the spool moves 1.8 m, a length $(0.03)\theta = 1.08$ m of thread is unwound. Hence the 4.4 N force acts through a distance of

$$d = 1.8 + 1.08 = 2.88 \text{ m}$$

and the work done by the force is

$$W = Fd = (4.4)(2.88) = 12.67 \text{ J}$$

The moment of inertia of the spool is

$$I = mk^2 = \left(\frac{1.1}{9.8}\right)(0.01)^2 = 1.122 \times 10^{-5} \text{ kg} \cdot \text{m}^2$$

The final kinetic energy of the spool is

$$K = \frac{1}{2} mv^2 + \frac{1}{2} I\omega^2 = \frac{1}{2}\left(\frac{1.1}{9.8}\right)v^2 + \frac{1}{2}(1.122 \times 10^{-5})\frac{v^2}{(0.05)^2} = 0.058 v^2$$

since $\omega = v/r$, and $r = 0.05$ m in this case. Equating the work done by the force of 4.4 N to the change in kinetic energy:

$$12.67 = 0.058 v^2 \qquad \text{or} \qquad v = 14.8 \text{ m/s}$$

11.9. Consider the two-part, unsymmetrical body in Fig. 11-6(a), which is initially at rest. Part A is one-fourth of a cylinder of radius $2R$; part B is one-half of a cylinder of radius R. The body rolls over, hitting the floor as shown in Fig. 11-6(b). In the initial position, the center of mass of A is at a height

$$y_1' = R\left(1 + \frac{8}{3\pi}\right)$$

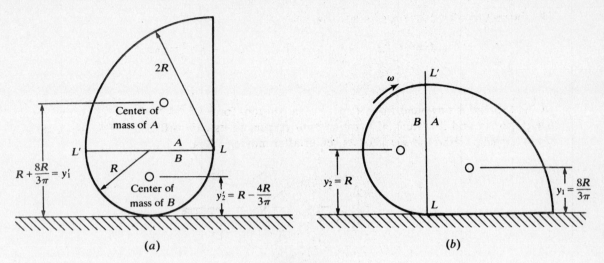

Fig. 11-6

(let $a = b = 2R$ in Problem 10.3) and the center of mass of B is at a height

$$y_2' = R\left(1 - \frac{4}{3\pi}\right)$$

(reason from Problem 10.3, with $a = b = R$).   Find the angular speed of the body about a perpendicular axis through point L, just before L comes in contact with the floor.

First we need the moment of inertia about L.   In Problem 11.15 it is shown that

$$I = I_A + I_B = \frac{7}{12} MR^2$$

where $M = M_A + M_B$ is the total mass.

Now apply conservation of energy.   Initially the body is at rest; at the moment of impact, it is rotating about an instantaneous axis through L.

$$\Delta U + \Delta K = 0$$

$$M_A g(y_1 - y_1') + M_B g(y_2 - y_2') + \frac{1}{2} I \omega^2 - 0 = 0$$

$$\frac{2M}{3} g(-R) + \frac{M}{3} g\left(\frac{4R}{3\pi}\right) + \frac{1}{2}\left(\frac{7}{12} MR^2\right)\omega^2 = 0$$

Solving,
$$\omega = 4\sqrt{\left(1 - \frac{2}{3\pi}\right)\frac{g}{7R}}$$

11.10. A solid homogeneous sphere, resting on top of another, fixed sphere, is slightly displaced and begins to roll down the fixed sphere.   Show that it will slip when the line of centers makes an angle with the vertical given by

$$2 \sin(\theta - \gamma) = 5(\sin \gamma)(3 \cos \theta - 2)$$

where $\mu = \tan \gamma$ is the coefficient of static friction between the two spheres.   (For a sphere, $I_{cm} = 2MR^2/5$.)

Figure 11-7 shows the situation at time t, up to which the motion is assumed to have been one of pure rolling.   The center of mass, C, of the rolling sphere is moving in a circle of radius $a + b$ about O; its equations of motion are

radial: $Mg \cos \theta - N = M(a + b)\dot{\theta}^2$ (1)

tangential: $Mg \sin \theta - f = M(a + b)\ddot{\theta}$ (2)

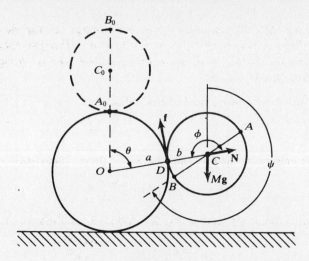

Fig. 11-7

The amount of rotation of the rolling sphere about its center of mass is measured by the angle ψ through which diameter $A_0C_0B_0$ has turned in time t. From the geometry it is seen that

$$\psi = \theta + \phi$$

(N makes angle θ with the vertical and CB makes angle ϕ with N). Moreover, since there has been no slipping,

$$\overset{\frown}{A_0D} = \overset{\frown}{AD} \qquad \text{or} \qquad a\theta = b\phi$$

Then the torque equation about C may be written as

$$fb = \frac{2Mb^2}{5}\ddot{\psi} = \frac{2Mb^2}{5}(\ddot{\theta} + \ddot{\phi}) = \frac{2Mb^2}{5}\left(\ddot{\theta} + \frac{a}{b}\ddot{\theta}\right)$$

or
$$f = \frac{2M}{5}(a+b)\ddot{\theta} \tag{3}$$

We can now solve (1), (2), and (3) for f and N as functions of θ. From (2) and (3),

$$\ddot{\theta} = \frac{5}{7}\frac{g}{a+b}\sin\theta$$

On integration of this we get

$$\ddot{\theta}\dot{\theta}\,dt = \frac{1}{2}d(\dot{\theta}^2) = \frac{5}{7}\frac{g}{a+b}\sin\theta\,d\theta$$

$$\int_0^{\dot{\theta}^2}\frac{1}{2}d(\dot{\theta}^2) = \int_0^\theta \frac{5}{7}\frac{g}{a+b}\sin\theta\,d\theta$$

$$\frac{\dot{\theta}^2}{2} = \frac{5}{7}\frac{g}{a+b}(1-\cos\theta)$$

Substituting the values for $\dot{\theta}^2$ and $\ddot{\theta}$ in (1) and (3), we find

$$f = \frac{2Mg}{7}\sin\theta \qquad N = \frac{Mg}{7}(17\cos\theta - 10)$$

The rolling sphere will start to slip when the friction becomes limiting ($f = N\tan\gamma$), i.e. when

$$\frac{2Mg}{7}\sin\theta = \frac{Mg}{7}(17\cos\theta - 10)(\tan\gamma)$$

This gives $(\cos\gamma)(2\sin\theta) = (\sin\gamma)(17\cos\theta - 10)$, or

$$2\sin(\theta - \gamma) = 5(\sin\gamma)(3\cos\theta - 2)$$

11.11. A floor polishing disk of diameter D is pushed against the floor with a force F. The coefficient of friction between the disk and floor is μ. The disk rotates with constant angular speed ω. Assuming that the pressure exerted on the floor is uniform, find the power to run the floor polisher (see Fig. 11-8).

The frictional force on the element of area dA is

$$df = \mu \frac{F}{A} dA$$

where F/A is the pressure of the polishing disk on the floor. Since $dA = r\, d\theta\, dr$ and $A = \pi D^2/4$,

$$df = \frac{4\mu F r\, d\theta\, dr}{\pi D^2}$$

The differential torque set up opposing the motion of the disk is

$$d\tau_{\text{ext}} = -r\, df = -\frac{4\mu F r^2\, d\theta\, dr}{\pi D^2}$$

and so

$$\tau_{\text{ext}} = -\frac{4\mu F}{\pi D^2}\int_0^{2\pi} d\theta \int_0^{D/2} r^2\, dr = -\frac{4\mu F}{\pi D^2}\int_0^{2\pi}\frac{D^3}{24}\, d\theta = -\frac{\mu F D}{3}$$

To keep the disk rotating at constant angular speed ($\alpha = 0$) an equal but opposite torque must be exerted by the motor of the polisher, which thus does work at the rate

$$(-\tau_{\text{ext}})\omega = \frac{\mu F D \omega}{3}$$

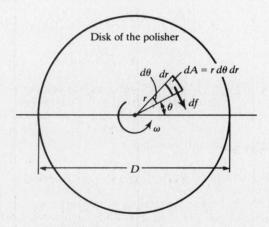

Fig. 11-8

11.12. A rotor weighs 430 N, and its radius of gyration is $k = 1.2$ m. A couple of magnitude $43\text{ N} \cdot \text{m}$ is needed to keep the shaft rotating at 32 rad/s. (a) Find the power required for this. (b) If an additional couple of $86\text{ N} \cdot \text{m}$ is applied to the shaft in the same sense as the previous one, find the angular acceleration of the shaft.

(a) $P = \tau\omega = (43)(32) = 1376$ W

(b)   The angular acceleration may be obtained from $\tau_{\text{ext}} = I\alpha$:

$$\alpha = \frac{\tau_{\text{ext}}}{I} = \frac{\tau_{\text{ext}}}{Mk^2} = \frac{86}{(430/9.8)(1.2)^2} = 1.36 \text{ rad/s}^2$$

11.13. If the belt shown in Fig. 11-9 transmits 33.557 kW to a pulley which is 0.41 m in diameter and rotates at 7.5 rev/s, find the difference in tension, $T_1 - T_2$, between the tight and slack sides of the belt.

The power transmitted to the shaft of the pulley is $P = \tau\omega = R(T_1 - T_2)\omega$. Hence,

$$T_1 - T_2 = \frac{P}{R\omega} = \frac{33.557 \times 10^3}{(0.41/2)(7.5 \times 2\pi)} = 3475 \text{ N}$$

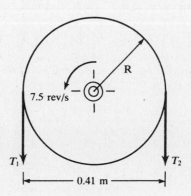

Fig. 11-9

Supplementary Problems

11.14. Repeat Problem 11.7 for a uniform solid sphere.

Ans. (a) $v^2 = v_0^2 + \frac{10}{7} gx \sin\theta$; (b) $f = \frac{2}{7} mg \sin\theta$

11.15. Refer to Problem 11.9. (a) Using the result of Problem 11.6 and the decomposition theorem, calculate the moment of inertia of part A about L. (b) Using the result of Problem 11.6, the parallel-axis theorem, and the decomposition theorem, calculate the moment of inertia of part B about L. *Ans.* (a) $I_A = MR^2/3$; (b) $I_B = MR^2/4$

11.16. A dumbbell consists of a 2.40 kg particle and a 1.60 kg particle connected by a rod which is 0.80 m long. Calculate the moment of inertia of this dumbbell about an axis perpendicular to the rod and passing through the rod at a point 0.20 m from the 1.60 kg particle. *Ans.* 0.9296 kg · m²

11.17. A sphere of mass M and radius R has a moment of inertia about a diameter given by $I = \frac{2}{5}MR^2$. Find the moment of inertia of the sphere about an axis that is tangent to the sphere.

Ans. $\frac{7}{5} MR^2$

11.18. A circumferential force of 1.2 N acts on the surface of a cylinder, which can turn on its axis. The mass of the cylinder is 2.5 kg and its radius is 0.1 m. Find (a) the torque acting on the cylinder, and (b) the angular acceleration of the cylinder. *Ans.* (a) 0.12 N · m; (b) 9.6 rad/s²

11.19. A homogeneous cylindrical wheel of radius 0.8 m and of mass 2.5 kg is free to rotate about its axis on frictionless bearings. Suppose that a force of 5 N is suddenly applied and then maintained tangent to the rim of the wheel. (*a*) What is the angular acceleration of the wheel? (*b*) What are the angular speed and the kinetic energy of the wheel at time $t = 3$ s? (*c*) How much work does the force do on the wheel during this 3 s interval?   *Ans.* (*a*) 5 rad/s^2; (*b*) 15 rad/s, 90 J; (*c*) 90 J

11.20. A 10 kg homogeneous sphere of radius 0.2 m is at a certain instant rotating about a shaft through its center at 10 rev/s. Assuming that a constant frictional torque acts so that the sphere comes to rest in 10 s, find the magnitude of the torque.   *Ans.* 0.2 N · m

11.21. A spool of thread of radius *r* and mass *m* has the shape of a uniform cylinder. Suppose that one end of the thread is attached to the ceiling and the spool is allowed to unwind under the action of gravity. Calculate the angular acceleration of the spool and the tension in the thread. Neglect the mass and thickness of the thread.   *Ans.* 2*g*/3*r*; *mg*/3

11.22. An automobile engine develops 1 MW of power while the crankshaft has an angular speed of 800 rad/s. Find the torque exerted on the crankshaft.   *Ans.* 1250 N · m

11.23. A flywheel is 2 m in diameter and weighs 4003 N; all of its mass may be considered as concentrated in its rim. What torque must be applied to the shaft to increase the angular speed uniformly from 5.2 rad/s to 6.3 rad/s in one revolution? Neglect friction.   *Ans.* 411.19 N · m

Chapter 12

Angular Momentum

12.1 ANGULAR MOMENTUM

If **r** is the position of a particle relative to a certain origin O, then its *angular momentum* **l** relative to this origin is defined as

$$\mathbf{l} = \mathbf{r} \times m\mathbf{v} = \mathbf{r} \times \mathbf{p}$$

where m is the mass of the particle, **v** is its velocity, and **p** is its linear momentum.   For a continuous distribution of mass,

$$\mathbf{L} = \int \mathbf{r} \times \mathbf{v}\, dm$$

The units of angular momentum are $\text{kg} \cdot \text{m}^2/\text{s}$, or, equivalently, $\text{J} \cdot \text{s}$.

12.2 PRINCIPLE OF ANGULAR MOMENTUM

The *principle of angular momentum* for a system of particles with a total angular momentum **L** is

$$\boldsymbol{\tau}_{\text{ext}} = \frac{d\mathbf{L}}{dt}$$

where $\boldsymbol{\tau}_{\text{ext}}$ is the resultant external torque (Section 10.1).   The torque and angular momentum are measured relative to the same origin.   The principle holds when the origin for **L** and $\boldsymbol{\tau}$ is either the center of mass or a point fixed in an inertial frame of reference.   This principle generalizes $\tau_{\text{ext}} = I\alpha$ (Section 11.4), which holds only when **L** is constrained to lie along the fixed axis of rotation of the rigid body, so that $L = I\omega$.

12.3 CONSERVATION OF ANGULAR MOMENTUM

The *law of conservation of angular momentum* states that if $\tau_{\text{ext}} = 0$, the system has a total angular momentum **L** that is constant.

Solved Problems

12.1.   The velocity of a particle of mass m is $\mathbf{v} = 5\mathbf{i} + 4\mathbf{j} + 6\mathbf{k}$ when at $\mathbf{r} = -2\mathbf{i} + 4\mathbf{j} + 6\mathbf{k}$.   Find the angular momentum of the particle about the origin.

$$\mathbf{l}_O = \mathbf{r} \times m\mathbf{v} = m \begin{vmatrix} \mathbf{i} & \mathbf{j} & \mathbf{k} \\ -2 & 4 & 6 \\ 5 & 4 & 6 \end{vmatrix} = m(42\mathbf{j} - 28\mathbf{k})$$

12.2 A light rigid axle of length D, rotating with a constant angular speed ω, has fixed on it a light rigid rod of length $2d$ that has two small spheres, each of mass m, attached to it. The rod makes an angle θ with the axle, as shown in Fig. 12-1. Find the torque exerted by the bearing reactions.

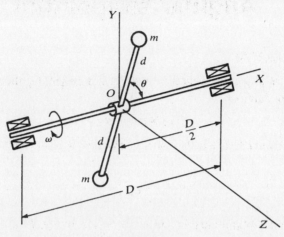

Fig. 12-1

The angular momentum of the spheres is

$$\mathbf{L}_O = \sum_{i=1}^{2} \mathbf{r}_i \times m_i \mathbf{v}_i$$

where the spheres are considered particles. Now, assuming that at $t = 0$ the rod was in the XY plane,

$$\mathbf{r}_1 = (d \cos \theta)\mathbf{i} + (d \sin \theta \cos \omega t)\mathbf{j} + (d \sin \theta \sin \omega t)\mathbf{k} = -\mathbf{r}_2$$

and

$$\mathbf{v}_1 = \dot{\mathbf{r}}_1 = (-\omega d \sin \theta \sin \omega t)\mathbf{j} + (\omega d \sin \theta \cos \omega t)\mathbf{k} = -\mathbf{v}_2$$

Further, $m_1 = m_2 = m$. Therefore,

$$\mathbf{L}_O = 2m\mathbf{r}_1 \times \mathbf{v}_1 = 2m \begin{vmatrix} \mathbf{i} & \mathbf{j} & \mathbf{k} \\ d \cos \theta & d \sin \theta \cos \omega t & d \sin \theta \sin \omega t \\ 0 & -\omega d \sin \theta \sin \omega t & \omega d \sin \theta \cos \omega t \end{vmatrix}$$

$$= \mathbf{i}(2m\omega d^2 \sin^2 \theta) - \mathbf{j}(2m\omega d^2 \cos \theta \sin \theta \cos \omega t) - \mathbf{k}(2m\omega d^2 \cos \theta \sin \theta \sin \omega t)$$

The torque is

$$\boldsymbol{\tau}_O = \frac{d\mathbf{L}_O}{dt} = \mathbf{j}(2m\omega^2 d^2 \cos \theta \sin \theta \sin \omega t) - \mathbf{k}(-2m\omega^2 d^2 \cos \theta \sin \theta \cos \omega t)$$

The magnitude of $\boldsymbol{\tau}_O$ is

$$|\boldsymbol{\tau}_O| = 2m\omega^2 d^2 \cos \theta \sin \theta = m\omega^2 d^2 \sin 2\theta$$

Notice that when $\theta = 90°$, giving a symmetrical system, $\boldsymbol{\tau}_O$ vanishes.

12.3. A rocket, of mass 10^6 kg, has a speed of 500 m/s in the horizontal direction. If its altitude (y) is 10 km and its horizontal distance (x) from the chosen origin is 10 km, what is its angular momentum with respect to this origin?

For this rocket, $\mathbf{r} = x\mathbf{i} + y\mathbf{j}$ and $\mathbf{v} = v\mathbf{i}$. Thus,

$$\mathbf{l} = \mathbf{r} \times m\mathbf{v} = m(x\mathbf{i} + y\mathbf{j}) \times v\mathbf{i} = -myv\mathbf{k} = -5 \times 10^{12} \text{ kg} \cdot \text{m}^2/\text{s}$$

12.4. Show that the angular momentum of a body may be expressed as the sum of two parts, one arising from the motion of the body's center of mass, and the other from the motion of the body with respect to its center of mass.

Relative to O, a fixed point in an inertial frame, the angular momentum is (see Fig. 12-2)

$$\mathbf{L}_O = \int \mathbf{r} \times \mathbf{v} \, dm = \int (\mathbf{r}_{cm} + \mathbf{r}') \times \mathbf{v} \, dm$$

$$= \mathbf{r}_{cm} \times \int \mathbf{v} \, dm + \int \mathbf{r}' \times \mathbf{v} \, dm$$

$$= \mathbf{r}_{cm} \times \mathbf{P} + \int \mathbf{r}' \times (\mathbf{v}' + \mathbf{v}_{cm}) \, dm$$

$$= \mathbf{r}_{cm} \times \mathbf{P} + \int \mathbf{r}' \times \mathbf{v}' \, dm + \left(\int \mathbf{r}' \, dm \right) \times \mathbf{v}_{cm}$$

$$= \mathbf{r}_{cm} \times \mathbf{P} + \mathbf{L}_C$$

since, by definition of the center of mass, $\int \mathbf{r}' \, dm = 0$.

The first term, $\mathbf{r}_{cm} \times \mathbf{P}$, is the angular momentum of the center of mass, with respect to O; this is called the *orbital angular momentum* of the body (with respect to O). The second term is just the angular momentum of the body about its center of mass; this is called the *spin angular momentum* of the body. In short,

$$\mathbf{L} = \mathbf{L}_{orb} + \mathbf{L}_{spin}$$

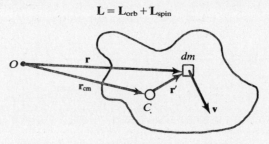

Fig. 12-2

12.5. A hoop of radius 0.10 m and mass 0.50 kg rolls across a table parallel to one edge, with a speed of 0.50 m/s. Refer its motion to a rectangular coordinate system with the origin at the left rear corner of the table. At a certain time t, a line drawn from the origin to the point of contact of the hoop with the table has length 1 m and makes an angle of 30° with the X axis (Fig. 12-3). What is the angular momentum of the hoop with respect to the origin at this time t?

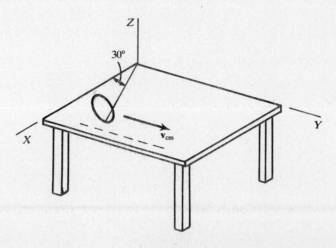

Fig. 12-3

Use the decomposition of Problem 12.4. The position vector of the center of mass at the time t is

$$\mathbf{r}_{cm} = \mathbf{i}(\cos 30°) + \mathbf{j}(\sin 30°) + \mathbf{k}(0.10) = 0.866\mathbf{i} + 0.5\mathbf{j} + 0.10\mathbf{k}$$

and the total momentum of the hoop is

$$\mathbf{P} = m\mathbf{v}_{cm} = (0.50)(0.50\mathbf{j}) = 0.25\mathbf{j}$$

Thus

$$\mathbf{L}_{orb} = \mathbf{r}_{cm} \times \mathbf{P} = (0.866\mathbf{i} + 0.5\mathbf{j} + 0.10\mathbf{k}) \times 0.25\mathbf{j}$$
$$= -0.025\mathbf{i} + 0.216\mathbf{k} \quad \text{kg} \cdot \text{m}^2/\text{s}$$

To find the spin angular momentum, note that every element of mass of the hoop is at the same distance from the center of mass, $r' = 0.10$ m, and every element rotates about the center of mass with a velocity $\mathbf{v}'$ (of magnitude 0.50 m/s) perpendicular to $\mathbf{r}'$. Thus,

$$\mathbf{L}_{spin} = \int \mathbf{r}' \times \mathbf{v}' \, dm = \int r'v'(-\mathbf{i}) \, dm = -mr'v'\mathbf{i} = -0.025\mathbf{i} \quad \text{kg} \cdot \text{m}^2/\text{s}$$

and

$$\mathbf{L} = \mathbf{L}_{orb} + \mathbf{L}_{spin} = -0.05\mathbf{i} + 0.216\mathbf{k} \quad \text{kg} \cdot \text{m}^2/\text{s}$$

12.6. The rigid pendulum shown in Fig. 12-4 is made up of two, nearly weightless bars carrying equal masses m. The pendulum swings in the vertical plane about a frictionless hinge O. Find the equation of motion of the pendulum.

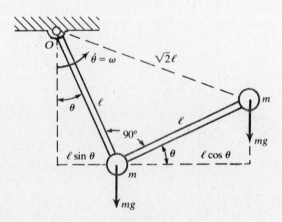

Fig. 12-4

The torque of the weights about the hinge is

$$\tau_{ext} = -mg\ell \sin\theta - mg\ell(\sin\theta + \cos\theta) = -mg\ell(2\sin\theta + \cos\theta)$$

The angular momentum of the pendulum about the hinge is

$$L = m\ell^2\dot\theta + m(\sqrt{2}\ell)^2\dot\theta = 3m\ell^2\dot\theta$$

and so the equation of motion, $\tau_{ext} = \dot L$, is

$$-mg\ell(2\sin\theta + \cos\theta) = 3m\ell^2\ddot\theta \qquad \text{or} \qquad \ddot\theta + \frac{g}{3\ell}(2\sin\theta + \cos\theta) = 0$$

For this system, $\mathbf{L}$ and $\boldsymbol{\omega}$ are both along the rotation axis, with $L = I\omega$ and $\dot L = I\dot\omega = I\alpha$. Hence,

$$\tau_{ext} = I\alpha$$

could also have been used as the equation of motion.

12.7. Refer to Problem 5.14 and Fig. 5-14(a). Find the direction of the force exerted on the particle by the tube.

Let us apply the principle of angular momentum to the particle, in an inertial frame with origin O. The angular momentum of the particle about O is

$$rp_\theta = r(mr\omega) = mr^2\omega$$

where r is the distance from the axis through O. The torque on the particle is rF_θ, where F_θ is the normal force that the tube exerts on the particle. Hence,

$$rF_\theta = \frac{d}{dt}(mr^2\omega) = 2m\omega r\frac{dr}{dt}$$

or
$$F_\theta = m\left(2\omega\frac{dr}{dt}\right)$$

Since r decreases with time (the particle is moving toward O), F_θ is negative; that is, the force opposes the rotation of the particle. (Is this the result you had predicted?)

This problem shows how considerations of angular momentum lead to the coriolis acceleration, $2\omega\dot{r}$ [see Problem 7.4(b)].

12.8. Assume that the center of mass of a girl crouching in a light swing has been raised to 1.2 m (see Fig. 12-5). The girl weighs 400 N, and her center of mass is 3.7 m from the pivot of the swing while she is in the crouched position. The swing is released from rest, and at the bottom of the arc the girl stands up instantaneously, thus raising her center of mass 0.6 m (returning it to its original level). Find the height of her center of mass at the top of the arc.

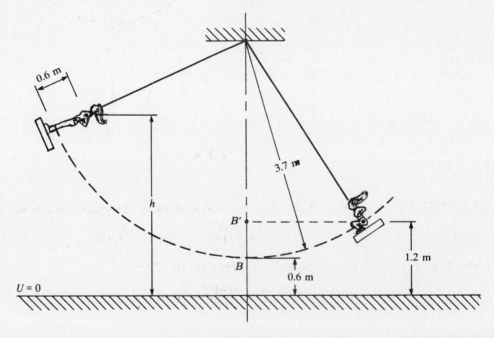

Fig. 12-5

The torque due to gravity is the only external torque acting about the pivot of the swing. This torque is zero at B and remains zero for the instant during which the girl stands up. Thus angular momentum is conserved at B; in fact, *orbital* angular momentum is conserved, since the girl straightens without rotating about her center of mass.

$$3.7mv_B = 3.1mv'_B \qquad \text{or} \qquad v'_B = 1.2v_B$$

where v_B and v'_B are the speeds of the swing at the bottom, before and after the girl stands up.

From the conservation of energy,

$$\tfrac{1}{2} m v_B^2 = mg(1.2 - 0.6) \qquad \text{or} \qquad v_B = 3.43 \text{ m/s}$$

Then $v_B' = 1.2\, v_B = 4.1$ m/s. Again we can use the conservation principle to write

$$\tfrac{1}{2} m v_B'^2 = mg(h - 1.2) \qquad \text{or} \qquad h = \frac{v_B'^2}{2g} + 1.2 = 2.1 \text{ m}$$

12.9. A bead of mass m is constrained by a light inextensible cord to move in a circular path of radius R on a frictionless horizontal plane (see Fig. 12-6). The angular speed of the section of string from O to the bead is ω_0, initially. The pull T exerted on the string is increased until the distance from O to the bead is $R/4$. Find (a) the ratio of the final angular speed to the initial angular speed and (b) the ratio of the final tension in the string to the initial tension.

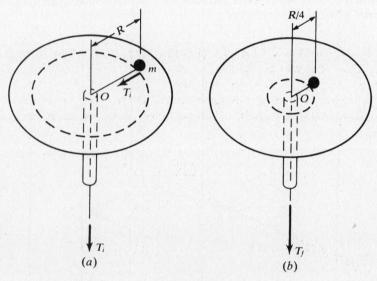

Fig. 12-6

(a) In this situation (a central force) there is no torque about O. Therefore, $l = \text{constant}$, or

$$m R^2 \omega_i = m\left(\frac{R}{4}\right)^2 \omega_f \qquad \text{or} \qquad \frac{\omega_f}{\omega_i} = 16$$

(b) Now, $T = mr\omega^2$, since T supplies the centripetal force. Thus,

$$\frac{T_f}{T_i} = \frac{(R/4)\omega_f^2}{R\omega_i^2} = \frac{1}{4}(16)^2 = 64$$

12.10. A man sits on a stool that rotates about a vertical axis. The moment of inertia of the man and the stool about the axis is $8 \text{ kg} \cdot \text{m}^2$. The man holds a 22 N weight in each hand. With his arms first outstretched, putting the weights a distance of 0.9 m from the axis of the stool, the man is set in rotation at 4 rad/s. The man then pulls the weights inward so that they are at the axis of rotation. What is the final angular speed of the man and stool?

Because the system is symmetric, its center of mass lies on the axis of rotation, and there is no gravitational torque about that axis. Hence, all other forces being internal,

$$\mathbf{L} = \text{constant} \qquad \text{or} \qquad (I_1 + I_2)\omega_0 = I_3\omega$$

where I_1 = moment of inertia of extended weights
         I_2 = moment of inertia of stool and man with arms outstretched
         I_3 = moment of inertia of stool and man with arms indrawn

We have

$$I_1 = 2mr^2 = 2\left(\frac{22}{9.8}\right)(0.9)^2 = 3.6 \text{ kg} \cdot \text{m}^2 \qquad I_2 \approx I_3 = 8 \text{ kg} \cdot \text{m}^2$$

and so

$$\omega = \frac{I_1 + I_2}{I_3}\,\omega_0 \approx \frac{3.6 + 8}{8}\,(4) = 5.8 \text{ rad/s}$$

Note that angular momentum would *not* be conserved if, for example, the man held 44 N in *one* hand.

12.11. A bead is set in motion along the inside rim of a hemispherical cup with a horizontal velocity v_0. If the bead slides without friction, will its path ever pass through the bottom of the cup (see Fig. 12-7)?

Relative to O, an external torque

$$\boldsymbol{\tau} = \mathbf{r} \times (\mathbf{N} + \mathbf{w})$$

acts on the bead, but this torque has no Z component. Now, the relation $\boldsymbol{\tau} = d\mathbf{l}/dt$ implies that $\tau_z = dl_z/dt$; hence

$$l_z = \text{constant}$$

for the motion of the bead.   At $t = 0$, $l_z = mv_0R > 0$; thus the bead cannot reach O, where l_z (and, in fact, $\mathbf{l}$) would vanish.

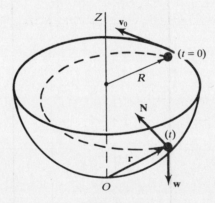

Fig. 12-7

12.12. Show that if $\mathbf{r}$, the position vector of a particle with respect to a certain origin, is at some instant parallel to its acceleration, $\ddot{\mathbf{r}}$, then the time rate of change of the angular momentum $\mathbf{l}$ with respect to the given origin is zero at that instant.

At the given instant, $\mathbf{r}$ and the force, $m\ddot{\mathbf{r}}$, are parallel, so that $\boldsymbol{\tau} = 0$.   Then, $\dot{\mathbf{l}} = 0$.

Supplementary Problems

12.13. A 0.2 kg ball, traveling at a speed of 5 m/s in a direction perpendicular to an upright door, strikes a nail in the door at a point 0.4 m from the axis of the door's hinges. The ball bounces off with a speed of 3 m/s at an angle of 30° with the plane of the door. Find the ball's angular momentum about the axis of the hinges before and after the impact.   *Ans.*   0.40 kg · m²/s; 0.12 kg · m²/s

12.14. Use the principle of angular momentum to find the acceleration of the center of mass of a sphere which rolls down a rough plane inclined at an angle θ with the horizontal. (Compare Problem 11.14.)
Ans. $a_{cm} = 5/7g \sin \theta$

12.15. Prove that the angular momentum of a system of particles has the same value for different reference points O and O' that are fixed in the center-of-mass frame. (*Hint:* Let O be the center of mass itself.)

12.16. (*a*) Calculate the orbital angular momentum of the earth with respect to an origin at the center of the sun. Take the earth's orbit as a circle of radius 1.5×10^8 km and assume that the earth travels uniformly at a speed of 30 km/s in the orbit. The earth's mass is 6.0×10^{24} kg. (*b*) Express the spin angular momentum of the earth as a fraction of its orbital angular momentum. Take the earth as a uniform sphere of radius 6400 km.   *Ans.*   (*a*) 2.7×10^{34} J · s; (*b*) 2.7×10^{-7}

12.17. A small body of mass m starts from rest at the top of a smooth inclined plane of angle θ. Let **N** represent the normal force acting on the body, and let **v** represent its velocity when it has moved a distance x down the plane. (*a*) Find the torque, with respect to an origin at the top of the inclined plane, produced by all the forces acting on the body. (*b*) Calculate the angular momentum **l** and its time derivative, with respect to the same origin. (*c*) What is the magnitude of **N**?
Ans. $\pm(N - mg \cos \theta)x$;   (*b*) 0, 0;   (*c*) $mg \cos \theta$

12.18. How much work is done by (*a*) the girl in Problem 12.8, (*b*) the man in Problem 12.10?
Ans.   (*a*) 240 J;   (*b*) 42 J

Chapter 13

Gravitation

13.1 GRAVITATIONAL FIELD

A particle of mass M is the source of a gravitational field which, at a distance r from M, has a *field intensity* **g** of magnitude

$$g = \frac{GM}{r^2}$$

and a direction toward M. The gravitational field exerts a force $\mathbf{F} = m\mathbf{g}$ on a particle of mass m at the given point.

$$G = 6.67 \times 10^{-11} \text{ N} \cdot \text{m}^2/\text{kg}^2$$

is called the *gravitational constant*. The units for **g** are N/kg (1 N/kg = 1 m/s^2).

The gravitational field caused by a spherically symmetrical distribution of matter is, at a point outside this distribution, the same as if all the matter were concentrated at the center.

13.2 GRAVITATIONAL FORCE

Newton's law of gravitation states that any two particles, of masses m and M and separated by a distance r, exert attractive forces on each other of magnitude

$$F = G \frac{Mm}{r^2}$$

The force is along the line connecting the particles.

13.3 GRAVITATIONAL POTENTIAL ENERGY

A particle of mass m, in an external gravitational field of intensity $g = GM/r^2$ (due to the presence of mass M), has a *gravitational potential energy*

$$U = -G \frac{Mm}{r}$$

Note that the reference level for the potential energy is $r = \infty$; for the approximation used in earlier chapters, $U \approx mgh$, the reference level could be chosen arbitrarily.

To escape the gravitational field (i.e. to reach $r = \infty$ with zero kinetic energy) the particle m must have the *escape speed*, v_e, given by

$$\frac{1}{2}mv_e^2 - G\frac{Mm}{r} = 0 + 0 \qquad \text{or} \qquad v_e = \sqrt{\frac{2GM}{r}}$$

The gravitational potential energy of any pair of particles due to their own gravitational interaction is

$$U = -G\frac{m_1 m_2}{r_{12}}$$

where m_1 and m_2 are the masses of the particles and r_{12} is their separation.

13.4 KEPLER'S LAWS. ORBITS

Kepler's laws: (1) The planets move around the sun in orbits which are ellipses, with the sun at one focus. (2) The radius vector from the sun to a planet sweeps out equal areas in equal times. (3) The ratio of the square of the orbital period to the cube of the ellipse's semimajor axis is the same for all planets.

The motion of an object of mass m under a central attractive force is characterized by two constants: E, the object's total energy, and L, the magnitude of the object's angular momentum about the center of force.

In particular, as is shown in Problem 13.5, the orbit of a particle of mass m, acted on by the attractive force GMm/r^2 directed toward a particle of mass M fixed at the origin of an inertial frame, is a conic section with one focus at the origin.

If the line $\theta = 0$ is chosen as the axis of symmetry of the conic section, its equation in polar coordinates may be written as

$$\frac{R}{r} = 1 + e \cos \theta \qquad (13.1)$$

The fundamental invariants E and L determine the parameters R (the *semi-latus rectum*) and e (the *eccentricity*) through

$$R = \frac{L^2}{GMm^2} \qquad e = \sqrt{1 + \frac{2EL^2}{G^2M^2m^3}} \qquad (13.2)$$

The orbit is an ellipse for $E < 0$ (a circle for $E < 0$ and $e = 0$), a parabola for $E = 0$, and a hyperbola for $E > 0$. Figure 13-1 indicates the geometrical significance of R and e in the case of an elliptical orbit.

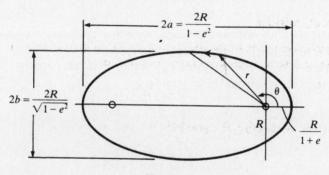

Fig. 13-1

The period in an elliptical orbit of semimajor axis

$$a = \frac{R}{1 - e^2} = \frac{GMm}{2(-E)}$$

is given by (see Problem 13.6)

$$T = \frac{2\pi a^{3/2}}{\sqrt{GM}} = 2\pi GM \left[\frac{m}{2(-E)} \right]^{3/2} \qquad (13.3)$$

Observe that the semimajor axis or the period is a measure of the particle's energy per unit mass, E/m.

13.5 GAUSS'S LAW

The gravitational flux into an arbitrary closed surface is $4\pi GM_i$, where M_i is the mass of the matter inside the surface. This law may be written in vector form as

$$\int_S \mathbf{g} \cdot d\mathbf{S} = -4\pi GM_i$$

Here, $d\mathbf{S} = \mathbf{n} \, dS$ is the directed element of area of the enclosing surface, $\mathbf{n}$ being the unit *outward* normal.

Solved Problems

13.1. What is the gravitational force exerted on a 40.0 gram stone by a 10.0 kg boulder which is 20.0 cm away?

$$F = G\,\frac{mM}{r^2} = \frac{(6.67 \times 10^{-11})(0.0400)(10.0)}{(0.200)^2} = 6.67 \times 10^{-10} \text{ N}$$

13.2. Find the gravitational field intensity (or gravitational acceleration) g on the surface of Mars, given that the radius of Mars is 3400 km and its mass is 0.11 the mass of the earth. The radius of the earth is 6400 km.

For earth, the surface gravitational acceleration (neglecting rotation) is

$$g_E = \frac{GM_E}{R_E^2}$$

and for Mars,

$$g_{\text{Mars}} = \frac{GM_{\text{Mars}}}{R_{\text{Mars}}^2}$$

Putting these two equations together,

$$g_{\text{Mars}} = g_E\,\frac{M_{\text{Mars}}R_E^2}{M_E R_{\text{Mars}}^2} = (9.8)\left(\frac{0.11\,M_E}{M_E}\right)\left(\frac{64}{34}\right)^2 = 3.8 \text{ m/s}^2$$

13.3. Consider the situation shown in Fig. 13-2, with the particle originally at the point $y = a$ $(a > 2R)$ and moving away from the planet of mass M at the speed v_0. Find (a) the maximum distance h from the center of the planet that the particle can reach, (b) its speed when it is at a distance $a/2$ from the center.

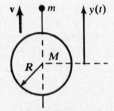

Fig. 13-2

(a) By conservation of energy,

$$\frac{1}{2}mv_0^2 - \frac{GMm}{a} = 0 - \frac{GMm}{h} \qquad \text{or} \qquad h = \frac{a}{1 - \dfrac{av_0^2}{2GM}}$$

(b) Again by conservation of energy,

$$\frac{1}{2}mv_0^2 - \frac{GMm}{a} = \frac{1}{2}mv^2 - \frac{GMm}{a/2} \qquad \text{or} \qquad v = -\sqrt{v_0^2 + \frac{2GM}{a}}$$

where the minus sign means that the particle is moving toward the planet.

13.4. A spaceship is at rest 10^8 m from the center of the earth. Calculate the speed with which it would hit the earth's surface if the earth had no atmosphere. Take the earth's mass as $M = 6.0 \times 10^{24}$ kg and its radius as $R = 6400$ km.

By conservation of energy,

$$0 - \frac{GMm}{r_0} = \frac{1}{2}mv^2 - \frac{GMm}{R}$$

or

$$v = -\sqrt{2GM\left(\frac{1}{R} - \frac{1}{r_0}\right)} = -\sqrt{2(6.7 \times 10^{-11})(6.0 \times 10^{24})\left(\frac{1}{6.4 \times 10^6} - \frac{1}{10^8}\right)}$$

$$= -1.1 \times 10^4 \text{ m/s} = -11 \text{ km/s}$$

13.5. Derive (*13.1*) for the path of a body moving under the influence of the gravitational force.

From Problem 7.4(*b*), the radial equation of motion is

$$F_r = -\frac{GMm}{r^2} = m\left[\frac{d^2r}{dt^2} - r\left(\frac{d\theta}{dt}\right)^2\right] \tag{1}$$

Now, for a central force, angular momentum is conserved:

$$mr^2\frac{d\theta}{dt} = L = \text{constant}$$

Use this relation to eliminate t from (*1*).

$$\frac{dr}{dt} = \frac{dr}{d\theta}\frac{d\theta}{dt} = \frac{L}{mr^2}\frac{dr}{d\theta} = -\frac{L}{m}\frac{d}{d\theta}\left(\frac{1}{r}\right)$$

Further,

$$\frac{d^2r}{dt^2} = \frac{d}{dt}\left(\frac{dr}{dt}\right) = \left[\frac{d}{d\theta}\left(\frac{dr}{dt}\right)\right]\frac{d\theta}{dt}$$

$$= \frac{L}{mr^2}\frac{d}{d\theta}\left(\frac{dr}{dt}\right) = \frac{L}{mr^2}\frac{d}{d\theta}\left[-\frac{L}{m}\frac{d}{d\theta}\left(\frac{1}{r}\right)\right] = -\frac{L^2}{m^2r^2}\frac{d^2}{d\theta^2}\left(\frac{1}{r}\right)$$

Substituting in the right-hand side (*1*), we obtain the following differential equation for $u = 1/r$:

$$\frac{d^2u}{d\theta^2} + u = \frac{1}{R} \tag{2}$$

where $R \equiv L^2/GMm^2$.

It is easy to see that a particular solution of (*2*) is $u_p = 1/R$, and that the general solution of the homogeneous equation is

$$u_h = A\cos\theta + B\sin\theta$$

where A and B are arbitrary constants. Thus,

$$u = u_p + u_h = \frac{1}{R} + A\cos\theta + B\sin\theta \tag{3}$$

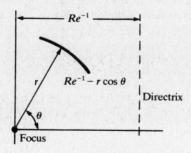

Fig. 13-3

Nothing is lost by setting $B = 0$; this merely rotates the coordinate system about the force center. Then, writing $A \equiv e/R$, we have from (3):

$$u = \frac{1}{R} + \frac{e}{R} \cos \theta \qquad \text{or} \qquad \frac{R}{r} = 1 + e \cos \theta \qquad\qquad (4)$$

Equation (4) is the standard equation of a conic section: rewritten as

$$\frac{r}{Re^{-1} - r \cos \theta} = e$$

it shows that the distance of a point on the curve from the focus (r) and its distance from the directrix $(Re^{-1} - r \cos \theta)$ are in constant ratio (e). See Fig. 13-3. The remaining arbitrary constant in (4), e, is determined by energy conservation. See Problem 13.16.

13.6. Derive (13.3) for the period of a body moving in an elliptical orbit under the influence of the gravitational force.

A triangular element of area of the ellipse (shown shaded in Fig. 13-1) is

$$dA = \frac{1}{2} r^2 \, d\theta$$

The rate of sweeping out area is then

$$\frac{dA}{dt} = \frac{1}{2} r^2 \frac{d\theta}{dt} = \frac{L}{2m} = \text{constant}$$

in accordance with Kepler's second law. The area of the ellipse is πab; hence, the period is given by

$$\int_0^{\pi ab} dA = \frac{L}{2m} \int_0^T dt \qquad \text{or} \qquad T = \frac{2\pi mab}{L}$$

Now, from Fig. 13-1, and from (13.2),

$$\frac{b}{a} = \sqrt{1 - e^2} = \sqrt{\frac{R}{a}} = \frac{L}{m\sqrt{GMa}}$$

whence

$$T = \frac{2\pi ma}{L} \frac{La}{m\sqrt{GMa}} = \frac{2\pi a^{3/2}}{\sqrt{GM}}$$

13.7. Is it possible for a particle moving under a central force to pass through the center of force?

Angular momentum about the center of force is conserved, so that the particle's angular momentum must always be zero if the particle once passes through the center of force. This is possible only if the particle's path is a straight line through the center.

13.8. A space capsule is to be returned to the earth from a satellite moving in a circular orbit at an altitude of $12\,600$ km above the earth's surface. It is desired that the capsule leave the satellite tangentially and make a tangential landing on earth. With what speed should the capsule be sent off, (a) measured relative to the center of the earth? (b) measured relative to the satellite?

(a) Choose an inertial frame with origin at the center of the earth, such that the capsule is at $\theta = 0$ at the moment of launching. It is seen from Fig. 13-4 that the desired conditions will be met if the capsule's path BB' is half of an ellipse of major axis $2a = r_0 + R_E$. But, from Fig. 13-1,

$$2a = \frac{2R}{1 - e^2} = \frac{GM_E m}{-E}$$

so that the capsule's energy per unit mass is determined as

$$\frac{E}{m} = -\frac{GM_E}{r_0 + R_E}$$

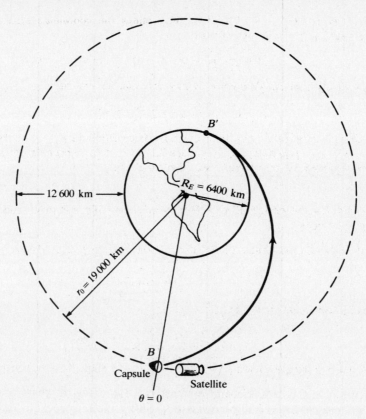

Fig. 13-4

Now, at B, the kinetic and potential energies per unit mass are $v_0^2/2$ and $-GM_E/r_0$, respectively.   Hence

$$-\frac{GM_E}{r_0+R_E}=\frac{v_0^2}{2}-\frac{GM_E}{r_0}$$

$$v_0=\sqrt{\frac{2GM_ER_E}{r_0(r_0+R_E)}}=\sqrt{\frac{2(6.67\times10^{-11})(6.0\times10^{24})(6.4\times10^6)}{(19\times10^6)(19\times10^6+6.4\times10^6)}}$$

$$=3.30\times10^3 \text{ m/s}$$

or 3.30 km/s.

(b)   Since the satellite is in a circular orbit of radius r_0, we can write, in the inertial frame,

$$F_{\text{centripetal}}=F_{\text{gravitational}}$$

$$\frac{mv^2}{r_0}=\frac{GmM_E}{r_0^2}$$

Thus

$$v=\sqrt{\frac{GM_E}{r_0}}=\sqrt{\frac{(6.67\times10^{-11})(6.0\times10^{24})}{1.9\times10^7}}=4.68 \text{ km/s}$$

and the initial speed of the capsule relative to the satellite is

$$v_0-v=3.30-4.68=-1.38 \text{ km/s}$$

13.9.   A space probe approaches the planet Venus along a parabolic path.   As the vehicle reaches the position B shown in Fig. 13-5, where it will be closest to Venus, retrorockets are fired to slow it down and put it into an elliptical orbit which will bring it to a tangential landing at position A.   At B the distance from the center of the planet to the probe is 16 090 km.   The

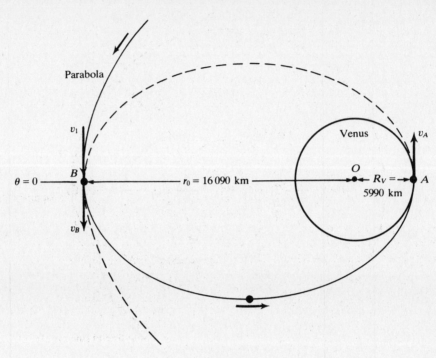

Fig. 13-5

radius of the planet is 5990 km and its mass is 0.815 times the mass of the earth.   Find the speed of the vehicle (*a*) as it approaches *B*, (*b*) after the retrorockets have been fired, (*c*) as it lands at *A*.

(*a*)   In the parabolic orbit, $E = 0$.   Thus

$$\frac{1}{2}mv_1^2 - \frac{GM_V m}{r_0} = 0 \quad \text{or} \quad v_1 = \sqrt{\frac{2GM_V}{r_0}} = \sqrt{\frac{2(6.67 \times 10^{-11})(0.815 \times 6 \times 10^{24})}{1.609 \times 10^7}} = 637 \text{ km/s}$$

(*b*)   Using the result of Problem 13.8,

$$v_B = \sqrt{\frac{2GM_V R_V}{r_0(r_0 + R_V)}} = \sqrt{\frac{2(6.67 \times 10^{-11})(0.815 \times 6 \times 10^{24})(5.99 \times 10^6)}{(16.09 \times 10^6)(16.09 \times 10^6 + 5.99 \times 10^6)}} = 3.31 \text{ km/s}$$

(By coincidence, this speed is very close to the speed v_0 found in Problem 13.8.)

(*c*)   By conservation of angular momentum,

$$r_0 m v_B = R_V m v_A \quad \text{or} \quad v_A = \frac{r_0}{R_V} v_B = \frac{16\,090}{5990}(3.31) = 8.87 \text{ km/s}$$

13.10. Two rocket ships, S_1 and S_2, are moving in coplanar, circular, counterclockwise orbits of radii r_1 and $r_2 = 6r_1$, respectively, about the earth.   A supply rocket is to be launched from S_1 in a direction tangent to its orbit and is to reach S_2 with a velocity tangent to the orbit of S_2; the supply rocket is to travel in a free-flight condition after launching.   (*a*) Find the launching speed of the supply rocket relative to S_1.   (*b*) What will have to be accomplished for the supply rocket to dock on S_2?   (*c*) Find the angle β giving the position of S_2 relative to S_1 at the time of launching the supply rocket.   See Fig. 13-6 for a sketch of the situation.

(*a*)   Choose an inertial frame fixed in the center of the earth and proceed as in Problem 13.8.   The major axis of the elliptical orbit of the supply rocket is $r_1 + r_2 = 7r_1$, and this gives

$$v_A = \sqrt{\frac{12GM_E}{7r_1}}$$

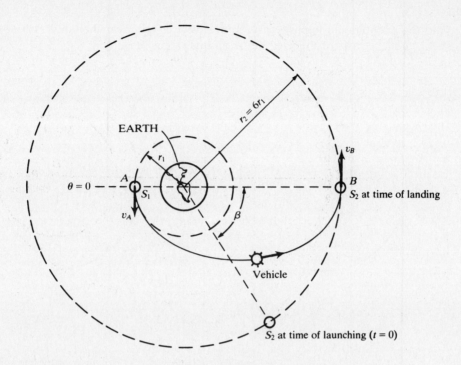

Fig. 13-6

The velocity of S_1 in its circular orbit is

$$v_1 = \sqrt{\frac{GM_E}{r_1}}$$

Thus the relative velocity of launching is

$$\sqrt{\frac{12GM_E}{7r_1}} - \sqrt{\frac{GM_E}{r_1}} = \left(\sqrt{\frac{12}{7}} - 1\right)v_1$$

(b) The supply rocket's speed at B is, by conservation of angular momentum,

$$v_B = \frac{r_1}{r_2}v_A = \frac{1}{6}\sqrt{\frac{12GM_E}{7r_1}} = \sqrt{\frac{GM_E}{21r_1}}$$

whereas the orbital speed of S_2 is

$$v_2 = \sqrt{\frac{GM_E}{r_2}} = \sqrt{\frac{GM_E}{6r_1}} > v_B$$

Therefore, on arrival at B, the supply rocket's engines must be fired to increase its speed up to that of S_2.

(c) The time taken for the supply rocket to travel from A to B is, by (13.3),

$$\frac{T}{2} = \frac{\pi a^{3/2}}{\sqrt{GM_E}} = \frac{\pi}{\sqrt{GM_E}}\left(\frac{7r_1}{2}\right)^{3/2}$$

while the time for S_2 to revolve through angle β is

$$\frac{r_2\beta}{v_2} = 6r_1\sqrt{\frac{6r_1}{GM_E}}\,\beta = \frac{1}{\sqrt{GM_E}}(6r_1)^{3/2}_\beta$$

Equating these two times,

$$\pi\left(\frac{7}{2}\right)^{3/2} = 6^{3/2}\beta \qquad \text{or} \qquad \beta = \left(\frac{7}{12}\right)^{3/2}\pi \text{ rad} = 61.2°$$

13.11. Find the gravitational intensity at a point inside a spherical shell and at a point outside the shell.   Let the mass of the shell be M and its radius R.

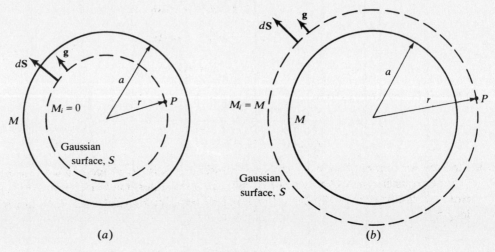

Fig. 13-7

Apply Gauss's theorem by drawing a concentric sphere S of radius r inside the spherical shell [Fig. 13-7(a)] through the point P.   By symmetry, **g** must be normal and of constant magnitude on S.

$$\int_S \mathbf{g} \cdot d\mathbf{S} = -4\pi G M_i$$

$$g(4\pi r^2) = -4\pi G(0)$$

$$g = 0 \qquad (r < a)$$

For an outside point, draw a concentric sphere S [Fig. 13-7(b)] around the spherical shell and passing through the point under consideration.   Again, **g** must be normal and of constant magnitude on S.

$$\int_S \mathbf{g} \cdot d\mathbf{S} = -4\pi G M_i$$

$$g(4\pi r^2) = -4\pi G M$$

$$g = -G\frac{M}{r^2} \qquad (r > a)$$

The negative sign shows that **g** was given the wrong direction in Fig. 13-7(b); the actual field is radially inwards.

13.12. Find the gravitational intensity at a point inside a uniform solid sphere of mass M and of radius a, and at a point outside the sphere.

Put a concentric sphere S of radius r inside the solid sphere [Fig. 13-7(a)] through P.   By Gauss's theorem:

$$\int_S \mathbf{g} \cdot d\mathbf{S} = -4\pi G M_i$$

$$g(4\pi r^2) = -4\pi G\left(\frac{r^3}{a^3}M\right)$$

$$g = -\frac{GMr}{a^3} \qquad (r < a)$$

i.e. $+GMr/a^3$, radially inwards.

For an outside point, draw a concentric sphere S [Fig. 13-7(b)] of radius r around the solid sphere running through the point under consideration.   By Gauss's theorem:

$$\int_S \mathbf{g} \cdot d\mathbf{S} = -4\pi G M_i$$

$$g(4\pi r^2) = -4\pi G M$$

$$g = -G\frac{M}{r^2} \qquad (r > a)$$

i.e. $+GM/r^2$, radially inwards.

13.13. Find the gravitational intensity at a point inside a shell in the shape of a long, uniform circular cylinder of radius a, and at a point outside the shell.   Assume that the points are far enough from either end of the cylinder so that $\mathbf{g}$ can be considered radial.   Let λ be the mass per unit length.

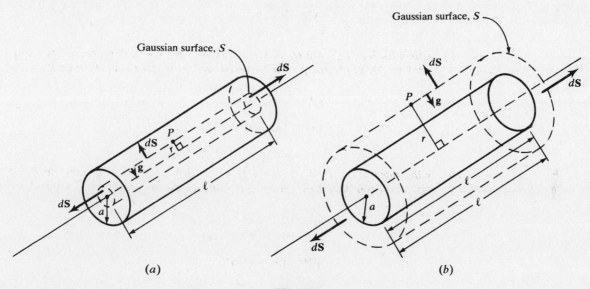

Fig. 13-8

For an interior point P, choose the Gaussian surface shown in Fig. 13-8(a).   Noting that $\mathbf{g} \cdot d\mathbf{S} = 0$ on the plane ends of S, we have

$$\int_S \mathbf{g} \cdot d\mathbf{S} = -4\pi G M_i$$

$$-g(2\pi r\ell) = -4\pi G(0)$$

$$g = 0 \qquad (r < a)$$

and for an exterior point P [Fig. 13-8(b)],

$$\int_S \mathbf{g} \cdot d\mathbf{S} = -4\pi G M_i$$

$$-g(2\pi r\ell) = -4\pi G \lambda \ell$$

$$g = 2G\frac{\lambda}{r} \qquad (r > a)$$

13.14. Find the gravitational intensity at a point on the axis of a circular disk of radius a, if the surface density at any distance r from the center is $\sigma = \lambda r$. The point is up the axis a distance z (Fig. 13-9).

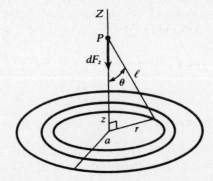

Fig. 13-9

If the disk is divided into a set of concentric rings, the total attraction on a unit mass at P—that is, **g**—can be found as the sum of the attractions of all the rings making up the disk. For a ring, the attraction at P will, by symmetry, have only a Z component, dF_z, given by

$$dF_z = \frac{G\,dm}{\ell^2}\cos\theta = \frac{Gz\,dm}{(r^2 + z^2)^{3/2}}$$

where $dm = \sigma(2\pi r\,dr) = 2\pi\lambda r^2\,dr$ is the mass of the ring. Thus

$$g = \int dF_z = 2\pi\lambda Gz \int_0^a \frac{r^2\,dr}{(r^2 + z^2)^{3/2}}$$

$$= 2\pi\lambda Gz\left[-\frac{r}{\sqrt{r^2 + z^2}} + \ln(r + \sqrt{r^2 + z^2})\right]_0^a$$

$$= 2\pi\lambda Gz\left(\ln\frac{a + \sqrt{a^2 + z^2}}{z} - \frac{a}{\sqrt{a^2 + z^2}}\right)$$

$$= 2\pi\lambda Gz\left(\sinh^{-1}\frac{a}{z} - \frac{a}{\sqrt{a^2 + z^2}}\right)$$

Supplementary Problems

13.15. Calculate the gravitational intensity due to the sun, of mass 1.99×10^{30} kg and radius 6.97×10^8 m, at (a) the surface of the sun and (b) the position of the earth ($d = 1.5 \times 10^{11}$ m).
    Ans.   (a) 270 N/kg;   (b) 5.9×10^{-3} N/kg

13.16. Assuming a circular orbit for the earth, determine the sun's gravitational intensity at the position of the earth in terms of the period of the earth. Use your result to check Problem 13.15(b).
    Ans. $4\pi^2 d/T^2$

13.17. Derive the expression (*13.2*) for e from

$$\frac{1}{2}m\left[\left(\frac{dr}{dt}\right)^2 + \left(r\frac{d\theta}{dt}\right)^2\right] - \frac{GMm}{r} = E$$

and

$$mr^2\frac{d\theta}{dt} = L$$

13.18. Express the semiminor axis of a planet's orbit in terms of its energy and angular momentum.

Ans. $b = L\sqrt{\dfrac{m}{2(-E)}}$

13.19. A cannon is fired horizontally. Neglecting atmospheric resistance and treating the earth as an inertial frame, find (*a*) the minimum and (*b*) the maximum muzzle velocity that will put the shell into periodic orbit about the earth.   *Ans.*   (*a*) 7.9 km/s;   (*b*) 11.2 km/s (the escape speed)

13.20. Prove that for a planet's elliptical orbit about the sun, of semimajor axis *a*,

$$\left(\frac{1}{r}\right)_{\text{avg}} = \frac{1}{a}$$

where the average is a *time*-average over one period of the motion.   (*Hint*: The definite integral

$$\int_0^{2\pi} \frac{d\theta}{1 + e\cos\theta} = \frac{2\pi}{\sqrt{1 - e^2}}$$

may be used.)

13.21. An important result for bound systems is that the total energy equals one-half the time-averaged potential energy (both energies being negative). Use Problem 13.20 to verify this result for the case of a planet orbiting the sun.

13.22. Find the gravitational attraction on a particle of unit mass at a point on the axis of a thin uniform circular cylinder. Assume that the point is a distance *c* above the top of the cylinder; the height of the cylinder is ℓ; the radius of the cylinder is *a*; and the density of the cylinder is ρ.

Ans. $2\pi G\rho[\ell - \sqrt{(\ell + c)^2 + a^2} + \sqrt{c^2 + a^2}]$

Chapter 14

Elasticity and Harmonic Motion

14.1 ELASTICITY AND HOOKE'S LAW

A body is said to be *elastic* if it returns to its original shape and dimensions upon the removal of a deforming force.

Stress is the ratio of the force exerted on a body to the area over which the force is exerted. *Linear strain* is the change in length per unit length of a deformed body, $(\Delta L)/L$. *Volume strain* is the change in volume per unit volume of the body, $(\Delta V)/V$. There may also be a *shearing strain*, given by $\tan \theta \approx \theta$, where θ is the angular change in shape of a body from its normal shape. The unit of stress is the N/m^2 or the *pascal* ($1\,Pa = 1\,N/m^2$); strain is a dimensionless quantity.

Hooke's law states that for a deformed elastic body, within the elastic limits, the stress in the body is proportional to the strain of the body:

$$\text{stress} = (\text{elastic modulus})(\text{strain})$$

For small deformations, the elastic moduli are constant.

14.2 SIMPLE HARMONIC MOTION

A body slightly displaced from equilibrium may be modeled by a particle of mass m attached to a spring. When the particle is let loose, the system undergoes oscillation. The restoring force in the spring when it is elongated is related to the elongation (x) as

$$F_{res} = -kx$$

where k is the *spring constant* (the force that will produce a unit elongation of the spring). The oscillating motion that occurs under F_{res} is called *simple harmonic motion* (SHM).

14.3 EQUATIONS FOR SHM

The *equation of motion* of a particle of mass m undergoing simple harmonic motion attached to a massless spring of constant k is

$$\ddot{x} = -\omega^2 x$$

where $\omega^2 = k/m$.

The *displacement* of the particle varies sinusoidally with time according to the equation

$$x = x_0 \cos(\omega t + \theta_0)$$

where x_0 is the *amplitude*, or maximum displacement, of the oscillating particle; θ_0 is the *phase constant*; and $f = \omega/2\pi$ is the *frequency* of the oscillatory motion.

The *period* is related to the frequency of the motion according to the equation

$$T = \frac{1}{f} = \frac{2\pi}{\omega} = 2\pi\sqrt{\frac{m}{k}}$$

The frequency and period have the units *hertz* ($1\,Hz = 1$ cycle/s) and s, respectively.

14.4 DAMPED HARMONIC MOTION

Simple harmonic motion is an idealization, since in reality the amplitude of the motion gradually becomes smaller because of frictional effects.   If a viscous force acts on a body attached to a spring undergoing vibratory motion, the equation of motion may be written

$$\ddot{x} = -\omega^2 x - \frac{b}{m}\dot{x}$$

where ω is the angular frequency of the corresponding undamped oscillator and b is a positive damping factor.   The minus sign in front of the last term reflects the fact that the viscous force is always opposite in sign to $\dot{x}$ (i.e. always opposes the motion).   For

$$b < 2m\omega = 2\sqrt{mk}$$

the motion is a damped oscillation.   At $b = 2\sqrt{mk}$ (critical damping), the motion becomes nonoscillatory, decaying exponentially to zero.

14.5 POTENTIAL ENERGY OF SIMPLE HARMONIC MOTION

The potential-energy function associated with the Hooke's-law force $F_{res} = -kx$ is

$$U = \frac{1}{2}kx^2$$

14.6 MOTION OF A SIMPLE PENDULUM

The equation of motion of a simple pendulum of length ℓ is

$$m\ell\ddot{\theta} = -mg\sin\theta$$

For small angles ($\theta \lesssim 5°$), the motion is simple harmonic, with period

$$T = 2\pi\sqrt{\frac{\ell}{g}}$$

Solved Problems

14.1. A steel wire is 4.0 m long and 2 mm in diameter.   How much is it elongated by a suspended body of mass 20 kg?   Young's modulus for steel is 196 GN/m^2.

Let ΔL be the elongation.   Then, by Hooke's law,

$$\frac{F}{A} = Y\frac{\Delta L}{L}$$

where Y is Young's modulus.   The elongation is

$$\Delta L = \frac{1}{Y}\frac{F}{A}L = \frac{mgL}{YA} = \frac{(20)(9.8)(4.0)}{(196 \times 10^9)\pi(0.001)^2} = 1.273 \times 10^{-3}\text{ m} = 1.273\text{ mm}$$

14.2. A copper wire 2.0 m long and 2 mm in diameter is stretched 1 mm. What tension is needed? Young's modulus for copper is 117.6 GN/m^2.

$$\frac{F}{A} = Y \frac{\Delta L}{L}$$

$$F = Y \frac{\Delta L}{L} A = (117.6 \times 10^9) \frac{0.001}{2} \pi(0.001)^2 = 184.73 \text{ N}$$

14.3. A wire is stretched 1 mm by a force of 1 kN. (a) How far would a wire of the same material and length but of four times that diameter be stretched? (b) How much work is done in stretching each wire?

(a) The elongation is inversely proportional to the cross-sectional area, and so

$$\Delta L = (1)\left(\frac{1}{4}\right)^2 = \frac{1}{16} \text{ mm}$$

(b) The work done in stretching the wire in the two cases is $W = \bar{F}x$.

$$W_1 = \frac{1000 + 0}{2}(0.001) = 0.5 \text{ J} \qquad W_2 = \frac{1}{16} W_1 = 0.0313 \text{ J}$$

14.4. A block of gelatin is 60 mm by 60 mm by 20 mm when unstressed. A force of 0.245 N is applied tangentially to the upper surface, causing a 5 mm displacement relative to the lower surface (Fig. 14-1). Find (a) the shearing stress, (b) the shearing strain, and (c) the shear modulus.

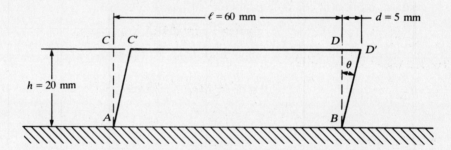

Fig. 14-1

(a)
$$\frac{F}{A} = \frac{0.245}{36 \times 10^{-4}} = 68.4 \text{ Pa}$$

(b)
$$\tan \theta = \frac{d}{h} = \frac{5}{20} = 0.25$$

(c)
$$\mu = \frac{F/A}{\tan \theta} = \frac{68.4}{0.25} = 273.6 \text{ N/m}^2$$

14.5. The pressure in an explosion chamber is 345 MPa. What would be the percent change in volume of a piece of copper subjected to this pressure? The bulk modulus for copper is 138 GPa.

The bulk modulus is defined as $B = -\Delta p/(\Delta V/V)$, where the minus sign is inserted because ΔV is negative when Δp is positive.

$$100 \left| \frac{\Delta V}{V} \right| = 100 \frac{\Delta p}{B} = 100 \frac{345 \times 10^6}{138 \times 10^9} = 0.25\%$$

14.6. A book sits on a horizontal board that is undergoing simple harmonic motion with an amplitude of 1 m. The coefficient of friction between the book and board is $\mu = 0.5$. Find the frequency of the motion of the board at which the book is just about to slip.

In the absence of slippage, the book partakes in the simple harmonic motion, according to the equation

$$x = x_0 \sin (2\pi f t + \theta_0)$$

where $x_0 = 1$ m. The horizontal force on the book, which can only be frictional, is therefore

$$m\ddot{x} = -4\pi^2 f^2 m x_0 \sin (2\pi f t + \theta_0)$$

When $m\ddot{x}$ exceeds the largest possible frictional force, $\mu N = \mu mg$, the book starts to slip. Now, $m\ddot{x}_{\text{max}} = 4\pi^2 f^2 m x_0$, and so

$$4\pi^2 f_{\text{max}}^2 m x_0 = \mu mg \qquad \text{or} \qquad f_{\text{max}} = \frac{1}{2\pi} \sqrt{\frac{\mu g}{x_0}} = \frac{1}{2\pi} \sqrt{\frac{(0.5)(9.8)}{1}} = 0.35 \text{ Hz}$$

14.7. The energy of recoil of a rocket launcher of mass $m = 4536$ kg is absorbed in a recoil spring. At the end of the recoil, a damping dashpot engages so that the launcher returns to the firing position without any oscillation (critical damping). The launcher recoils 3 m with an initial speed of 10 m/s. Find the recoil spring constant (k) and the dashpot's coefficient for critical damping ($b = 2\sqrt{mk}$).

To find the spring constant we may make use of the conservation of energy: $K + U_{\text{elastic}} = \text{constant}$.

$$\frac{1}{2} m v_0^2 + 0 = 0 + \frac{1}{2} k x_{\text{max}}^2$$

$$k = \frac{m v_0^2}{x_{\text{max}}^2} = \frac{(4536)(10)^2}{(3)^2} = 50\,400 \text{ N/m}$$

The coefficient of critical damping is

$$b = 2\sqrt{mk} = 2\sqrt{(4536)(50\,400)} = 30\,240 \text{ N} \cdot \text{s/m}$$

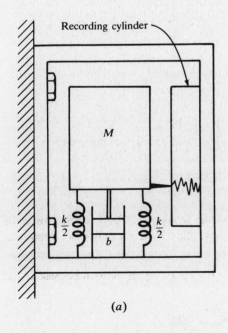

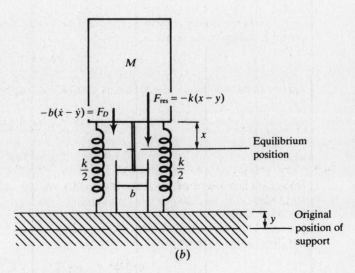

Fig. 14-2

14.8. A seismograph used to record vertical oscillations of the earth is shown in Fig. 14-2(a). When the support for the instrument is displaced a distance y, the body of mass M is displaced a distance x from the equilibrium position. The relative displacement between the inertial member (mass M) and the support is $s = x - y$. Let the oscillation of the earth be given by $y = y_0 \sin \omega t$. Find the differential equation for the relative movement of the inertial member.

Newton's second law for the inertial member of the seismograph is, from Fig. 14-2(b),

$$\sum F = -k(x - y) - b(\dot{x} - \dot{y}) = M\ddot{x}$$

(Why is the weight Mg not included among the acting forces?) With $s = x - y$,

$$-ks - b\dot{s} = M\ddot{s} + M\ddot{y}$$

or
$$M\ddot{s} + b\dot{s} + ks = -M\ddot{y} = My_0\omega^2 \sin \omega t$$

The solution of this differential equation will be made up of a transient component, corresponding to the undriven oscillator ($y_0 = 0$), and a steady-state component, of the form $A \sin(\omega t + \phi)$, which is determined by the given motion of the earth.

14.9. From energy considerations, discuss the small motions of the system shown in Fig. 14-3. Find the natural frequency of the system, assuming no frictional effects.

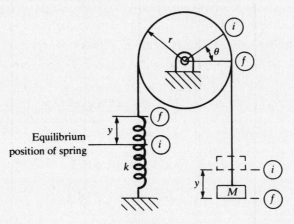

Fig. 14-3

Choose the (small) angle θ to represent the configuration of the system. Since $y = r\theta$, the elastic potential energy of the spring is

$$U = \frac{1}{2} ky^2 = \frac{1}{2} kr^2\theta^2$$

relative to its equilibrium length (*not* its unstretched length). The kinetic energy of the system is

$$K = K_M + K_{\text{disk}} = \frac{1}{2} M\dot{y}^2 + \frac{1}{2} I_{\text{disk}}\dot{\theta}^2 = \frac{1}{2}(Mr^2 + I_{\text{disk}})\dot{\theta}^2 = \frac{1}{2} I\dot{\theta}^2$$

where
$$I = Mr^2 + I_{\text{disk}} = \left(M + \frac{1}{2} M_{\text{disk}}\right)r^2$$

is the total moment of inertia of the disk and the weight about the axle.

Since the oscillations are very small, we are neglecting gravitational potential-energy changes. When the spring is stretched its maximum length, the kinetic energy will be zero and the potential energy a maximum; in the equilibrium position, the kinetic energy will be a maximum and the potential energy zero. Energy conservation thus gives:

$$U_{\text{max}} = K_{\text{max}} \qquad \text{or} \qquad \frac{1}{2} kr^2\theta_{\text{max}}^2 = \frac{1}{2} I\dot{\theta}_{\text{max}}^2$$

To find the natural frequency ω of the system, assume a solution of the form

$$\theta = \theta_0 \sin \omega t$$

Then $\theta_{max} = \theta_0$, $\dot{\theta}_{max} = \omega \theta_0$, and

$$\frac{1}{2} k r^2 \theta_0^2 = \frac{1}{2} I \omega^2 \theta_0^2 \qquad \text{or} \qquad \omega = \sqrt{\frac{k r^2}{I}} = \sqrt{\frac{k}{M + \frac{1}{2} M_{disk}}}$$

This result, when compared with $\omega^2 = k/m$, shows that the effective mass of the system is

$$m = M + \tfrac{1}{2} M_{disk}$$

14.10. A body of weight 27 N hangs on a long spring of such stiffness that an extra force of 9 N stretches the spring 0.05 m. If the body is pulled downward and released, what is its period?

The spring constant is

$$k = \frac{9}{0.05} = 180 \text{ N/m}$$

and so

$$T = 2\pi \sqrt{\frac{m}{k}} = 2\pi \sqrt{\frac{27/9.8}{180}} = 0.78 \text{ s}$$

14.11. A particle attached to a spring undergoes SHM. The maximum acceleration of the particle is 18 m/s^2 and the maximum speed is 3 m/s. Find (a) the frequency of the particle's motion, and (b) the amplitude.

(a) The equation of motion of the particle is $x = x_0 \cos(\omega t + \theta_0)$. Thus,

$$\dot{x} = -x_0 \omega \sin(\omega t + \theta_0) \qquad \text{and} \qquad \ddot{x} = -x_0 \omega^2 \cos(\omega t + \theta_0)$$

from which $\qquad\qquad \dot{x}_{max} = x_0 \omega \qquad$ and $\qquad \ddot{x}_{max} = x_0 \omega^2$

Thus, $\omega = \ddot{x}_{max}/\dot{x}_{max}$, and

$$f = \frac{\omega}{2\pi} = \frac{\ddot{x}_{max}}{2\pi \dot{x}_{max}} = \frac{18}{2\pi(3)} = 0.95 \text{ Hz}$$

(b)

$$x_0 = \frac{\dot{x}_{max}}{\omega} = \frac{3}{6} = 0.5 \text{ m}$$

14.12. A bob is attached to a string of length 1.8 m and is released from a rest angle $\theta_A = 3°$, as shown in Fig. 14-4. Suppose that $d = 0.9$ m. Find the time for the bob to get back to A after a complete swing.

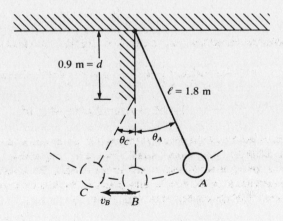

Fig. 14-4

From A to B, $\ell = 1.8$ m, so that

$$t_{AB} = \frac{1}{4}T = \frac{1}{4}\left(2\pi\sqrt{\frac{\ell}{g}}\right) = \frac{1}{4}\left(2\pi\sqrt{\frac{1.8}{9.8}}\right) = 0.67 \text{ s}$$

From B to C, $\ell' = \ell/2$, so that

$$t_{BC} = \frac{1}{4}T' = \frac{1}{4}\frac{T}{\sqrt{2}} = \frac{0.67}{\sqrt{2}} = 0.47 \text{ s}$$

Therefore, the time from A back to A is

$$t = 2(t_{AB} + t_{BC}) = 2(0.67 + 0.47) = 2.28 \text{ s}$$

14.13. Write the equations of motion for the systems shown in Fig. 14-5. Each mass must move vertically.

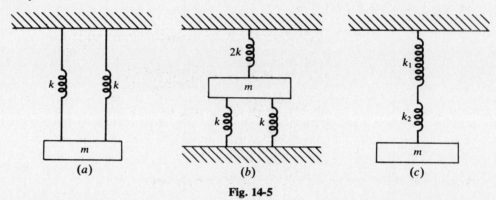

Fig. 14-5

In each case choose the equilibrium position of the mass m as $x = 0$, thereby eliminating the weight mg from the equation of motion, which will be of the form

$$m\ddot{x} = -k_{\text{eff}}x$$

(a)   A displacement Δx of mass m sets up a restoring force $-2k(\Delta x)$; $k_{\text{eff}} = 2k$.

(b)   A displacement Δx of mass m sets up a restoring force $-4k(\Delta x)$; $k_{\text{eff}} = 4k$.

(c)   A displacement Δx of mass m causes elongations $\Delta\ell_1$, $\Delta\ell_2$ of the two springs, such that

$$\Delta x = \Delta\ell_1 + \Delta\ell_2$$

The restoring force on m is $-k_2(\Delta\ell_2)$. But the force must be continuous at the junction of the springs:

$$k_1(\Delta\ell_1) = k_2(\Delta\ell_2)$$

Hence

$$k_{\text{eff}} = \frac{k_2(\Delta\ell_2)}{\Delta x} = \frac{k_2(\Delta\ell_2)}{\frac{k_2}{k_1}(\Delta\ell_2) + \Delta\ell_2} = \frac{k_1 k_2}{k_1 + k_2}$$

14.14. Determine whether vertical SHM is possible for the system shown in Fig. 14-6(a). If so, find the natural frequency ω.

The criterion for SHM is that the restoring force be proportional to the displacement *from the equilibrium position*. Because of the weight mg, the equilibrium configuration is as shown in Fig. 14-6(b), with the springs stretched to a distance

$$d' = \sqrt{d^2 + h^2}$$

If the mass m is displaced a distance Δx below the equilibrium position, where Δx is small compared to h and to d', the force in each spring is

$$F = k(\sqrt{d^2 + (h + \Delta x)^2} - d')$$

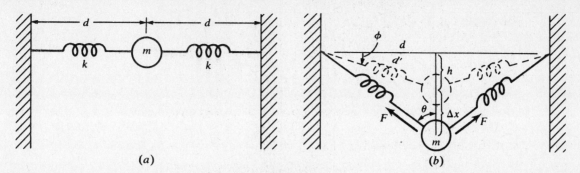

Fig. 14-6

and so the restoring force on mass m is

$$F_{res} = -2F\cos\theta = -2k(\sqrt{d^2 + (h + \Delta x)^2} - d')\frac{h + \Delta x}{\sqrt{d^2 + (h + \Delta x)^2}}$$

$$= -2kh\left(1 + \frac{\Delta x}{h}\right)\left[1 - \frac{1}{\sqrt{1 + \frac{2h(\Delta x) + (\Delta x)^2}{d'^2}}}\right]$$

$$\approx -2kh\left(1 + \frac{\Delta x}{h}\right)\left[\frac{1}{2}\frac{2h(\Delta x) + (\Delta x)^2}{d'^2}\right]$$

$$\approx -2\frac{kh^2}{d'^2}(\Delta x) = -(2k\sin^2\phi)\Delta x$$

where, in the third line, we have used the binomial expansion

$$(1 + u)^{-1/2} = 1 - \frac{1}{2}u + \cdots$$

and then retained only terms of first order in small quantities.

Thus, SHM can occur, the effective spring constant being $k_{eff} = 2k\sin^2\phi$, and

$$\omega = \sqrt{\frac{k_{eff}}{m}} = \sqrt{\frac{2k}{m}}\sin\phi$$

14.15. The cranking apparatus shown in Fig. 14-7 moves with very small oscillations. Find the equation of motion of the system, if the rods and dashpot are of negligible mass.

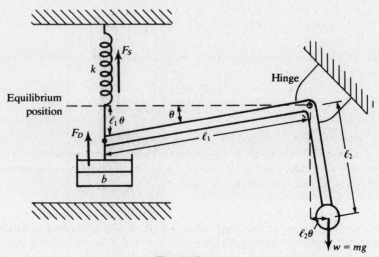

Fig. 14-7

The torques of the three forces about the hinge are, for small θ,

$$\tau_S = -F_S\ell_1 = -k\ell_1^2\theta$$

$$\tau_D = -F_D\ell_1 = -b\ell_1^2\dot{\theta}$$

$$\tau_w = -mg\ell_2\theta$$

The moment of inertia about the hinge is $I = m\ell_2^2$. Thus the equation of motion is

$$\sum \tau = I\ddot{\theta}$$

$$-(k\ell_1^2 + mg\ell_2)\theta - b\ell_1^2\dot{\theta} = m\ell_2^2\ddot{\theta}$$

Because of the dashpot term, this equation represents a damped angular harmonic motion.

Supplementary Problems

14.16. A steel piano wire ($Y = 196$ GN/m^2), of radius 1 mm, is stretched between two points 1.0 m apart, causing a tension in the wire of 39.3 N. The midpoint of the wire is pulled laterally a distance of 0.04 m. (a) By how much is the tension increased in the wire? (b) What is the lateral force applied to the wire?   Ans. (a) 2462 N; (b) 392 N

14.17. A particle of mass M is suddenly attached to the end of a suspended spring of natural length L, and immediately released.   Relative to a Y axis directed vertically downward and with origin at the point of suspension of the spring, determine the position of the particle at any time t.
Ans. $y = L + \dfrac{Mg}{k}\left(1 - \cos\sqrt{\dfrac{k}{M}}\,t\right)$

14.18. The energy of a body of mass m on the end of a spring undergoing SHM is

$$\frac{1}{2}mv^2 + \frac{1}{2}kx^2 = \frac{1}{2}kx_0^2$$

where k is the spring constant, x_0 is the amplitude of the motion, v is the velocity, and x is the displacement from the equilibrium position.   Make use of the integral

$$\int \frac{dx}{\sqrt{a^2 - x^2}} = -\cos^{-1}\frac{x}{a}$$

to show directly that $x = x_0 \cos(\omega t + \theta_0)$.

14.19. For the damped harmonic oscillator, energy is not conserved, but steadily decreases with time.   Show that

$$\frac{d}{dt}(K + U) = -bv^2$$

14.20. A particle undergoes SHM with an angular frequency (ω) of 4.0 rad/s.   Initially the particle is 6.0 mm to the right of its equilibrium position and is moving to the right with a speed of 0.032 m/s.   Where will the particle be after 0.40 s has elapsed?   Ans.   7.2 mm to the right of equilibrium position

14.21. The spring constant k of a certain spring is measured by observing that the spring is stretched 0.200 m by a force of 20.0 N applied at one end.   A body of mass 0.250 kg is attached to one end of the spring, while the other end is fixed.   The spring is compressed by moving the body a distance 0.400 m to the left of its equilibrium position; the body is then released.   Describe the resulting motion of the body.
Ans. $x = 0.200 \cos(20.0t + \pi)$ (m) $= 0.200 \cos(20.0t)$ (m)

14.22. (a) At what position will the tension in an oscillating simple pendulum reach its maximum value?   (b) Find the maximum tension in a simple pendulum of mass M and length ℓ, for small oscillations of amplitude a.   *Ans.*   (a) the vertical position;   (b) $Mg[1 + (a/\ell)^2]$

14.23. A body of mass 0.4 kg vibrates at the end of a spring with a frequency of 0.56 Hz.   (a) If its maximum displacement from the equilibrium position is 0.20 m, what is the spring constant?   (b) If its motion starts when it is at its maximum displacement, what is its phase angle?   (c) Find the equation for its position at time t.   *Ans.*   (a) 4.95 N/m;   (b) 0;   (c) $x = 0.20 \cos{(3.52t)}$ (m)

14.24. A bar in the form of a circular cylinder is of length ℓ, radius R, and shear modulus μ.   One end of the bar is twisted through a small angle $d\alpha$ relative to the other end, as indicated in Fig. 14-8.   Find the restoring torque in the bar.

Ans. $\tau_{\text{res}} = -\dfrac{\pi\mu}{\ell}\dfrac{R^4}{2}\,d\alpha$

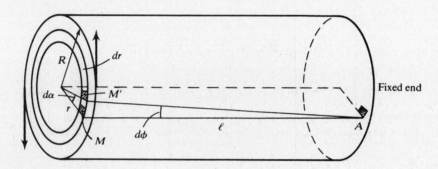

Fig. 14-8

Chapter 15

Fluid Statics

15.1 PRESSURE IN A FLUID

The term *fluid* applies to a substance that does not have a fixed shape but is able to flow and take the shape of its container: a liquid or a gas.   Fluid statics is the study of fluids at rest.

The *pressure* at a given location in a fluid is the ratio of the magnitude dF of the normal force exerted by the fluid to the area dA of a small plane surface passing through the point in question:

$$p = \frac{dF}{dA}$$

The normal force on an extended plane surface is then

$$F = \int p\, dA$$

If p is constant over the plane surface, $F = pA$.

The SI unit of pressure is the pascal (Pa); from Section 14.1, 1 Pa = 1 N/m^2.

15.2 PASCAL'S PRINCIPLE

Pressure applied to an enclosed fluid is transmitted in the fluid undiminished to every portion of the fluid and to the walls of the containing vessel.

15.3 DENSITY

The *density* of a homogeneous body is defined as its mass per unit volume:

$$\rho = \frac{m}{V}$$

For a nonhomogeneous body, the density is defined pointwise by $\rho = dm/dV$.   The units for density are kg/m^3.   The *specific gravity* of a homogeneous body is

$$S = \frac{\rho}{\rho_{\text{water}}}$$

where ρ is the density of the body, and $\rho_{\text{water}} = 1000$ kg/m^3.

15.4 LAWS OF FLUID STATICS

If $p_2 - p_1$ is the pressure difference between any two points with a difference in elevations $y_2 - y_1$,

$$p_2 - p_1 = -\rho g(y_2 - y_1)$$

provided that the two points can be connected by a path lying within a fluid of constant density ρ and that there is a constant gravitational acceleration **g** directed vertically downward.   The differential form, which is applicable when ρ or **g** is variable, is:

$$dp = -\rho g\, dy$$

Archimedes' principle says that a fluid acts on a foreign body immersed in it with a net force that is vertically upward and equal in magnitude to the weight of the fluid displaced by the body. This upward force is called the *buoyant* force.

Solved Problems

15.1. Calculate the pressure at a depth of 100 m below the surface of the ocean. Take atmospheric pressure as 100 kPa and the density of seawater as $\rho = 1030$ kg/m³.

$$p = p_{atm} + \rho g h = 10^5 + (1030)(9.8)(100) = 1.11 \times 10^6 \text{ Pa}$$

or 1.11 MPa.

15.2. What is the weight of 3 m³ of copper, whose specific gravity is 8.8?

$$w = mg = \rho_{copper} V g = S_{copper} \rho_{water} V g$$
$$= (8.8)(1000)(3)(9.8) = 259 \text{ kN}$$

15.3. Water stands at a depth h behind the vertical face of a dam, Fig. 15-1(a). It exerts a resultant horizontal force on the dam, tending to slide it along its foundation, and a torque tending to overturn the dam about the point O. Find (a) the horizontal force, (b) the torque about O, and (c) the height at which the resultant force would have to act to produce the same torque.

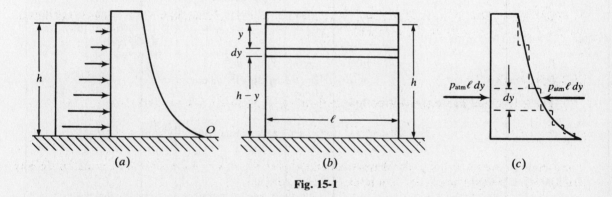

Fig. 15-1

(a) Figure 15-1(b) is a view of the face of the dam from upstream. The pressure at depth y is

$$p = \rho g y$$

We may neglect the atmospheric pressure since it acts on the other side of the dam also. [The construction shown in Fig. 15-1(c) may be used to justify the neglect of atmospheric pressure.] The force against the shaded strip is

$$dF = p \, dA = \rho g y \ell \, dy$$

The total force is

$$F = \rho g \ell \int_0^h y \, dy = \frac{\rho g \ell h^2}{2}$$

(b) The torque of the force dF about an axis through O is, in magnitude,

$$d\tau = (h - y) \, dF = \rho g \ell y (h - y) \, dy$$

The total torque about O is

$$\tau = \rho g \ell \int_0^h y(h-y)\,dy = \frac{\rho g \ell h^3}{6}$$

(c)   If H is the height above O at which the total force F would have to act to produce this torque,

$$HF = \tau \qquad \text{or} \qquad H = \frac{\tau}{F} = \frac{\rho g l h^3/6}{\rho g l h^2/2} = \frac{h}{3}$$

15.4.   A conical cup, $r = (b - z)\tan\alpha$, rests open-end-down on a smooth flat surface, as shown in Fig. 15-2.   The cup is to be filled to a height h with liquid of density ρ.   What will be the lifting force on the cup?

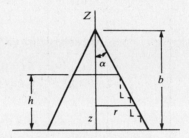

Fig. 15-2

Imagine the inside surface of the cup to consist of an infinite number of infinitesimal ring-shaped steps (Fig. 15-2).   The pressure $p(z)$ acting on the vertical face of a step does not contribute to the lifting force, since it acts horizontally.   Thus, the infinitesimal lifting force is just the pressure force on the horizontal face of the step:

$$dF_z = p(z)\,dA = \rho g(h-z)(2\pi r\,dr)$$
$$= \rho g(h-z)[2\pi(b-z)\tan\alpha](-dz\tan\alpha)$$

Integrating to obtain the total lifting force,

$$F_z = -2\pi\rho g \tan^2\alpha \int_h^0 (h-z)(b-z)\,dz = \pi\rho g \left(bh^2 - \frac{h^3}{3}\right)\tan^2\alpha$$

Another Method
The total pressure force exerted by a static fluid on its container is equal to the fluid's weight (why?).   Hence, for this problem,

$$F_b - F_z = w$$

where F_b is the downward force on the plane surface; F_z is the lifting force on the cup (by symmetry, the horizontal pressure forces on the cup cancel); and w is the weight of the liquid.   Now,

$$F_b = p(0)\,A = \rho g h(\pi b^2 \tan^2\alpha)$$

and
$$w = \rho g V = \rho g \int_0^h \pi r^2\,dz = \rho g \pi \tan^2\alpha \int_0^h (b-z)^2\,dz$$

$$= (\rho g \pi \tan^2\alpha)\left(b^2 h - bh^2 + \frac{h^3}{3}\right)$$

Consequently,

$$F_z = F_b - w = (\rho g \pi \tan^2\alpha)\left(bh^2 - \frac{h^3}{3}\right)$$

15.5. Find the pressure at a depth of 10 m in water when the atmospheric pressure is that corresponding to a mercury column of height 760 mm. The densities of water and mercury are 10^3 kg/m^3 and 13.6×10^3 kg/m^3, respectively.

$$p = p_{atm} + \rho g y = \rho_{Hg} g h_{Hg} + \rho g y$$
$$= (13.6 \times 10^3)(9.8)(0.760) + (10^3)(9.8)(10)$$
$$= 1.99 \times 10^5 \text{ Pa} = 199 \text{ kPa}$$

15.6. A mercury barometer stands at 762 mm. A gas bubble, whose volume is 33 cm^3 when it is at the bottom of a lake 45.7 m deep, rises to the surface. What is its volume at the surface of the lake?

In terms of the weight density, ρg, of water,

$$p_{bottom} = \rho g y + p_{atm} = \rho g y + \rho \left(\frac{\rho_{Hg}}{\rho}\right) g h_{Hg} = \rho g[45.7 + (13.6)(0.762)] = 56.1 \rho g$$

For the bubble, *Boyle's law* states that $pV = $ constant, assuming that the temperature stays fixed. Then,

$$V_{surface} = \frac{p_{bottom}}{p_{surface}} V_{bottom} = \frac{56.1\rho g}{10.4\rho g}(33) = 180 \text{ cm}^3$$

15.7. A small block of wood, of density 0.4×10^3 kg/m^3, is submerged in water at a depth of 2.9 m. Find (a) the acceleration of the block toward the surface when the block is released, and (b) the time for the block to reach the surface.

(a) By Archimedes' principle, the net upward force on the block is

$$F = \rho_{water} g V - \rho_{wood} g V$$

where V is the volume of the block. Then

$$a = \frac{F}{m} = \frac{\rho_{water} g V - \rho_{wood} g V}{\rho_{wood} V} = \left(\frac{\rho_{water}}{\rho_{wood}} - 1\right) g$$
$$= \left(\frac{1}{0.4} - 1\right)(9.8) = 14.7 \text{ m/s}^2$$

(b)
$$s = \frac{1}{2} at^2 \qquad \text{or} \qquad t = \sqrt{\frac{2s}{a}} = \sqrt{\frac{2(2.9)}{14.7}} = 0.63 \text{ s}$$

15.8. A cylindrical wooden buoy, of height 3 m and mass 80 kg, floats vertically in water. If its specific gravity is 0.80, how much will it be depressed when a body of mass 10 kg is placed on its upper surface?

By Archimedes' principle, the submerged height, h, of the unloaded buoy is given by

$$\rho_{water} g A h = \rho_{wood} g A(3) \qquad \text{or} \qquad h = \frac{\rho_{wood}}{\rho_{water}}(3) = (0.80)(3) = 2.40 \text{ m}$$

Under loading, the submerged height is directly proportional to the total weight or mass.

$$\frac{h + \Delta h}{h} = \frac{80 + 10}{80} \qquad \text{or} \qquad \Delta h = \frac{10}{80} h = \frac{10}{80}(2.40) = 0.30 \text{ m}$$

15.9. A man whose weight is 667 N and whose density is 980 kg/m^3 can just float in water with his head above the surface with the help of a life jacket which is wholly immersed. Assuming that the volume of his head is 1/15 of his whole volume and that the specific gravity of the life jacket is 0.25, find the volume of the life jacket.

The man's volume is

$$V = \frac{w}{\rho g} = \frac{667}{(980)(9.8)} = 0.07 \text{ m}^3$$

Equating the buoyant force to the weight of the man plus the weight of the life jacket,

$$\rho_{\text{water}}g\left[\frac{14}{15}V + V_{lj}\right] = 667 + (0.25\,\rho_{\text{water}})gV_{lj}$$

Solving,

$$V_{lj} = \frac{(667/\rho_{\text{water}}) - (14/15)gV}{0.75g}$$

$$= \frac{0.667 - (14/15)(9.8)(0.07)}{(0.75)(9.8)} = 0.004 \text{ m}^3$$

or 4 liters.

15.10. The weight of a balloon and the gas it contains is 11.12 kN. If the balloon displaces 1132 m³ of air and the weight of 1 m³ of air is 12.3 N, find the acceleration with which the balloon begins to rise.

The equation of motion of the balloon is

$$\sum F = F_B - w = ma$$

or

$$a = \frac{F_B - w}{m} = \frac{(1132)(12.3) - (1.112 \times 10^4)}{(1.112 \times 10^4)/9.8} = 2.57 \text{ m/s}^2$$

15.11. A slender homogeneous rod of length 2ℓ floats partly immersed in water, being supported by a string fastened to one of its ends, as pictured in Fig. 15-3. If the specific gravity of the rod is 0.75, find the fraction of the length of the rod that extends out of the water.

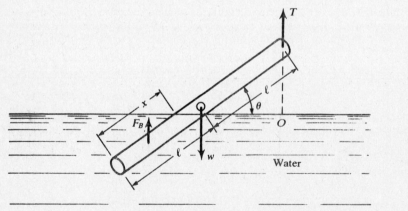

Fig. 15-3

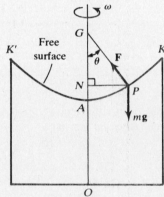

Fig. 15-4

Since the buoyant force acts through the center of gravity of the displaced water, the condition for rotational equilibrium is (A = area of cross section)

$$0 = \sum \tau_O = w\ell \cos\theta - F_B\left(\ell + \frac{x}{2}\right)\cos\theta$$

$$= \rho_{\text{rod}}gA(2\ell)(\ell \cos\theta) - \rho_{\text{water}}gAx\left(\ell + \frac{x}{2}\right)\cos\theta$$

$$= \left(-\frac{1}{2}\rho_{\text{water}}gA \cos\theta\right)\left(x^2 + 2\ell x - 4\frac{\rho_{\text{rod}}}{\rho_{\text{water}}}\ell^2\right)$$

From this,

$$x^2 + 2\ell x - 3.00\,\ell^2 = 0 \qquad \text{or} \qquad x = \ell, -3\ell$$

Discarding the negative root, we see that one-half the rod extends out of the water.

Strictly speaking, the above solution is only approximate, since the water surface does not cut the rod perpendicularly. However, the error will be negligible if A is small.

15.12. If a vessel and the liquid contained in it rotate uniformly about a vertical axis, show that the free surface of the liquid is a paraboloid (the surface formed by the rotation of a parabola about its axis).

As shown in Fig. 15-4, let the surface of the liquid take the shape generated by the rotation of the curve APK about the axis of rotation OA. This surface is one of equal pressure, since it is in contact with the air, which exerts essentially the same pressure everywhere on the surface. Therefore, the force $\mathbf{F}$ exerted on a surface element at P by the rest of the liquid is normal to the surface at P. (Otherwise there would be a tangential force along the surface.) The only other force on the element is its weight, $m\mathbf{g}$.

The element moves in a circle of radius $\overline{NP}$ with angular speed ω. The equations of motion in the vertical and horizontal directions are then

$$\sum F_{\text{ver}} = F\cos\theta - mg = 0$$

$$\sum F_{\text{hor}} = F\sin\theta = m\omega^2\,\overline{PN}$$

Combine these equations to get:

$$\tan\theta = \frac{\omega^2\,\overline{PN}}{g} \qquad \text{or} \qquad \overline{NG} = \frac{\overline{PN}}{\tan\theta} = \frac{g}{\omega^2} = \text{constant}$$

The subnormal of the curve AP is $\overline{NG}$ and is constant. Thus, the curve AP is a parabola, since a constant subnormal is a defining property of the parabola.

15.13. A small uniform tube is bent into a circle of radius r whose plane is vertical. Equal volumes of two fluids whose densities are ρ and σ ($\rho > \sigma$) fill half the circle (see Fig. 15-5). Find the angle that the radius passing through the common surface makes with the vertical.

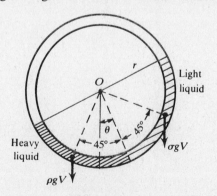

Fig. 15-5

Of the external forces acting on the two fluid segments only the two weights, $\rho g V$ and $\sigma g V$, have torques about the center O; the forces exerted by the container are purely radial. Thus, for equilibrium,

$$0 = \rho g V r \sin(45° - \theta) - \sigma g V r \sin(45° + \theta)$$

$$0 = \rho(\sin 45° \cos\theta - \cos 45° \sin\theta) - \sigma(\sin 45° \cos\theta + \cos 45° \sin\theta)$$

$$0 = \rho(1 - \tan\theta) - \sigma(1 + \tan\theta)$$

$$\tan\theta = \frac{\rho - \sigma}{\rho + \sigma}$$

Supplementary Problems

15.14. Consider a gas confined to a container by means of a piston of area 40 cm². If a perpendicular force of 20 N is exerted on the piston to keep the gas from expanding, find the gas pressure.   *Ans.* 5 kPa

15.15. A square board has dimension 0.15 m; its upper edge is horizontal and at a depth of 0.30 m below the surface of the ocean. Find the force that the water exerts on one side when the board is inclined at 30° to the horizontal. Ocean water weighs 10.1 kN/m³.   *Ans.* 77 N

15.16. Find the angle θ that the surface of water in a bucket makes with the horizontal if the entire system is caused to move in the horizontal direction with an acceleration **a**.   *Ans.* $\tan \theta = a/g$

15.17. A block of wood weighing 71.2 N and of specific gravity 0.75 is tied by a string to the bottom of a tank of water in order to have the block totally immersed. What is the tension in the string?
Ans. 23.6 N

15.18. A tank contains water on top of mercury. A cube of iron, 60 mm along each edge, is sitting upright in equilibrium in the liquids. Find how much of it is in each liquid. The densities of iron and mercury are 7.7×10^3 kg/m³ and 13.6×10^3 kg/m³, respectively.   *Ans.* 32 mm in mercury, 28 mm in water

15.19. A body of density ρ' is dropped from rest at a height h into a lake of density ρ, where $\rho > \rho'$. Neglect all dissipative effects and calculate (*a*) the speed of the body just before entering the lake, (*b*) the acceleration of the body while it is in the lake, and (*c*) the maximum depth to which the body sinks before returning to float on the surface.

Ans. (*a*) $\sqrt{2gh}$;   (*b*) $g\left(\dfrac{\rho}{\rho'} - 1\right)$, upward;   (*c*) $h\dfrac{\rho'}{\rho - \rho'}$

15.20. Rework Problem 15.13 by requiring that the fluid pressures be equal at the interface.

Chapter 16

Fluid Dynamics

16.1 SOME PROPERTIES OF FLUID FLOW

A *streamline* is an imaginary line in a fluid, taken at an instant of time, such that the velocity vector at each point of the line is tangential to it. A *stream tube* is a tube whose surface is made up of streamlines, across which there is no transport of fluid.

In *steady flow*, the fluid velocity at a given location is independent of time. (However, the velocity will in general vary from point to point.) Streamlines and stream tubes are fixed in steady flow.

A flow is *incompressible* if the fluid density ρ is constant; it is *irrotational* if there is no swirling (i.e. if the line segment defined by two neighboring fluid particles maintains a fixed direction).

16.2 THE CONTINUITY EQUATION

The conservation of mass requires that the net rate of flow of mass inward across any closed surface equal the rate of increase of the mass within the surface, assuming that there are no sources or sinks of matter within the surface. Applying this to a stream tube in steady flow, we obtain the *continuity equation* in the form

$$\rho_1 A_1 v_1 = \rho_2 A_2 v_2$$

where ρ is the density, assumed uniform over the cross section of area A, and v is the average velocity over the cross section (and normal to it).

If, besides being steady, the flow is incompressible, the continuity equation reduces to

$$A_1 v_1 = A_2 v_2$$

16.3 BERNOULLI'S EQUATION

For the steady flow of a nonviscous incompressible fluid, *Bernoulli's equation* relates the pressure p, the fluid speed v, and the height y at any two points *on the same streamline*, as follows:

$$p_1 + \frac{1}{2}\rho v_1^2 + \rho g y_1 = p_2 + \frac{1}{2}\rho v_2^2 + \rho g y_2$$

Note that each term in Bernoulli's equation has the dimensions of energy per unit volume. In fact, the equation simply states that the work done by pressure forces along a streamline is equal to the change in kinetic and potential energy (all per unit volume).

Solved Problems

16.1. A tank 10 m^3 in volume has an intake valve through which a gas is being pumped at the rate of 1 m^3/s. The gas escapes through an outlet at the same volume rate, but at the density existing in the tank, which is one-third the intake density. Find an expression for the density inside the tank.

Let ρ be the density in the tank. The net rate of mass flow into the tank is

$$(1)(3\rho) - (1)(\rho) = 2\rho \ \text{(kg/s)}$$

By conservation of mass, this must equal the rate of increase of mass in the tank:

$$\frac{dm}{dt} = \frac{d\rho}{dt}(10) = 2\rho$$

Integrating,

$$\int_{\rho_0}^{\rho} \frac{d\rho}{\rho} = 0.2 \int_0^t dt$$

or

$$\rho = \rho_0 e^{0.2t}$$

Obviously, this process could not continue very long.

16.2. Find the rate of change of density in a tank 0.28 m^3 in volume if a gas is escaping through an outlet 0.13 m in diameter at a speed of 305 m/s. The density in the tank at the start of the flow was 16.1 kg/m^3.

The conservation of mass equation is

$$\frac{dm}{dt} = \frac{d\rho}{dt} V = -\rho A v$$

where V, A, v are the volume of the tank, the area of the outlet, and the escape speed. Thus,

$$\frac{d\rho}{dt} = -\frac{Av}{V}\rho = -\frac{\pi(0.13/2)^2(305)}{0.28}\rho = -14.46\rho$$

Integrating,

$$\int_{16.1}^{\rho} \frac{d\rho}{\rho} = -14.46 \int_0^t dt$$

or

$$\rho = 16.1 e^{-14.46t} \ \text{(kg/m}^3)$$

and finally,

$$\frac{d\rho}{dt} = -14.46\rho = -232.8 e^{-14.46t} \ \text{(kg/m}^3 \cdot \text{s)}$$

16.3. A conical smokestack 97.6 m tall is 30.5 m in diameter at its base and 6.1 m in diameter at its top. Waste gas of density 0.64 kg/m^3 enters the stack at the bottom and condenses as it moves upward; it leaves the stack at a speed of 12.2 m/s. Assuming that the density increases linearly with height to a final value of 1.28 kg/m^3, find the speed at any height h. See Fig. 16-1.

The continuity equation, applied between cross sections at height h and at the top of the stack, gives

$$\rho A v = \rho_2 A_2 v_2$$

$$\rho\left(\frac{\pi d^2}{4}\right)v = (1.28)\left[\frac{\pi(6.1)^2}{4}\right](12.2)$$

$$v = \frac{581.0}{\rho d^2} \ \text{(m/s)}$$

Now, the diameter of a cross section varies linearly with height:

$$d = 30.5 - \frac{30.5 - 6.1}{97.6} h = 30.5 - 0.25 h \quad \text{(m)}$$

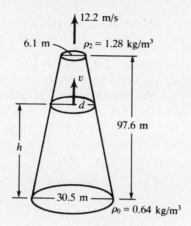

Fig. 16-1

and, by assumption, so does the density:

$$\rho = 0.64 + \frac{1.28 - 0.64}{97.6}\, h = 0.64 + 0.0066\,h \quad (\text{kg/m}^3)$$

Then

$$v = \frac{581.0}{(0.64 + 0.0066\,h)(30.5 - 0.25\,h)^2} \quad (\text{m/s})$$

16.4. In the case of a static fluid, show that Bernoulli's equation reduces to the hydrostatic equation,

$$\frac{dp}{dy} = -\rho g$$

Bernoulli's equation may be written as

$$\int \frac{dp}{\rho} + \frac{1}{2} v^2 + gy = C$$

where C is constant along a streamline, but generally varies from one streamline to another.   However, *if the flow is irrotational* (and this includes the case of the static fluid), it can be shown that C is constant over the whole fluid.   Then, with $v = 0$,

$$\int \frac{dp}{\rho} + gy = \text{constant}$$

and differentiation of this equation with respect to y gives the hydrostatic equation.

16.5. The overflow of a dam occurs in the form of an isosceles triangular notch of angle θ, as shown in Fig. 16-2.   Find an expression for the mass rate of flow through the notch in terms of θ and y_0.

An element of the mass rate of flow Q is given by

$$dQ = \rho v\, dA = \rho v \left(2y \tan \frac{\theta}{2}\right) dy$$

Assuming irrotational flow, Bernoulli's equation states that $p_{\text{atm}} + \frac{1}{2}\rho v^2 + \rho gy$ has the same value at all points in the face of the notch, whence

$$v^2 + 2gy = C \qquad \text{or} \qquad v = \sqrt{C - 2gy}$$

where C is a constant.   Substituting and integrating:

$$Q = \left(2\rho \tan \frac{\theta}{2}\right) \int_0^{y_0} y\sqrt{C - 2gy}\, dy$$

$$= \left(\frac{2\rho}{15g^2} \tan \frac{\theta}{2}\right)[C^{5/2} - (C + 3gy_0)(C - 2gy_0)^{3/2}]$$

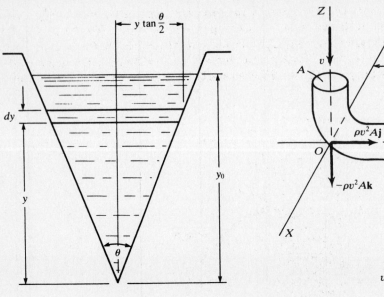

Fig. 16-2   **Fig. 16-3**

It is seen that the mass flow depends not only on the geometry of the notch (θ and y_0) but also on the constant C. To evaluate C we should have to know the flow speed at some particular height. For instance, if $v = 0$ at $y = y_0$, then

$$Q = \frac{2^{7/2}\rho g^{1/2} y_0^{5/2}}{15} \tan \frac{\theta}{2}$$

16.6. An incompressible fluid of density ρ flows through a uniform tube with two $90°$ bends, as shown in Fig. 16-3. The bends are in perpendicular planes. For static equilibrium of the tube, find (*a*) the force F, applied as shown, to prevent the tube from rotating around the pivot axis, Z, and (*b*) the reaction force at the pivot, O.

(*a*) The magnitude of the momentum flux through the tube is $\rho v^2 A$. At O, the momentum flux changes in direction from $-\mathbf{k}$ to $\mathbf{j}$; the force on the fluid is thus

$$\rho v^2 A\mathbf{j} + \rho v^2 A\mathbf{k}$$

and so the force on the tube at O is

$$\mathbf{F}_O = -\rho v^2 A\mathbf{j} - \rho v^2 A\mathbf{k}$$

Similarly, the force on the tube at P is

$$\mathbf{F}_P = -\rho v^2 A\mathbf{i} + \rho v^2 A\mathbf{j}$$

Taking torques about Z,

$$0 = d\mathbf{j} \times F\mathbf{i} + L\mathbf{j} \times \mathbf{F}_P = -dF\mathbf{k} + L\rho v^2 A\mathbf{k}$$

or $$F = \frac{L\rho v^2 A}{d}$$

(*b*) If $\mathbf{R}$ is the reaction force at O, then the force condition for equilibrium is

$$0 = \mathbf{R} + \mathbf{F}_O + F\mathbf{i} + \mathbf{F}_P$$

which gives

$$\mathbf{R} = -\rho v^2 A\left(\frac{L}{d} - 1\right)\mathbf{i} + \rho v^2 A\mathbf{k}$$

16.7. A cylindrical tank 0.9 m in radius rests atop a platform 6 m high, as shown in Fig. 16-4. Initially the tank is filled with water ($\rho = 1 \times 10^3$ kg/m³) to a depth $h_0 = 3$ m. A plug whose area is 6.3 cm² is removed from an orifice in the side of the tank at the bottom. (a) With what speed does the water flow initially from this orifice? (b) What is the speed of the stream initially as it strikes the ground? (c) How long will it take to empty the tank entirely?

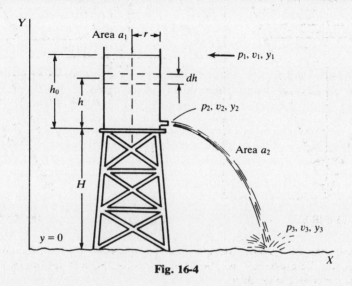

Fig. 16-4

(a) At the very top surface of the liquid, p, v, and y have the values p_1, v_1, and y_1, and in the stream which is just outside the orifice at the bottom of the tank, they have the values p_2, v_2, and y_2. According to Bernoulli's equation,

$$\rho g y_1 + \frac{1}{2}\rho v_1^2 = \rho g y_2 + \frac{1}{2}\rho v_2^2$$

since $p_1 = p_2 = p_{atm}$.

The speeds v_1 and v_2 are related by the continuity equation:

$$a_1 v_1 = a_2 v_2 \qquad \text{or} \qquad v_2 = \frac{a_1}{a_2} v_1 = \frac{\pi(0.9)^2}{6.3 \times 10^{-4}} v_1 = 4039 v_1$$

It follows that the term $\frac{1}{2}\rho v_1^2$ will be $(4039)^2 \approx 1.6 \times 10^7$ times smaller than the term $\frac{1}{2}\rho v_2^2$. Thus, $\frac{1}{2}\rho v_1^2$ may be considered effectively zero in Bernoulli's equation, which then gives

$$v_2 = \sqrt{2g(y_1 - y_2)} = \sqrt{2gh_0} = \sqrt{2(9.8)(3)} = 7.6 \text{ m/s}$$

The fact that v_2 equals the free-fall speed is known as *Torricelli's law*.

(b) Upon assigning values for p, v, and y just outside the tank orifice and where the stream hits the ground, we find according to Bernoulli's equation:

$$\rho g y_2 + \frac{1}{2}\rho v_2^2 = \rho g y_3 + \frac{1}{2}\rho v_3^2$$

since $p_2 = p_3 = p_{atm}$. Then,

$$\rho g H + \rho g h_0 = 0 + \frac{1}{2}\rho v_3^2$$

$$v_3 = \sqrt{2g(h_0 + H)} = \sqrt{2(9.8)(9)} = 13.28 \text{ m/s}$$

just as if the liquid had fallen from rest at the top of the tank.

(c) By conservation of mass

$$-\rho a_1 \frac{dh}{dt} = \rho a_2 v_2$$

As in (a) we find that $v_2 = \sqrt{2gh}$, again neglecting the motion of the top surface.   Thus,

$$dt = -\frac{a_1}{a_2}\frac{dh}{\sqrt{2gh}}$$

$$t = -\frac{a_1}{a_2\sqrt{2g}}\int_{h_0}^{0}\frac{dh}{\sqrt{h}} = \frac{a_1}{a_2}\sqrt{\frac{2h_0}{g}}$$

$$= 4039\sqrt{\frac{2(3)}{9.8}} = 3160.5 \text{ s} = 52.67 \text{ min}$$

16.8.   A fixture is attached to the orifice of the tank in Problem 16.7 to cause the stream to move upward at an angle θ, without any effect on its speed or cross-sectional area.   What is the maximum height h' achieved by the stream?

It was seen in Problem 16.7 that the stream undergoes projectile motion upon leaving the orifice, where its speed is $v_2 = \sqrt{2gh_0}$.   Then, applying the appropriate constant-acceleration formula in the Y direction,

$$0^2 = (v_2 \sin\theta)^2 - 2g(h' - H) \qquad \text{or} \qquad h' = H + h_0 \sin^2\theta$$

Supplementary Problems

16.9.   A Venturi flow meter introduces a constriction of cross-sectional area a_2 in a pipe of cross-sectional area a_1.   The meter records the difference in pressure, $p_1 - p_2$, between the ordinary fluid pressure, p_1, and the pressure at the constriction, p_2.   From this, infer the fluid speed in the unconstricted pipe.

Ans. $v_1 = a_2\sqrt{\dfrac{2(p_1 - p_2)}{\rho(a_1^2 - a_2^2)}}$

16.10.   Using Problem 16.9, find the fluid volume per unit time passing any cross section of the pipe.

Ans. $V = a_1 a_2\sqrt{\dfrac{2(p_1 - p_2)}{\rho(a_1^2 - a_2^2)}}$

16.11.   Consider the flow of a fluid at speed v_0 through a cylindrical pipe of radius r.   What would be the speed of this fluid at a point where, because of a constriction in the pipe, the fluid flow is confined to a cylindrical opening of radius $r/4$?   *Ans.* $16v_0$

16.12.   Suppose that the gas in the explosion chamber of a rocket ship is kept at a density ρ_1 and a pressure p_1, and that it exudes from the chamber into empty space through an opening of area a at one end of the rocket ship.   Find (a) the exhaust speed of the gas relative to the rocket, in terms of p_1 and ρ_1; (b) the thrust produced on the rocket ship.   *Ans.*   (a) $\sqrt{2p_1/\rho_1}$;   (b) $2p_1 a$

16.13.   What is the initial thrust on the tank in Problem 16.7?   *Ans.* $2\rho gh_0 a_2 = 37.0$ N

16.14.   A water barrel stands on a table of height h.   If a small hole is punched in the side of the barrel at its base, it is found that the resultant stream of water strikes the ground at a horizontal distance R from the barrel.   What is the depth of water in the barrel?   *Ans.* $R^2/4h$

16.15.   A flat plate moves normally toward a discharging jet of water at the rate of 3 m/s.   The jet discharges water at the rate of 0.1 m³/s and at a speed of 18 m/s.   (a) Find the force on the plate due to the jet and (b) compare it with that if the plate were stationary.   *Ans.*   (a) 2450 N;   (b) 1800 N

Chapter 17

Gases, Thermal Motion, and the
First Law of Thermodynamics

17.1 EQUATION OF STATE

The *equation of state of a gas* in thermal equilibrium (uniform temperature throughout the system) relates the pressure, the volume, and the temperature of the gas. At sufficiently low densities, all gases have the same equation of state, called the *ideal gas law*:

$$pV = NkT = nRT \qquad (17.1)$$

where N is the number of molecules in the gas; n is the number of moles of the gas; and T is the Kelvin temperature of the gas, related to the Celsius temperature by

$$T_{\text{Kelvin}} = T_{\text{Celsius}} + 273.15$$

The values of k, *Boltzmann's constant*, and R, the *universal gas constant*, are

$$k = 1.38 \times 10^{-23} \text{ J/K} \qquad R = 8.314 \text{ J/mol} \cdot \text{K}$$

The ratio of these constants is *Avogadro's number*,

$$N_0 = \frac{R}{k} = 6.02 \times 10^{23} \text{ mol}^{-1}$$

which is the number of molecules in a mole.

17.2 THERMAL MOTION

Viewed in the center-of-mass reference frame, the molecules of a gas are in random motion, with a wide distribution of kinetic energies. The *root-mean-square speed* (or *thermal speed*) v_{rms} may be defined as the speed of a hypothetical molecule whose translational kinetic energy equals the average translational kinetic energy over the whole ensemble of gas molecules:

$$\tfrac{1}{2}mv_{\text{rms}}^2 \equiv \overline{\tfrac{1}{2}mv^2} \qquad \text{or} \qquad v_{\text{rms}} = \sqrt{\overline{v^2}} \qquad (17.2)$$

For a dilute gas, obeying the ideal gas law,

$$\tfrac{1}{2}mv_{\text{rms}}^2 = \tfrac{3}{2}kT \qquad (17.3)$$

Thus, the absolute temperature of a macroscopic object is a measure of the average translational kinetic energy of its molecules, as determined in the object's center-of-mass frame. (See Problem 17.11.)

The total energy E of an ideal gas (as measured in the center-of-mass frame) is the same as its total kinetic energy. Thus, from (17.2) and (17.3),

$$E = N(\overline{\tfrac{1}{2}mv^2}) = \tfrac{3}{2}NkT \qquad \text{(monatomic gas)} \qquad (17.4)$$

A diatomic molecule possesses rotational, as well as translational, kinetic energy. According to the *equipartition theorem*, each molecular *degree of freedom* (independent coordinate) has the same amount of energy, $\tfrac{1}{2}kT$, associated with it. Hence,

$$E \approx \tfrac{5}{2}NkT \qquad \text{(diatomic gas)} \qquad (17.5)$$

153

The expression (17.5) is only approximate, because a diatomic molecule, besides translating and rotating as a rigid body, can undergo internal vibrations, the energies of which are quantized. With the vibrational contribution included, E ceases to be simply proportional to T.

The average distance traveled by a gas molecule between random collisions is called the *mean free path*, ℓ. For an ideal gas of spherical molecules with diameter d,

$$\ell = \frac{1}{(N/V)\pi\sqrt{2}\,d^2} \tag{17.6}$$

N/V being the number of molecules per unit volume.

17.3 THE FIRST LAW OF THERMODYNAMICS

For a process in which the internal energy of a system changes from an initial value E to a final value $E + \Delta E$,

$$\Delta E = Q - W \tag{17.7}$$

where W is the work done *by* the system on the surroundings during the process and Q is the heat *into* the system from the surroundings. (Work done on the system and heat out of the system are counted as negative.) Equation (17.7) expresses the first law of thermodynamics in the center-of-mass reference frame; it may be applied in another reference frame if E is defined as the *total* energy (the internal energy plus any macroscopic kinetic or potential energy possessed by the system).

The work W and the heat Q in general depend on the particular process, whereas ΔE is the same for all processes that take the system from a given initial state to a given final state. For this reason, (17.7) is written for infinitesimal processes as

$$dE = \dbar Q - \dbar W \tag{17.8}$$

The bar is a warning that $\dbar Q$ and $\dbar W$ are not exact differentials; they cannot be integrated to yield unique functions Q and W.

If the system is an ideal gas, the internal energy E is given by (17.4) or (17.5). If the system is a gas (ideal or not), and the only work done in the process is that due to the expansion of the system from volume V_1 to volume V_2, then

$$W = \int_{V_1}^{V_2} p\,dV$$

On a p-V diagram, this work is equal to the area under the curve describing the expansion.

An *adiabatic* process is one for which $Q = 0$. For an ideal gas, an adiabatic change in volume is represented by

$$pV^\gamma = \text{constant} \qquad \text{or} \qquad TV^{\gamma-1} = \text{constant}$$

where $\gamma > 1$ is the ratio of the heat capacities (see Chapter 18). For a monatomic ideal gas, $\gamma = 5/3$ (see Problem 17.3).

Solved Problems

17.1. An ideal gas exerts a pressure of 1.52 MPa when its temperature is 298.15 K (25 °C) and its volume is 10^{-2} m^3 (10 liters). (*a*) How many moles of gas are there? (*b*) What is the mass density if the gas is molecular hydrogen, H_2? (*c*) What is the mass density if the gas is oxygen, O_2?

(a)
$$n = \frac{pV}{RT} = \frac{(1.52 \times 10^6)(10^{-2})}{(8.31)(298.15)} = 6.135 \text{ mol}$$

(b) The atomic mass of hydrogen is 1.008, so that one mole of hydrogen (H_2) contains 2.016 g, or 2.016×10^{-3} kg. The density of the hydrogen is then

$$\rho = \frac{nM}{V} = \frac{(6.13)(2.016 \times 10^{-3})}{10^{-2}} = 1.24 \text{ kg/m}^3$$

(c) The atomic mass of oxygen is 16, so that one mole of O_2 contains 32 g, or 32×10^{-3} kg. The density of the oxygen is then

$$\rho = \frac{nM}{V} = \frac{(6.13)(32 \times 10^{-3})}{10^{-2}} = 19.6 \text{ kg/m}^3$$

17.2. One mole of helium gas, initially at STP ($p_1 = 1$ atm $= 101.3$ kPa, $T_1 = 0\,°C = 273.15$ K), undergoes an isovolumetric process in which its pressure falls to half its initial value. (a) What is the work done by the gas? (b) What is the final temperature of the gas? (c) The helium gas then expands isobarically to twice its volume; what is the work done by the gas? (d) Suppose that the gas undergoes a process from the initial state in (a) to the final state in (c) by an isothermal expansion; what is the work done by the gas?

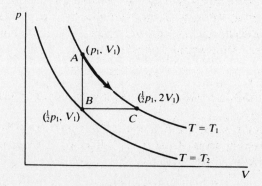

Fig. 17-1

(a) Refer to Fig. 17-1. $W_{AB} = 0$, since $dW = p\,dV = 0$.

(b) By the ideal gas law at constant volume,

$$\frac{T_2}{T_1} = \frac{p_2}{p_1} = \frac{1}{2} \qquad \text{or} \qquad T_2 = \frac{T_1}{2} = 136.58 \text{ K}$$

(c) The constant-pressure process returns the gas to the original temperature, $T_1 = 273.16$ K.

$$W_{BC} = \int_{V_1}^{2V_1} \frac{1}{2} p_1 \, dV = \frac{1}{2} p_1 V_1 = \frac{1}{2} R T_1$$

$$= \frac{1}{2}(8.31)(273.15) = 1135 \text{ J}$$

or 1.14 kJ.

(d) Along the isotherm $T = T_1$, $p = RT_1/V$; hence

$$W_{AC} = \int_{V_1}^{2V_1} p \, dV = RT_1 \int_{V_1}^{2V_1} \frac{dV}{V} = RT_1 \ln 2$$

$$= (8.31)(273.15)(0.693) = 1573 \text{ J}$$

Note that $W_{AC} \neq W_{AB} + W_{BC}$; the work between states A and C is path-dependent.

17.3. Determine the p-V relation for a monatomic ideal gas undergoing an adiabatic process such that only expansion work is done.

The first law of thermodynamics, (17.8), gives

$$dE = -\bar{d}W$$

for the adiabatic process.   The energy of a monatomic ideal gas is given by (17.4), whence

$$dE = \frac{3}{2} Nk\, dT$$

Moreover, using the ideal gas law,

$$\bar{d}W = p\, dV = \frac{NkT}{V}\, dV$$

Thus, $\dfrac{3}{2} Nk\, dT = -\dfrac{NkT}{V}\, dV$ or $\dfrac{dT}{T} + \dfrac{2}{3}\dfrac{dV}{V} = 0$

Integrating, $\ln T + \dfrac{2}{3} \ln V = \text{constant}$ or $TV^{2/3} = \text{constant}$

Finally, substituting for T from the ideal gas law,

$$pV^{5/3} = \text{constant}$$

17.4. Find the work done by an ideal gas in expanding adiabatically from a state (p_1, V_1) to a state (p_2, V_2).

For the adiabatic expansion,

$$pV^\gamma = \text{constant} = C \qquad \text{or} \qquad p = CV^{-\gamma}$$

Then $W = \displaystyle\int_{V_1}^{V_2} p\, dV = C \int_{V_1}^{V_2} V^{-\gamma}\, dV = \dfrac{CV_1^{1-\gamma} - CV_2^{1-\gamma}}{\gamma - 1}$

But $C = p_1 V_1^\gamma = p_2 V_2^\gamma$, and so

$$W = \frac{p_1 V_1 - p_2 V_2}{\gamma - 1}$$

17.5. Calculate the rms speed of hydrogen molecules (H_2) at 373.15 K (100 °C).

The mass of an H_2 molecule may be calculated from the molecular weight as

$$m = \frac{M}{N_0} = \frac{2.016 \times 10^{-3} \text{ kg/mol}}{6.02 \times 10^{23} \text{ mol}^{-1}} = 3.35 \times 10^{-27} \text{ kg}$$

Then $\dfrac{1}{2} m v_{rms}^2 = \dfrac{3}{2} kT$

$$v_{rms} = \sqrt{\frac{3kT}{m}} = \sqrt{\frac{3(1.38 \times 10^{-23})(373.15)}{3.35 \times 10^{-27}}} = 2.15 \text{ km/s}$$

17.6. Derive *Dalton's law of partial pressures.*

Consider a mixture of nonreactive gases in thermal equilibrium in a container of volume V.   The rms molecular speed for the mixture is the same as for each component gas, since it depends on temperature alone.   The pressure of the mixture is, from (17.1) and (17.3),

$$p = \frac{1}{3} \rho v_{rms}^2$$

Now, if the mixture consists of N_1 molecules of gas 1, each of mass m_1; N_2 molecules of gas 2, each of mass m_2; . . . ; then the density ρ of the mixture is given by

$$\rho = \frac{\text{mass}}{\text{volume}} = \frac{N_1 m_1 + N_2 m_2 + \cdots}{V} \qquad \text{and} \qquad p = \frac{1}{3}\left(\frac{N_1 m_1}{V}\right) v_{rms}^2 + \frac{1}{3}\left(\frac{N_2 m_2}{V}\right) v_{rms}^2 + \cdots$$

But $N_i m_i / V = \rho_i$, the density of the ith gas if it alone occupied the container. Thus

$$p = \frac{1}{3}\rho_1 v_{rms}^2 + \frac{1}{3}\rho_2 v_{rms}^2 + \cdots = p_1 + p_2 + \cdots$$

where p_i is the pressure of the ith gas if it alone occupied the container. This is Dalton's law.

17.7. According to the *law of atmospheres*, atmospheric pressure decreases exponentially with altitude, assuming that the temperature is uniform. Obtain this result from the ideal gas law.

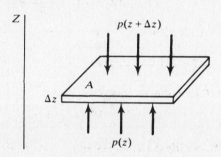

Fig. 17-2

Figure 17-2 shows a thin slab of air at altitude z. For equilibrium of the slab,

$$[p(z) - p(z + \Delta z)]A = \rho g A\, \Delta z$$

or

$$\frac{dp}{dz} = -\rho g$$

(the hydrostatic equation; see Section 15.4). From the ideal gas law,

$$p = \frac{\rho}{m} kT \qquad \text{or} \qquad \rho = \frac{m}{kT}p$$

where m is the average molecular mass for air. Thus

$$\frac{dp}{dz} = -\frac{mg}{kT}p$$

$$\int_{p_0}^{p}\frac{dp}{p} = -\frac{mg}{kT}\int_0^z dz$$

$$p = p_0 e^{-(mg/kT)z}$$

In the integration, we assumed that g, as well as T, is independent of z.

17.8. Molecular speeds may be measured with the device shown in Fig. 17-3. In one experiment it is found that molecules will pass through the velocity selector with the disks separated by 0.50 m and with an angular displacement of 180° between the two slits, when the disks turn at the rate of 600 rev/s. What are the possible speeds of the molecules?

A molecule that has passed through the first slit will pass through the second if the second disk turns through π, 3π, 5π, etc., while the molecule is in transit between the two disks. Thus

$$t = \frac{\theta}{\omega} = \frac{d}{v} \qquad \text{or} \qquad v = \frac{\omega d}{\theta} = \frac{(600 \times 2\pi)(0.50)}{(2n+1)\pi} = \frac{600}{2n+1} \text{ m/s}$$

where $n = 0, 1, 2, \ldots$.

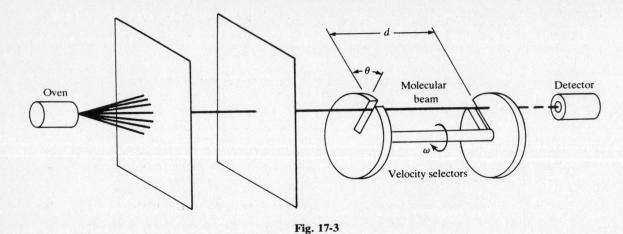

Fig. 17-3

17.9. The moving system of a D'Arsonval galvanometer consists of a coil of wire and a mirror suspended by a fine fiber and capable of rotation about a vertical axis. Random collisions of air molecules with the suspended system produce torques which are not equal and opposite at all instants. The result is that the angular position is continuously fluctuating and the system exhibits an unsteady "zero" (an example of Brownian motion). See Fig. 17-4. If the rms angular displacement of the system is $\theta_{rms} = 2 \times 10^{-4}$ rad and the torque constant for a fine quartz fiber is $K = 1 \times 10^{-13}$ N · m/rad, find the temperature of the air.

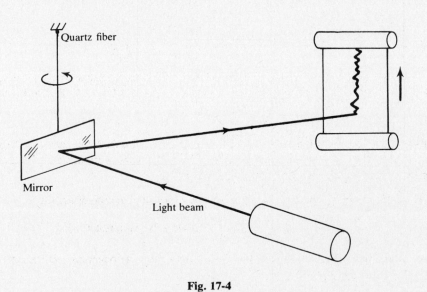

Fig. 17-4

The equipartition theorem applies to both types of disordered energy, kinetic and potential. In its random oscillations the mirror-wire system has rotational kinetic energy, $\frac{1}{2}I\omega^2$, and elastic potential energy of the twist of the wire, $\frac{1}{2}K\theta^2$. By the theorem, the mean value of each energy is $\frac{1}{2}kT$. Hence,

$$\frac{1}{2}K\theta_{rms}^2 = \frac{1}{2}kT$$

$$T = \frac{K\theta_{rms}^2}{k} = \frac{(1 \times 10^{-13})(2 \times 10^{-4})^2}{1.38 \times 10^{-23}} = 290 \text{ K} = 17 \text{ °C}$$

This is the temperature of the system, which is also the temperature of the air.

17.10. A satellite sent into space samples the density of matter within the solar system and gets a value of 2.5 hydrogen atoms per cm³. What is the mean free path of the hydrogen atoms? Take the diameter of a hydrogen atom as 2.4×10^{-10} m.

$$\ell = \frac{1}{(N/V)\pi\sqrt{2}d^2} = \frac{1}{(2.5 \times 10^6)\pi\sqrt{2}(2.4 \times 10^{-10})^2} = 1.56 \times 10^{12} \text{ m}$$

(which is about ten times the distance of the earth from the sun).

17.11. Imagine a thermometer in an ideal gas whose temperature is T. Let the thermometer move through the gas with speed v_0 in the $+Y$ direction. What "temperature" will the thermometer then read?

Relative to the moving thermometer, a gas molecule has velocity components v_x, $v_y - v_0$, v_z. The rms speed, v'_{rms}, relative to the thermometer is then given by

$$(v'_{rms})^2 = \overline{v_x^2 + (v_y - v_0)^2 + v_z^2} = \overline{v_x^2} + \overline{(v_y - v_0)^2} + \overline{v_z^2}$$
$$= \overline{v_x^2} + (\overline{v_y^2} - 2\overline{v_y}v_0 + v_0^2) + \overline{v_z^2}$$

But $\overline{v_y} = 0$, since for each gas molecule having a certain component of speed in the $+Y$ direction there will be another molecule having the same speed in the $-Y$ direction. Therefore,

$$(v'_{rms})^2 = \overline{v_x^2} + \overline{v_y^2} + \overline{v_z^2} + v_0^2 = (v_{rms})^2 + v_0^2$$

The thermometer will convert v'_{rms} to a temperature, T', by use of the ideal gas relationship $\frac{1}{2}mv_{rms}^2 = \frac{3}{2}kT$. Thus,

$$\tfrac{1}{2}m(v'_{rms})^2 = \tfrac{3}{2}kT'$$

$$\tfrac{1}{2}m(v_{rms})^2 + \tfrac{1}{2}mv_0^2 = \tfrac{3}{2}kT'$$

$$\tfrac{3}{2}kT + \tfrac{1}{2}mv_0^2 = \tfrac{3}{2}kT'$$

$$T' = T + \frac{mv_0^2}{3k}$$

The moving thermometer gives the false reading T' because, relative to it, the gas molecules possess *ordered* kinetic energy, in the amount $\frac{1}{2}mv_0^2$ per molecule, in addition to the *disordered* kinetic energy which alone determines the true temperature.

17.12. A spring having a spring constant 5 N/m is compressed 0.04 m, clamped in this configuration, and dropped into a container of acid in which the spring dissolves. By how much is the internal energy of the spring–vat system increased?

The spring in being compressed acquired elastic potential energy of amount

$$U_{elastic} = \frac{1}{2}kx^2 = \frac{1}{2}(5)(0.04)^2 = 0.004 \text{ J}$$

When the spring is dissolved, this ordered potential energy is converted into disordered potential and kinetic energy of the system.

17.13. A tank contains a fluid that is stirred by a paddle wheel. The power input to the paddle wheel is 2.24 kW. Heat is transferred from the tank at the rate of 0.586 kW. Considering the tank and the fluid as the system, determine the change in the internal energy of the system per hour.

$$\Delta E = Q - W = -0.586 - (-2.24) = 1.654 \text{ kW} = 1.654 \text{ kJ/s} = 5954 \text{ kJ/h}$$

or about 6 MJ/h.

17.14. Consider a system made up of a rock having a mass $m_r = 14.6$ kg and a bucket containing $m_w = 14.6$ kg of water. Initially the rock is 23.7 m above the water, and the rock and water are at the same temperature. The rock falls into the water. Find ΔE_{int}, ΔK, ΔU, and W (a) when the rock is about to enter the water, (b) just after the rock comes to rest at the bottom of the bucket, (c) after enough heat has been transferred so that the rock and water are at the same temperature they were at initially.

We apply the first law in the ground frame of reference, so that E includes the macroscopic kinetic and potential energies, K and U, of the system, as well as its internal energy, E_{int}.

$$\Delta E_{int} + \Delta K + \Delta U = Q - W$$

(a) We have $Q = W = \Delta E_{int} = 0$, whence

$$\Delta K + \Delta U = 0$$

i.e. the conservation of mechanical energy. Then,

$$\Delta K = -\Delta U = m_r g h = (14.6)(9.8)(23.7) = 3391 \text{ J}$$

(b) Just after the rock comes to rest at the bottom of the bucket,

$$Q = W = \Delta K = 0$$

and, neglecting the depth of the bucket, $\Delta U = -3391$ J. Then,

$$\Delta E_{int} = -\Delta U = 3391 \text{ J}$$

(c) After enough heat has been lost so that the rock and water are at the original temperature, $\Delta E_{int} = 0$. Also, $\Delta K = 0$, $W = 0$, and $\Delta U = -3391$ J. Therefore,

$$Q = \Delta U = -3391 \text{ J}$$

Supplementary Problems

17.15. If 2.1212 grams of a monatomic gas occupies a volume of 1.49 liters when the temperature is 0 °C and the pressure is 810.6 kPa, what is the gas? *Ans.* He

17.16. An ideal gas has been placed in a tank at 40 °C. The gauge pressure is initially 608 kPa. One-fourth of the gas is then released from the tank and thermal equilibrium is established. What will be the gauge pressure if the temperature is 315 °C? Take standard atmospheric pressure as 101 kPa.
Ans. 897.7 kPa

17.17. A spherical balloon of diameter 2 m is filled with helium. Assume STP and neglect the mass of the balloon. (a) How much helium is contained in the balloon if it just floats in air? What are (b) the density and (c) the pressure, of the helium gas? *Ans.* (a) 5.45 kg; (b) 1.3 kg/m^3; (c) 738 kPa

17.18. The statistical energy-distribution underlying the ideal gas law is the *Maxwell-Boltzmann distribution*. It gives the number of molecules with kinetic energies between E and $E + dE$ as

$$\frac{2N}{\sqrt{\pi}(kT)^{3/2}} \sqrt{E}\, e^{-E/kT}\, dE$$

Find the average kinetic energy over the collection of N molecules.

Ans. $\bar{E} = \frac{3}{2}kT$

17.19. Show that the molecular collision frequency for a given type of gas in a container of constant volume varies as the square root of the absolute temperature.

17.20. At what temperature is v_{rms} of H_2 molecules equal to the escape speed from the earth's surface ($v_e = \sqrt{2gR}$)? (b) What is the corresponding temperature for escape of hydrogen from the moon's surface ($g_M = 1.6$ m/s^2, $R_M = 1750$ km)? *Ans.* (a) 10 100 K; (b) 449 K

17.21. A nuclear fusion reaction will occur in a gas of deuterium nuclei when the nuclei have an average kinetic energy of at least 0.72 MeV. What is the temperature required for nuclear fusion to occur with deuterium? ($1 \text{ eV} = 1.6 \times 10^{-19}$ J.)   *Ans.* 5.57×10^9 K

17.22. Two gases occupy two containers, A and B. The gas in A, of volume 0.11 m^3, exerts a pressure of 1.38 MPa. The gas in B, of volume 0.16 m^3, exerts a pressure of 0.69 MPa. The two containers are united by a tube of negligible volume and the gases are allowed to intermingle. What is the final pressure in the container if the temperature remains constant?   *Ans.*   0.965 MPa

17.23. Five molecules have speeds of 12, 16, 32, 40, and 48 m/s. Find (a) the average speed and (b) the root-mean-square speed, for these molecules. (c) Show that for any distribution of speeds, $v_{rms} \geq \bar{v}$.
    Ans.   (a) 29.6 m/s;   (b) 32.6 m/s;   (c) $(v_{rms})^2 = (\bar{v})^2 + \overline{(v - \bar{v})^2}$

17.24. One mole of hydrogen gas, initially at STP, expands isobarically to twice its initial volume. (a) What is the final temperature of the gas? (b) What is the work done by the gas in expanding? (c) By how much does the internal energy of the gas change? (d) Does thermal energy leave or enter the gas and, if so, how much?   *Ans.*   (a) 546.30 K;   (b) 2.27 kJ: (c) 3.41 kJ;   (d) 5.68 kJ enters

17.25. A cylinder fitted with a piston contains 0.10 mol of air at room temperature (20 °C). The piston is pushed so slowly that the air within the cylinder remains essentially in thermal equilibrium with the surroundings. Find the work done by the air within the cylinder if the final volume is one-half the initial volume.   *Ans.*   −169 J

Chapter 18

Thermal Properties of Matter

18.1 THERMAL EXPANSION

The change in length, ΔL, in the length L_0 of a solid, when its temperature is changed a small amount ΔT, is

$$\Delta L = \alpha L_0 \Delta T$$

where α is the *coefficient of linear expansion* of the solid. The area and volume changes for a solid are given by

$$\Delta A = 2\alpha A_0 \Delta T \qquad \Delta V = 3\alpha V_0 \Delta T$$

The volume change for a liquid is

$$\Delta V = \beta V_0 \Delta T$$

where β is the *coefficient of volume expansion*.

The unit for α and β is $°C^{-1}$ (or K^{-1}, since a unit temperature interval is the same on the Celsius and Kelvin scales).

18.2 HEAT CAPACITY

A system's *heat capacity* C for some process (assumed reversible; see Section 19.1) is defined as

$$C = \lim_{\Delta T \to 0} \frac{Q}{\Delta T}$$

where Q is the heat into the system during the process and ΔT is the change in the temperature of the system. For n moles or a mass m of a homogeneous substance, the *molar heat capacity* is $c' = C/n$ and the *specific heat capacity* (or, simply, *specific heat*) is $c = C/m$. The units for C, c', and c are J/K, J/mol $\cdot$ K, and J/kg $\cdot$ K, respectively.

Ordinarily, heat capacities are defined for only two types of processes.

Constant-Volume Processes

For these processes the first law of thermodynamics, *(17.7)*, gives $Q = \Delta E$, so that the heat capacity can be obtained from the system's internal energy:

$$C_v = \lim_{\substack{\Delta T \to 0 \\ V = \text{const}}} \frac{\Delta E}{\Delta T} = \left(\frac{\partial E}{\partial T}\right)_V$$

For an ideal gas (see Chapter 17), E is the disordered kinetic energy of the molecules, which depends only on temperature. For point-molecules, *(17.4)* gives

$$C_v = \frac{3}{2} Nk = \frac{3}{2} nR \qquad \text{(monatomic ideal gas)}$$

or $c'_v = \frac{3}{2}R = 12.47$ J/mol $\cdot$ K. For real gases, C_v will not be a constant but will vary with temperature; the more complex the molecule, the greater the variation.

In liquids and solids, random intramolecular vibrations and atomic lattice oscillations contribute to E.   Only approximations to C_v can be given.   For metals at high temperatures,

$$c_v' \approx 3R$$

(the empirical *law of Dulong and Petit*).

Constant-Pressure Processes

It is convenient to introduce the *enthalpy* of a system, defined as

$$H = E + pV$$

Like E, H is a state function whose change is independent of the path of the process.   In a constant-pressure process,

$$\Delta H = \Delta E + p\,\Delta V = \Delta E + W = Q$$

by (*17.7*), so that the heat capacity can be written as

$$C_p = \lim_{\substack{\Delta T \to 0 \\ p=\text{const}}} \frac{\Delta H}{\Delta T} = \left(\frac{\partial H}{\partial T}\right)_p$$

For an ideal gas, $H = E + nRT$ is a function of T only; therefore,

$$C_p = \frac{dE}{dT} + nR = C_v + nR \qquad \text{(ideal gas)}$$

or $c_p' = c_v' + R$.   In particular, for a monatomic ideal gas,

$$C_p = \frac{5}{2}nR \qquad \text{(monatomic ideal gas)}$$

For solids and liquids, the difference between C_p and C_v is small and is often neglected.

18.3   HEAT TRANSFER

The rate dQ/dt at which heat is conducted (in a steady state) through a slab of cross-sectional area A and thickness dx is

$$\frac{dQ}{dt} = -kA\,\frac{dT}{dx}$$

where dT/dx is the temperature gradient in the slab and k is its *thermal conductivity*.   The minus sign reflects the fact that the conduction is in the direction of decreasing temperature.   The units for k are $\text{W/m} \cdot \text{K}$.

Also thermal energy may be transferred by the *emission* or *absorption* of electromagnetic radiation.   The absorption process is temperature independent, but the emission of radiation from a blackbody (a perfect radiator or absorber) of surface area A is governed by the *Stefan-Boltzmann law*,

$$\text{power} = \sigma A T^4$$

where $\sigma = 5.669\,61 \times 10^{-8}\ \text{W/m}^2 \cdot \text{K}^4$.

Solved Problems

18.1. What is the change in length of 2 km of steel wire when the temperature changes from 0 °C to 40 °C? $\alpha_{steel} = 12 \times 10^{-6}$ °C^{-1}.

$$\Delta L = L_0 \alpha \, \Delta T = (2 \times 10^3)(12 \times 10^{-6})(40 - 0) = 0.96 \text{ m}$$

18.2. A clock with a metal pendulum keeps correct time when the temperature is 0 °C. How many seconds does it gain (or lose) per day when the temperature is 40 °C? $\alpha = 16 \times 10^{-6}$ °C^{-1}.

The period of a simple pendulum is (Section 14.6)

$$P = 2\pi \sqrt{\frac{L}{g}} = 2\pi \sqrt{\frac{L_0(1 + \alpha \, \Delta T)}{g}} = \left(2\pi \sqrt{\frac{L_0}{g}}\right) \sqrt{1 + \alpha \, \Delta T} = P_0 \sqrt{1 + \alpha \, \Delta T}$$

where P_0 is the correct period. Thus,

$$\Delta t = P_0 - P = P_0(1 - \sqrt{1 + \alpha \, \Delta T}) \approx P_0\left(-\frac{1}{2} \alpha \, \Delta T\right)$$

$$= (24 \times 3600)\left[-\frac{1}{2}(16 \times 10^{-6})(40)\right] = -27.65 \text{ s}$$

The clock loses 27.65 s per day.

18.3. If a brass rod ($Y = 88.2$ kPa, $\alpha = 18.8 \times 10^{-6}$ K^{-1}) could be prevented from changing length when its temperature is changed, what stress would be produced in the rod when the temperature is changed by 50 K?

The stress produced in the rod is the stress that must be exerted externally to prevent the change in length ΔL caused by the temperature change of 50 K:

$$\frac{F}{A} = Y \frac{\Delta L}{L_0} = Y \frac{L_0 \alpha \, \Delta T}{L_0} = Y \alpha \, \Delta T$$

$$= (88.2 \times 10^3)(18.8 \times 10^{-6})(50) = 82.9 \text{ Pa}$$

18.4. A glass beaker holds exactly one liter (1 L) at 0 °C. (a) What is its volume at 50 °C? (b) If the beaker is filled with mercury at 0 °C, what volume of mercury overflows when the temperature is raised to 50 °C? $\alpha_{glass} = 6.9 \times 10^{-6}$ K^{-1} and $\beta_{mercury} = 1.82 \times 10^{-4}$ K^{-1}.

(a) The volume of the beaker after the temperature change is

$$V_{beaker} = V_0(1 + 3\alpha \, \Delta T) = (1)[1 + 3(6.9 \times 10^{-6})(50)] = 1.001 \text{ L}$$

(b) For the mercury expansion,

$$V_{mercury} = V_0(1 + \beta \, \Delta T) = (1)[1 + (1.82 \times 10^{-4})(50)] = 1.009 \text{ L}$$

The overflow is thus $1.009 - 1.001 = 0.008$ L, or 8 mL.

18.5. Consider two parallel bars [Fig. 18-1(a)] of different metals, having linear expansion coefficients α', α'' and fastened together so as to keep them at a fixed distance d apart. A change of temperature will cause their bending into two circular arcs intercepting an angle θ [Fig. 18-1(b)]. Find their mean radius of curvature, R.

Assume as the scale zero-point that temperature at which the bars are straight. If their common length at this temperature is L_0, their lengths at any other temperature T will be

$$L' = \theta R' = L_0(1 + \alpha' T) \qquad L'' = \theta R'' = L_0(1 + \alpha'' T)$$

where R', R'' are the radii of curvature of the bars and θ is the angle subtended at the center of curvature by the joined bars. Also, $R' - R'' = d$.

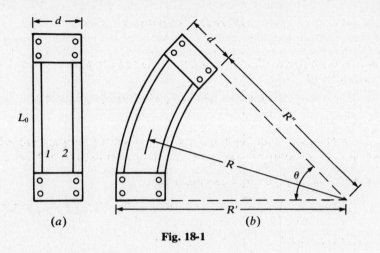

Fig. 18-1

Subtract the second of the above equations from the first to get

$$\theta(R' - R'') = (\alpha' - \alpha'')L_0 T \qquad \text{or} \qquad \theta = \frac{(\alpha' - \alpha'')L_0 T}{d}$$

Now add the two equations to obtain

$$\theta(R' + R'') = 2L_0 + (\alpha' + \alpha'')L_0 T$$

The mean radius of curvature is

$$R = \frac{R' + R''}{2} = \frac{2L_0 + (\alpha' + \alpha'')L_0 T}{2\theta} = \frac{2L_0 + (\alpha' + \alpha'')L_0 T}{2(\alpha' - \alpha'')L_0 T/d} = \frac{2 + (\alpha' + \alpha'')T}{2(\alpha' - \alpha'')T}\,d$$

But, $(\alpha' + \alpha'')T \ll 2$; so,

$$R \approx \frac{2d}{2(\alpha' - \alpha'')T} = \frac{d}{(\alpha' - \alpha'')T}$$

18.6. Consider the all-solid pendulum clock shown in Fig. 18-2. This pendulum is compensated for temperature change by using differential expansion so as to keep the center of oscillation of the pendulum at a fixed distance below the point of suspension. Find the dimensions of this pendulum if it is to tick off seconds.

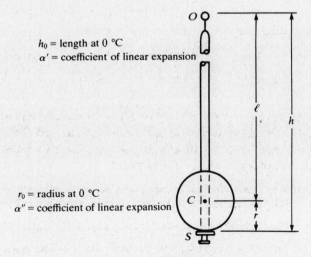

h_0 = length at 0 °C
α' = coefficient of linear expansion

r_0 = radius at 0 °C
α'' = coefficient of linear expansion

Fig. 18-2

The light rod OS supports the heavy bob by means of the regulating screw S. Regarding the center of the bob C as approximately the center of oscillation of the pendulum (conjugate to the center of suspension O), the effective length of the pendulum is

$$\ell = h - r$$

Now, $$h = h_0(1 + \alpha' \Delta T) \qquad r = r_0(1 + \alpha'' \Delta T)$$

where $\alpha'' \gg \alpha'$. Then,

$$\ell = h_0 - r_0 + (h_0\alpha' - r_0\alpha'') \Delta T$$

which will be independent of temperature if

$$h_0\alpha' - r_0\alpha'' = 0$$

Also $$\ell_0 = h_0 - r_0$$

where ℓ_0 is the length that must be maintained. Solve these last two equations for h_0 and r_0:

$$h_0 = \frac{\alpha'' \ell_0}{\alpha'' - \alpha'} \qquad r_0 = \frac{\alpha' \ell_0}{\alpha'' - \alpha'}$$

If the clock is to tick off seconds, the period (2 s) of the pendulum is

$$P = 2\pi \sqrt{\frac{\ell_0}{g}} \qquad \text{or} \qquad 2 = 2\pi \sqrt{\frac{\ell_0}{g}} \qquad \text{or} \qquad \ell_0 = \frac{g}{\pi^2}$$

For the dimensions of the pendulum, we have:

$$h_0 = \frac{\alpha'' \ell_0}{\alpha'' - \alpha'} = \frac{\alpha''(g/\pi^2)}{\alpha'' - \alpha'} \qquad \text{and} \qquad r_0 = \frac{\alpha'(g/\pi^2)}{\alpha'' - \alpha'}$$

18.7. Prove that the change in the moment of inertia of a solid body is given by $\Delta I = 2\alpha I_0 \Delta T$, when the temperature changes by ΔT.

The moment of inertia is

$$I_0 = \int r_0^2 \, dm$$

Corresponding to a temperature change ΔT,

$$I_0 + \Delta I = \int (r_0 + \alpha r_0 \Delta T)^2 \, dm \approx \int (r_0^2 + 2\alpha r_0^2 \Delta T) \, dm = I_0 + (2\alpha \Delta T)I_0$$

and the result follows.

18.8. If an anisotropic solid has coefficients of linear expansion α_x, α_y, and α_z for three mutually perpendicular directions in the solid, what is the coefficient of volume expansion for the solid?

Consider a cube, with edges parallel to X, Y, Z, of dimension L_0 at $T = 0$. After a change in temperature ΔT, the dimensions change to

$$L_x = L_0(1 + \alpha_x T)$$
$$L_y = L_0(1 + \alpha_y T)$$
$$L_z = L_0(1 + \alpha_z T)$$

and the volume of the parallelopiped is

$$V = V_0(1 + \alpha_x T)(1 + \alpha_y T)(1 + \alpha_z T) \approx V_0[1 + (\alpha_x + \alpha_y + \alpha_z)T]$$

where $V_0 = L_0^3$. Therefore, the coefficient of volume expansion is given by

$$\beta = \frac{1}{V_0} \frac{dV}{dT} = \alpha_x + \alpha_y + \alpha_z$$

18.9. A copper container of mass 0.30 kg contains 0.45 kg of water. Container and water are initially at room temperature, 20 °C. A 1 kg block of metal is heated to 100 °C and placed in the water in the calorimeter. The final temperature of the system is 40 °C. Find the specific heat of the metal. The specific heat of water is 4.2 kJ/kg · K and that of copper, 0.39 kJ/kg · K.

$$\left\{ \begin{array}{c} \text{heat lost by} \\ \text{metal block} \end{array} \right\} = \left\{ \begin{array}{c} \text{heat gained} \\ \text{by water} \end{array} \right\} + \left\{ \begin{array}{c} \text{heat gained by} \\ \text{copper calorimeter} \end{array} \right\}$$

$$(mc\,\Delta T)_{\text{metal}} = (mc\,\Delta T)_{\text{water}} + (mc\,\Delta T)_{\text{copper}}$$

$$(1)c(100-40) = (0.45)(4.2)(40-20) + (0.30)(0.39)(40-20)$$

$$c = 0.67 \text{ kJ/kg} \cdot \text{K}$$

18.10. If 5 kg of iron at 1812.16 K, which is just above the melting point of iron, is dropped into 3 kg of water at 273.16 K (just above the freezing point), how much water is evaporated if no heat is lost? The heat of fusion (i.e. the enthalpy change) of iron is 272 kJ/kg and the specific heat of solid iron is 0.50 kJ/kg · K. The heat of vaporization (enthalpy change) of water is 2260 kJ/kg and the specific heat of the liquid is 4.2 kJ/kg · K.

heat (or enthalpy) lost by hot bodies = heat (or enthalpy) gained by cold bodies

To solve this problem, an educated guess is required as to the final equilibrium state of the system. Let us assume that the mixture ends up at 373.16 K (≈ 100 °C), with all the iron solidified and a certain mass, x, of the water in the form of steam.

$$(mH)_{\text{iron}} + (mc\,\Delta T)_{\text{iron}} = (mc\,\Delta T)_{\text{water}} + (mH)_{\text{water}}$$

$$5(272) + 5(0.50)(1812.16 - 373.16) = 3(4.2)(373.16 - 273.16) + x(2260)$$

$$x = 1.64 \text{ kg}$$

Our guess was correct; had it not been so, x would have come out negative or greater than 3 kg.

18.11. Show how the equipartition theorem leads to the law of Dulong and Petit.

At high temperatures, it may be supposed that essentially all the internal energy of a metal is due to the vibrations of the atoms about their equilibrium positions in the crystalline lattice. If we picture each atom as connected to its neighbors by springs, then the atom will have kinetic and potential energies along three mutually perpendicular directions—six modes in all. By the equipartition theorem, its total energy will be $6(\frac{1}{2}kT) = 3kT$, giving a molar energy of

$$E' = N_0(3kT) = 3RT$$

and a molar heat capacity of

$$c'_v = \frac{dE'}{dT} = 3R$$

18.12. An ordinary refrigerator is thermally equivalent to a box of corkboard 90 mm thick and 5.6 m² in inner surface area. When the door is closed, the inside wall is kept, on the average, 22.2 °C below the temperature of the outside wall. If the motor of the refrigerator runs 15% of the time the door is closed, at what rate must heat be taken from the interior while the motor is running? The thermal conductivity of corkboard is $k = 0.05$ W/m · K.

Consider a time interval Δt during which the door is closed. As an approximation, take the heat conduction to be steady over Δt. Then the rate of heat into the box is

$$\frac{Q}{\Delta t} = kA\left(\frac{\Delta T}{\Delta x}\right) = (0.05)(5.6)\left(\frac{22.2}{0.090}\right) = 69.1 \text{ W}$$

To remove this heat, the motor must, since it runs only for a time $(0.15)\,\Delta t$, cause heat to leave at the rate

$$\frac{69.1}{0.15} = 460 \text{ W}$$

18.13. A very small hole in an electric furnace used for treating metals acts nearly as a black-body. If the hole has an area 100 mm², and it is desired to maintain the metal at 1100 °C, how much power travels through the hole?

$$\text{power} = \sigma A T^4 = (5.67 \times 10^{-8})(10^{-4})(1373.15)^4 = 20.2 \text{ W}$$

18.14. Find the rate of heat flow through the hollow sphere shown in cross section in Fig. 18-3.

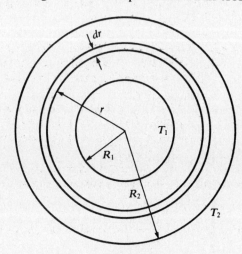

Fig. 18-3

The steady-state heat flow (which is independent of the radius r) through a spherical shell of radius r and thickness dr is

$$\frac{Q}{t} = -k(4\pi r^2)\left(\frac{dT}{dr}\right) \qquad \text{or} \qquad \frac{Q}{t}\frac{dr}{r^2} = -4\pi k\, dT$$

Integrate this from the inner surface of the hollow sphere to the outer surface:

$$\frac{Q}{t}\int_{R_1}^{R_2}\frac{dr}{r^2} = -4\pi k\int_{T_1}^{T_2} dT$$

$$\frac{Q}{t}\left(\frac{1}{R_1} - \frac{1}{R_2}\right) = 4\pi k(T_1 - T_2)$$

$$\frac{Q}{t} = \frac{4\pi k(T_1 - T_2)}{(1/R_1) - (1/R_2)}$$

18.15. A 200 W heater is turned on in an insulated garage of dimensions 8 m by 6 m by 5 m. How much time does it take for the heater to raise the temperature of the air from 0 °C to 10 °C? $c'_{\text{air}} = 20.93$ J/mol · K.

In the time interval Δt, the heat into the garage is $(200\,\text{W})\Delta t$. This is related to the temperature rise by

$$200\,\Delta t = nc'\,\Delta T$$

where n is the number of moles of air in the garage. Treating air as an ideal gas,

$$n = \frac{V_{\text{garage}}}{0.0224} = \frac{(8)(6)(5)}{0.0224} = 10\,714 \text{ mol}$$

since 1 mol, at the initial standard conditions, occupies 0.0224 m³. Thus

$$\Delta t = \frac{nc'\,\Delta T}{200} = \frac{(10\,714)(20.93)(10)}{200} = 11\,212 \text{ s} = 3.115 \text{ h}$$

Supplementary Problems

18.16. A brass rod is 0.70 m long at 40 °C. If the coefficient of linear expansion of brass is 18.8×10^{-6} °C^{-1}, find the length of this rod at 50 °C.   *Ans.*   0.7002 m

18.17. A nickel-steel rod at 21 °C is 0.624 06 m in length. Raising the temperature to 31 °C produces an elongation of 121.6 μm. Find the length at 0 °C and the coefficient of linear expansion.
Ans.   0.623 805 m; 19.5×10^{-6} °C^{-1}

18.18. The mercury in a thermometer has a volume of 210 mm^3 at 0 °C, at which temperature the diameter of the bore is 0.2 mm. If the volume coefficients of mercury and the glass are respectively 1.82×10^{-4} °C^{-1} and 2.4×10^{-5} °C^{-1}, how far apart are the degree marks on the stem?   *Ans.*   1.06 mm

18.19. How many kilograms of ice at 0 °C must be added to 0.6 kg of water at 100 °C in an insulated, 0.1 kg copper container in order to cool the container and its contents to 30 °C? The specific heats of water and copper are 4.2 and 0.39 kJ/kg · K, respectively; the heat of fusion of ice is 335 kJ/kg.   *Ans.*   0.39 kg

18.20. If all the heat produced were used, how many cubic meters of natural gas at STP (the heat of combustion is 37.3 MJ/m^3) are needed to heat 4.54 kg of water in a 0.45 kg copper cup ($c = 0.39$ kJ/kg · K) from 297.59 K to 373.15 K?   *Ans.*   0.039 m^3

18.21. Find the steady-state heat flow through the hollow circular cylinder shown in Fig. 18-4.
Ans. $\dfrac{Q}{t} = 2\pi k \ell \left[\dfrac{T_1 - T_2}{\ln (R_2 / R_1)} \right]$

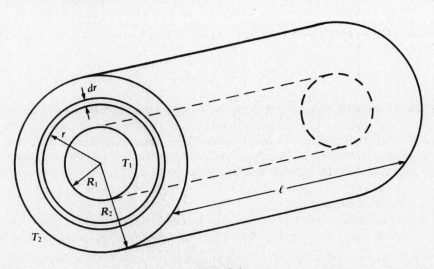

Fig. 18-4

Chapter 19

Entropy and the Second Law of Thermodynamics

19.1 REVERSIBLE PROCESSES

In a *reversible process* a system goes from an initial equilibrium state to a final equilibrium state through a continuous sequence of equilibrium states. This means that at every instant during the process the system is in thermal and mechanical equilibrium with its surroundings. The direction of such a process can be reversed (hence, "reversible") at any instant by an infinitesimal change in external conditions.

Reversibility is an idealization that is never realized in macroscopic experiments. An actual process approaches reversibility to the degree that it is quasistatic (i.e. extremely slow) and that dissipative effects (e.g. friction) are absent.

19.2 ENTROPY

Any thermodynamic system has a state function S, called the *entropy*, which may be defined as follows. Let the system undergo an infinitesimal *reversible* process in which it absorbs heat $đQ$. Then the change in entropy of the system is given by

$$dS = \frac{đQ}{T}$$

where T is the Kelvin temperature. Entropy will have the units J/K.

The *Clausius equation*, $dS = đQ/T$, holds only for reversible processes. However, since S is a state function, the entropy change accompanying an irreversible process can be calculated by integrating $đQ/T$ along the path of an arbitrary *reversible* process connecting the initial and final states.

The importance of the entropy function is exhibited in the *second law of thermodynamics*: *In any process, the total entropy of the system and its surroundings increases or* (*in a reversible process*) *does not change.* The second law applies to the system alone if the system is isolated; that is, if it in no way interacts with its surroundings.

The Clausius equation, together with the first law of thermodynamics, allows a new definition of the Kelvin temperature of a system, in terms of the system's internal energy E and volume V:

$$\frac{1}{T} = \left(\frac{\partial S}{\partial E}\right)_V$$

Statistical Definition of Entropy

The second law of thermodynamics indicates that entropy is a measure of irreversibility. Now, irreversibility is associated, on the molecular level, with the increase of disorder. Molecular systems tend, as time passes, to become chaotic, and it is extremely unlikely that a more organized state, once left, will ever be regained. For a given macroscopic equilibrium state that corresponds to Ω distinct microscopic states, the entropy may be defined as

$$S = k \ln \Omega$$

where k is Boltzmann's constant. This definition of entropy, due to Boltzmann, agrees in its predictions with that afforded by the Clausius equation. (See Problem 19.14.)

19.3 HEAT ENGINES AND REFRIGERATORS

A *heat engine* is a device or system that converts heat into work. Heat engines operate by absorbing heat from a reservoir at a high temperature, performing work, and giving off heat to a reservoir at a lower temperature. The *efficiency* η of a cyclic heat engine is

$$\eta = \frac{W}{Q_{\text{hot}}} = 1 - \frac{Q_{\text{cold}}}{Q_{\text{hot}}}$$

where Q_{hot}, Q_{cold}, and W (see Fig. 19-1) represent, respectively, the heat absorbed per cycle from the higher-temperature reservoir, the heat rejected per cycle to the lower-temperature reservoir, and the work carried out per cycle. The greatest possible thermal efficiency of an engine operating between two heat reservoirs is that of a *Carnot engine*, one that operates in a reversible cycle between the two fixed temperatures. This maximal efficiency is

$$\eta^* = 1 - \frac{T_{\text{cold}}}{T_{\text{hot}}}$$

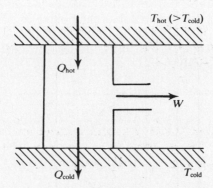

Fig. 19-1

EXAMPLE 19.1. For a heat engine, over one cycle, $\Delta S = \Delta E = 0$, since the engine returns to its original state. The first law of thermodynamics then gives for the work done by the engine per cycle

$$W = Q_{\text{hot}}\left(1 - \frac{T_{\text{cold}}}{T_{\text{hot}}}\right) - T_{\text{cold}}\,\Delta S_{\text{total}} = W^* - T_{\text{cold}}\,\Delta S_{\text{total}}$$

where W^* is the work that would be done by a Carnot engine operating between the same two temperatures, and ΔS_{total} is the entropy change of the universe (in this case, the entropy change of the hot and cold reservoirs) during one cycle. In general, *when an irreversible process produces an increase* $\Delta S_{\text{total}} > 0$ *in the entropy of the universe, an amount of energy* $T\,\Delta S_{\text{total}}$ *becomes unavailable for work, where T is the lowest temperature available for the rejection of heat.*

A *refrigerator* (or *heat pump*) is a heat engine operated backwards; it takes heat from a low-temperature reservoir, is supplied work, and rejects heat to a high-temperature reservoir. When a refrigerator in one cycle takes out heat Q_{cold} from a cold reservoir, the work input W_i to the refrigerator satisfies

$$W_i \geq Q_{\text{cold}}\left(\frac{T_{\text{hot}}}{T_{\text{cold}}} - 1\right)$$

The equality holds only for a Carnot heat pump.

19.4 ALTERNATIVE STATEMENTS OF THE SECOND LAW OF THERMODYNAMICS

Kelvin-Planck: It is impossible to construct an engine that operates in a cycle and produces no effect other than the extraction of heat from a reservoir and the performance of an equal amount of work.

Clausius: It is impossible to construct an engine that, operating in a cycle, will produce no effect other than the transfer of heat from a cooler to a hotter body.

The equivalence of these two formulations is demonstrated in Problem 19.7.

Solved Problems

19.1.   Calculate the entropy change associated with the boiling of 10 kg of water at 100 °C to steam at the same temperature.   What is the entropy change associated with the reverse process (the condensation of 10 kg of steam at 100 °C to water at the same temperature)?   The heat of vaporization of water is 2260 kJ/kg.

    The temperature of the water remains constant while it boils.   The entropy change is

$$\Delta S = \frac{Q}{T} = \frac{(10)(2260 \times 10^3)}{373} = 60\ 600 \text{ J/K}$$

    When steam condenses, it is necessary to extract 2260 kJ/kg of heat.   The change in entropy of 10 kg of steam in condensing to water at 100 °C is $\Delta S = -60\ 600$ J/K.

19.2.   One mole of water at 290 K is poured into a container of negligible heat capacity, which contains 1 mol of water originally at 310 K.   (*a*) Find the final temperature of the mixture.   (*b*) What is the entropy change, ΔS_1, of the originally cold water?   (*c*) What is the entropy change, ΔS_2, of the other mole of water?   (*d*) What is the total entropy change of the system?   For water, $c_v' = 75.4$ J/mol $\cdot$ K.

(*a*)   By symmetry, the final temperature T of the mixture must be the arithmetic mean of the two:

$$T = \frac{1}{2}(290 + 310) = 300 \text{ K}$$

(*b*)   Now, $dQ = nc_v' \, dT$, where n is the number of moles and c_v' is the molar specific heat.   Thus,

$$\Delta S_1 = \int_{290}^{300} nc_v' \frac{dT}{T} = (1)(75.4)[\ln T]_{290}^{300} = 2.56 \text{ J/K}$$

(*c*)   Similarly,
$$\Delta S_2 = \int_{310}^{300} nc_v' \frac{dT}{T} = (1)(75.4)[\ln T]_{310}^{300} = -2.47 \text{ J/K}$$

(*d*)
$$\Delta S_{\text{total}} = \Delta S_1 + \Delta S_2 = 2.56 + (-2.47) = 0.09 \text{ J/K}$$

19.3.   Figure 19-2 shows a dilute gas confined to a cylinder of cross-sectional area A and length ℓ, with a movable piston at one end.   Assume that the walls are diathermal (heat-conducting) and that the system is in thermal contact with a heat reservoir which is so large that its temperature remains essentially constant at T_0 as it gives up or absorbs heat.   The gas is suddenly compressed by a large force from a volume $A\ell$ to a volume $A\ell/2$.   When thermal equilibrium (all parts and the environment are at the same temperature) is restored, find (*a*) the entropy change, ΔS_1, of the heat reservoir; (*b*) the entropy change, ΔS_2, of the gas; and (*c*) the total entropy change of heat reservoir and gas.

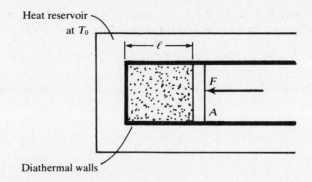

Fig. 19-2

(a) The final temperature of the gas is the same as its initial value, T_0.   The internal energy of an ideal gas depends only on temperature, so that the change in internal energy is $\Delta E = 0$ in this process.   The work done by the gas is the negative of the work done *on* the gas: $W = -F\ell/2$.   Then, by the first law of thermodynamics, Q for the gas is

$$Q = W = -\frac{F\ell}{2}$$

where the minus indicates that heat is extracted from the gas.

This heat is taken in by the heat reservoir while the gas is *relaxing* to thermal equilibrium.   The temperature of the heat reservoir does not change, so

$$\Delta S_1 = \frac{-Q}{T_0} = \frac{F\ell}{2T_0}$$

(b) The entropy change of a system undergoing some process depends only on the entropy difference between the initial and final states.   Suppose then that the entropy change ΔS_2 is accomplished by an isothermal compression of the gas from volume $V_0 = A\ell$ to $V_0/2$.   For the process,

$$p = \frac{nRT_0}{V}$$

and so the work done by the gas is

$$W = \int_{V_0}^{V_0/2} p\, dV = nRT_0 \int_{V_0}^{V_0/2} \frac{dV}{V} = -nRT_0 \ln 2$$

Since $\Delta E = 0$ for the gas, $Q = W$ and

$$\Delta S_2 = \frac{W}{T_0} = -nR \ln 2$$

(c) $$\Delta S_{\text{total}} = \Delta S_1 + \Delta S_2 = \frac{F\ell}{2T_0} - nR \ln 2$$

This may be rewritten as

$$\Delta S_{\text{total}} = \frac{V_0}{2T_0} \left(\frac{F}{A} - p_0\, 2 \ln 2 \right)$$

which, in view of the second law, gives the lower limit $p_0\, 2 \ln 2 = 1.39 p_0$ for the applied pressure.

19.4. A hot rock is thrown into a pond which has a temperature of 285 K.   As the rock cools, it gives off 295 kJ of heat.   Find the change in entropy of the pond.

The pond may be considered a heat reservoir, so that its temperature remains fixed at 285 K.   The entropy increase of the pond is

$$\Delta S = \frac{\text{heat absorbed by the pond}}{T} = \frac{295\,000}{285} = 1035 \text{ J/K}$$

19.5. Find the expression for the entropy change of a system undergoing an isochoric process (one in which the volume remains constant).

If two thermodynamic states that are involved in the process are infinitesimally close, $đQ = T\,dS$, so that

$$C_v = \left(\frac{đQ}{dT}\right)_V = T\left(\frac{dS}{dT}\right)_V \qquad \text{or} \qquad \left(\frac{dS}{dT}\right)_V = \frac{C_v}{T}$$

If, then, C_v is constant,

$$\Delta S = \int_{T_0}^{T} \frac{dS}{dT}\,dT = C_v \int_{T_0}^{T} \frac{dT}{T} = C_v \ln \frac{T}{T_0}$$

19.6. Ten kilograms of aluminum ($c_v = 0.91$ J/kg $\cdot$ K) at 250 K is placed in contact with 30 kg of copper ($c_v = 0.39$ J/kg $\cdot$ K) at 375 K. (a) If there can be no transfer of energy to the surroundings, what will be the final temperature of the metals? (b) Calculate the change in entropy of the system when the two blocks of metal are placed in contact.

(a) In the constant-volume process, the changes in internal energy of the aluminum and copper are

$$\Delta E_{Al} = (mc_v\,\Delta T)_{Al} = (10)(0.91)(T - 250)$$

$$\Delta E_{Cu} = (mc_v\,\Delta T)_{Cu} = (30)(0.39)(T - 375)$$

where T is the final temperature of the system. Since there is no energy transfer,

$$\Delta E = \Delta E_{Al} + \Delta E_{Cu} = 10(0.91)(T - 250) + 30(0.39)(T - 375) = 0$$

or $T = 321$ K.

(b) The change in entropy of each metal in the constant-volume process is given by the result of Problem 19.5:

$$\Delta S = C_v \ln \frac{T}{T_0} = mc_v \ln \frac{T}{T_0}$$

Thus,

$$\Delta S_{Al} = (10)(0.91) \ln \frac{321}{250} = 2.27 \text{ J/K}$$

$$\Delta S_{Cu} = (30)(0.39) \ln \frac{321}{375} = -1.82 \text{ J/K}$$

The net entropy change is

$$\Delta S = \Delta S_{Al} + \Delta S_{Cu} = 0.46 \text{ J/K}$$

19.7. Show that the Clausius and Kelvin-Planck statements of the second law of thermodynamics are equivalent.

Statements A and B are equivalent if a violation of A implies a violation of B, and conversely.

Shown in Fig. 19-3(a) is an engine that would violate the Kelvin-Planck statement by extracting heat Q_2 from the reservoir at temperature T_2 and doing work $W = Q_2$. This engine could be used to drive an ordinary refrigerator between reservoirs at $T_1 < T_2$ and T_2. As shown in Fig. 19-3(b), the composite engine would transfer a net amount of heat Q_1 from T_1 to T_2 without any external work being supplied. It would therefore violate the Clausius statement.

The Clausius statement would be violated by the refrigerator shown in Fig. 19-4(a). Using this refrigerator in conjunction with an ordinary heat engine, as in Fig. 19-4(b), one could convert an amount of heat $Q_2 - Q_1$ completely into work. This would violate the Kelvin-Planck statement.

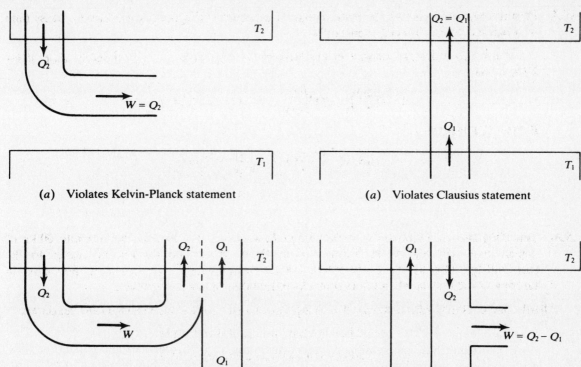

(a) Violates Kelvin-Planck statement

(a) Violates Clausius statement

(b) Violates Clausius statement
Fig. 19-3

(b) Violates Kelvin-Planck statement
Fig. 19-4

19.8. For a gasoline engine undergoing the cycle shown in Fig. 19-5, an air-standard Otto cycle, determine the thermal efficiency.

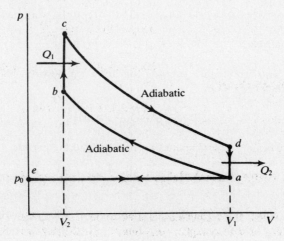

Fig. 19-5

We assume that the cycle is reversible and that C_v is constant for the working substance (air). Then, along the line $b \rightarrow c$, the heat into the system is

$$Q_1 = \int_{T_b}^{T_c} C_v \, dT = C_v(T_c - T_b)$$

Similarly, for process $d \to a$, the heat out of the system is

$$Q_2 = -\int_{T_d}^{T_a} C_v \, dT = C_v(T_d - T_a)$$

The thermal efficiency is therefore

$$\eta = 1 - \frac{Q_2}{Q_1} = 1 - \frac{T_d - T_a}{T_c - T_b}$$

Now, assuming an ideal gas, the two adiabatic processes may be described by the equations

$$T_d V_1^{\gamma-1} = T_c V_2^{\gamma-1} \qquad \text{and} \qquad T_a V_1^{\gamma-1} = T_b V_2^{\gamma-1}$$

(see Section 17.3). These give, upon subtraction,

$$(T_d - T_a)V_1^{\gamma-1} = (T_c - T_b)V_2^{\gamma-1} \qquad \text{or} \qquad \frac{T_d - T_a}{T_c - T_b} = \left(\frac{V_2}{V_1}\right)^{\gamma-1}$$

Thus

$$\eta = 1 - \left(\frac{V_2}{V_1}\right)^{\gamma-1}$$

The Otto cycle is an example of a reversible cycle that differs from the Carnot cycle (see Fig. 19-6).

19.9. A room is kept at 27 °C by air conditioning while the outside temperature is 42 °C. The refrigerating unit has compression cylinders operating at 57 °C (outside) and expansion coils inside the house operating at 17 °C. (*a*) If the unit operates reversibly, how much work must be done for each transfer of 4 kJ of heat from the house? (*b*) What entropy changes occur inside and outside the house for this amount of refrigeration?

(*a*) The unit operates between reservoirs at $273 + 17 = 290$ K and $273 + 57 = 330$ K. Since the unit operates reversibly, its thermal efficiency is maximal:

$$\eta^* = 1 - \frac{T_{\text{cold}}}{T_{\text{hot}}} = 1 - \frac{290}{330}$$

But also,

$$\eta^* = 1 - \frac{Q_{\text{cold}}}{Q_{\text{hot}}} = 1 - \frac{4000}{Q_{\text{hot}}}$$

and the two expressions yield $Q_{\text{hot}} = 4552$ J. Then the work input is $W_i = Q_{\text{hot}} - Q_{\text{cold}} = 552$ J.

Alternatively,

$$W_i = Q_{\text{cold}}\left(\frac{T_{\text{hot}}}{T_{\text{cold}}} - 1\right) = 4000\left(\frac{330}{290} - 1\right) = 552 \text{ J}$$

(*b*) There are two irreversible transfers of heat: (1) between the inside of the house and the low-temperature reservoir,

$$\Delta S' = \frac{Q_{\text{cold}}}{290} - \frac{Q_{\text{cold}}}{300} = 4000\left(\frac{1}{290} - \frac{1}{300}\right) = 0.46 \text{ J/K}$$

and (2) between the high-temperature reservoir and the outside,

$$\Delta S'' = \frac{Q_{\text{hot}}}{315} - \frac{Q_{\text{hot}}}{330} = 4552\left(\frac{1}{315} - \frac{1}{330}\right) = 0.66 \text{ J/K}$$

19.10. A heat engine, which operates with an efficiency of 17%, absorbs 100 kJ of heat per cycle from the high-temperature reservoir. (*a*) How much heat is rejected per cycle to the low-temperature reservoir? (*b*) How much work does the engine carry out during each cycle?

(*a*)

$$\eta = 1 - \frac{Q_{\text{cold}}}{Q_{\text{hot}}}$$

or

$$Q_{\text{cold}} = Q_{\text{hot}}(1 - \eta) = 100(1 - 0.17) = 83 \text{ kJ}$$

(*b*)

$$W = Q_{\text{hot}} - Q_{\text{cold}} = 17 \text{ kJ}$$

19.11. Show that for a *Carnot cycle*, Fig. 19-6,

$$\frac{Q_2}{Q_1} = \frac{T_2}{T_1}$$

Assume 1 mol of ideal gas as the working substance.

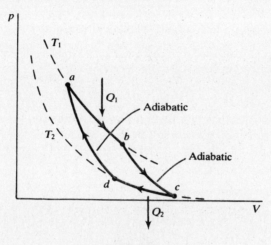

Fig. 19-6

During the stage $a \to b$, while the gas expands isothermally, its internal energy does not change, and the heat Q_1 absorbed from the T_1-reservoir is the same as the work carried out on it. Proceeding as in Problem 19.3(*b*), we find:

$$Q_1 = RT_1 \ln \frac{V_b}{V_a}$$

Similarly, in the stage $c \to d$, the heat rejected, Q_2, is

$$Q_2 = RT_2 \ln \frac{V_c}{V_d}$$

Along the paths $b \to c$ and $d \to a$ the processes are adiabatic, so

$$T_1 V_b^{\gamma-1} = T_2 V_c^{\gamma-1} \qquad T_2 V_d^{\gamma-1} = T_1 V_a^{\gamma-1}$$

Upon division of these equations,

$$\frac{V_b}{V_a} = \frac{V_c}{V_d}$$

Thus

$$\frac{Q_2}{Q_1} = \frac{RT_2 \ln(V_c/V_d)}{RT_1 \ln(V_b/V_a)} = \frac{T_2}{T_1}$$

19.12. The number Ω of states accessible to N atoms of a monatomic ideal gas with a volume V, when the energy of the gas is between E and $E + dE$, can be shown to be

$$\Omega = A(N) \, V^N E^{3N/2}$$

where the factor $A(N)$ depends only on N. (*a*) Find the entropy S as a function of V and E. (*b*) Using this entropy function and the definition of the Kelvin temperature, show that

$$E = \frac{3}{2} NkT$$

(*a*) The entropy is given by

$$S = k \ln \Omega = k \ln A(N) + Nk \ln V + \frac{3}{2} Nk \ln E$$

(b) The Kelvin temperature is given by

$$\frac{1}{T} = \left(\frac{\partial S}{\partial E}\right)_V = \frac{3}{2} Nk \frac{d(\ln E)}{dE} = \frac{3}{2} \frac{Nk}{E}$$

whence
$$E = \frac{3}{2} NkT$$

19.13. Is it possible to devise a heat engine which will cause no thermal pollution?

No; heat must be exhausted to a cold reservoir to provide an entropy increase that will more than compensate for the entropy decrease of the hot reservoir.

19.14. An insulated container has two sections, each of volume V_0, separated by a wall. Suppose that initially each section is occupied by N atoms of a gas at temperature T_0. If the wall is removed, calculate the change in entropy if (a) both gases are the same, (b) one gas is different from the other.

Substituting $E = \frac{3}{2}NkT$ into the expression for S found in Problem 19.12(a) yields

$$S = k \ln A(N) + Nk \ln V + \frac{3}{2} Nk \ln \frac{3}{2} Nk + \frac{3}{2} Nk \ln T$$

$$= B(N) + Nk \ln T^{3/2} V$$

where (for a given gas) $B(N)$ depends on N alone.

(a) After the wall is removed, each section of volume V_0 contains N atoms of the same species at temperature T_0, just as it did before the removal of the wall. Thus, $\Delta S = 0$.

(b) The initial total entropy is

$$S_i = B_1(N) + Nk \ln T_0^{3/2} V_0 + B_2(N) + Nk \ln T_0^{3/2} V_0$$

[$B(N)$ depends parametrically on the atomic mass], and the final total entropy is

$$S_f = B_1(N) + Nk \ln T_0^{3/2} 2V_0 + B_2(N) + Nk \ln T_0^{3/2} 2V_0$$
$$= S_i + 2Nk \ln 2$$

since each gas occupies a final volume $2V_0$ and the temperature stays the same.

Hence
$$\Delta S = 2Nk \ln 2$$

Recall from Problem 19.3(b) that the entropy change of either gas in an isothermal volume-doubling should be

$$nR \ln 2 = Nk \ln 2$$

Thus the statistical concept of entropy is in agreement with the thermodynamical.

Supplementary Problems

19.15. Derive the equation given in Example 19.1.

19.16. What is the change in the entropy of 2.00 kg of H_2O molecules when transformed at a constant pressure of 1 atm from water at 100 °C to steam at the same temperature?   *Ans.*   12 120 J/K

19.17. A pot contains 2.00 kg of hot water at 40 °C. This is poured into a kettle containing 2.00 kg of cold water at 0 °C. After mixing, the water has a uniform temperature of 20 °C throughout. Calculate the change in entropy in the process.   *Ans.*   20 J/K

19.18. What is the minimum power that must be supplied to a refrigerator that freezes 1.00 kg of water at 0 °C into ice at 0 °C in a time interval of 10 min?   The room temperature is 20 °C.   *Ans.*   40.7 W

19.19. Prove that any heat engine which operates cyclically between a hot reservoir and a cold reservoir has a thermal efficiency less than or equal to the Carnot efficiency, $1 - (T_{cold}/T_{hot})$.

19.20. On a hot day a student leaves the refrigerator door open in an effort to cool his apartment.   Will this work?
  Ans.   No.   Although the refrigerator takes out heat Q_{cold}, it gives off heat $Q_{hot} = Q_{cold} + W$, where W is the energy supplied to operate the refrigerator.

19.21. Compare the Carnot efficiency of a steam engine which has a boiler temperature of 180 °C and a condenser temperature of 50 °C to the Carnot efficiency of a gasoline engine which has a combustion temperature of 1510 °C and an exhaust temperature of 410 °C.
  Ans. $\eta^{*}_{steam} = 0.287$, $\eta^{*}_{gas} = 0.617$

19.22. A system, initially in a macroscopic state that can occur in Ω_1 microscopically different ways, undergoes a process resulting in a macroscopic state that can occur in Ω_2 microscopically different ways.   If the initial entropy is S_1 and final entropy is S_2, show that

$$\frac{\Omega_2}{\Omega_1} = e^{(S_2 - S_1)/k}$$

19.23. Find an expression for the entropy of an ideal gas having constant heat capacities, C_p and C_v.
  Ans. $S = C_v \ln pV^{\gamma} + \text{constant}$, where $\gamma = C_p/C_v$.

19.24. From Problem 19.23 infer that during a reversible adiabatic process the pressure and volume of an ideal gas are related through $pV^{\gamma} = \text{constant}$.

Chapter 20

Wave Phenomena

20.1 WAVE FUNCTION

A wave $y(x, t)$ traveling along X with an unchanging form and speed v is given by

$$y = f(x \pm vt) \qquad (20.1)$$

where the minus and plus signs refer to wave propagation in the positive and negative X-directions, respectively.

These functions satisfy the one-dimensional *wave equation*

$$\frac{\partial^2 y}{\partial x^2} = \frac{1}{v^2} \frac{\partial^2 y}{\partial t^2} \qquad (20.2)$$

20.2 WAVE ON A STRETCHED STRING

The *speed* of a wave traveling on a stretched uniform string, causing small transverse displacements $y(x, t)$, is given by

$$v = \sqrt{\frac{F}{\rho}} \qquad (20.3)$$

where F is the tension in the string and ρ is the linear density of the string.

The *instantaneous power transmitted* by the wave is

$$P = -F \frac{\partial y}{\partial x} \frac{\partial y}{\partial t} \qquad (20.4)$$

20.3 THE SINUSOIDAL WAVE

The sinusoidal traveling wave

$$y = A \cos (kx - \omega t) \qquad (20.5)$$

has a wave speed

$$v = \frac{\omega}{k} \qquad (20.6)$$

For a particular x, y is a periodic function of t, with period

$$T = \frac{2\pi}{\omega} \qquad (20.7)$$

For a particular t, y is a periodic function of x, with spatial period, called the *wavelength*, given by

$$\lambda = \frac{2\pi}{k} \qquad (20.8)$$

The average power transmitted by a sinusoidal wave can be calculated from (20.4):

$$P_{\text{avg}} = \frac{1}{2}\,\omega^2 A^2 \rho v \qquad\qquad (20.9)$$

That is to say, an amount of energy $P_{\text{avg}}T$ passes a fixed location x during each temporal period.

20.4　SUPERPOSITION PRINCIPLE FOR WAVES

When two waves are simultaneously present, the displacement $y(x, t)$ at any instant is the algebraic sum of the displacements $y_1(x, t)$ and $y_2(x, t)$ that would occur if each individual wave were present alone.

20.5　STANDING WAVES

The superposition of two sinusoidal waves having the same amplitude and the same wavelength λ, but traveling in opposite directions along a string, is a *standing wave* with nodes (locations of permanently zero displacement) at a distance $\lambda/2$ apart.

The *normal modes* of vibration of a string of length L, fixed at both ends, are standing waves with nodes at each end of the string. In the nth normal mode, each internodal segment of the string oscillates with the *natural frequency* $f_n = nf_1$, where $f_1 = v/2L$. Adjacent segments are out of step by a half-period, $1/2f_n$.

Solved Problems

20.1.　A rope under a tension of 20 N has a wave traveling in it at a speed of 5 m/s. What is the linear density of the rope?

From (20.3),

$$\rho = \frac{F}{v^2} = \frac{20}{5^2} = 0.8 \text{ kg/m}$$

20.2.　A uniform, flexible wire 20 m long and weighing 50 N hangs vertically under its own weight. What is the speed of a transverse pulse at (a) the midpoint of the wire and (b) a point very close to the upper end?

(a)　At the wire's midpoint, the tension is $F = w/2$. The linear density of the wire is

$$\rho = \frac{w}{g\ell}$$

Then

$$v = \sqrt{\frac{F}{\rho}} = \sqrt{\frac{g\ell}{2}} = \sqrt{\frac{(9.8)(20)}{2}} = 9.9 \text{ m/s}$$

(b)　At the upper end, $F = w$, and

$$v = \sqrt{g\ell} = 14.0 \text{ m/s}$$

20.3.　Two wires, with linear densities ρ_1 and ρ_2, are joined together and are under the same tension. (a) If a wave's frequency is 125 Hz in wire #1, what is its frequency in wire #2? (b) If a wave has a wavelength of 0.03 m in wire #1, what is its wavelength in wire #2?

(a)　125 Hz.

(b)　The wavelength is $\lambda = v/f$. In wire #1 the wavelength is $\lambda_1 = v_1/f$ and in wire #2, $\lambda_2 = v_2/f$. Thus,

$$\lambda_2 = \lambda_1 \frac{v_2}{v_1} = \lambda_1 \sqrt{\frac{F/\rho_2}{F/\rho_1}} = \lambda_1 \sqrt{\frac{\rho_1}{\rho_2}} = 0.03\sqrt{\frac{\rho_1}{\rho_2}} \quad \text{(m)}$$

20.4. Figure 20-1 shows a sinusoidal traveling wave at three different instants. What are (a) the wavelength, (b) the wave speed, (c) the amplitude, and (d) the frequency? (e) In what direction does the wave travel? (f) Write the wave function.

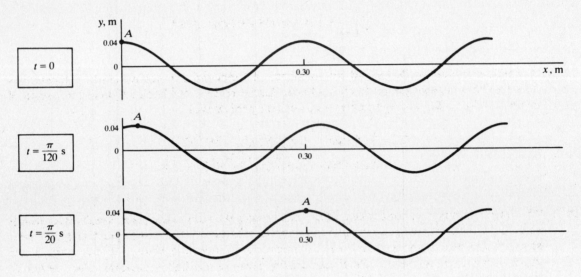

Fig. 20-1

(a) Any of the three graphs shows that the spatial period, or wavelength, is

$$\lambda = 0.30 \text{ m}$$

(b) Inspection of the three graphs shows that at any fixed location (say, $x = 0$) the disturbance *first* repeats itself after $(\pi/20)$ s. Hence,

$$T = \frac{\pi}{20} \text{ s}$$

and

$$v = \frac{\lambda}{T} = \frac{0.30}{\pi/20} = 1.9 \text{ m/s}$$

(c)

$$A = 0.04 \text{ m}$$

(d)

$$f = \frac{1}{T} = \frac{1}{\pi/20} = 6.37 \text{ Hz}$$

(e) The wave moves to the right.

(f)

$$y = A \cos 2\pi \left(\frac{t}{T} - \frac{x}{\lambda} \right) = 0.04 \cos (0.16t - 3.33x) \quad \text{(m)}$$

20.5. A traveling transverse wave is given by

$$y = 0.004 \sin (25x + 250t) \quad \text{(m)}$$

Find (a) the wave speed, (b) the wavelength, (c) the frequency, (d) the amplitude, and (e) the transverse particle speed of the medium at $x = x_0$, $t = t_0$.

(a) From the given expression we can write $\omega = 250 \text{ s}^{-1}$ and $k = 25 \text{ m}^{-1}$. The wave speed is then

$$v = \lambda f = \frac{\omega}{k} = \frac{250}{25} = 10 \text{ m/s}$$

(b)

$$\lambda = \frac{2\pi}{k} = \frac{2\pi}{25} = 0.251 \text{ m}$$

(c)
$$f = \frac{\omega}{2\pi} = \frac{250}{2\pi} = 39.8 \text{ Hz}$$

(d) By inspection, $A = 0.004$ m.

(e) The transverse particle speed at x_0, t_0 is

$$\left.\frac{\partial y}{\partial t}\right|_{x_0, t_0} = 1.000 \cos (25x_0 + 250t_0) \quad \text{(m/s)}$$

20.6. A simple harmonic oscillator with an amplitude of 4 mm and a frequency of 450 Hz generates a sinusoidal wave along a string with a linear density of 15×10^{-3} kg/m and under a tension of 225 N. What is the average power transmitted by this wave?

$$P_{\text{avg}} = \frac{1}{2}\omega^2 A^2 \rho v = \frac{1}{2}(2\pi f)^2 A^2 \rho \sqrt{\frac{F}{\rho}} = 2\pi^2 f^2 A^2 \sqrt{F\rho}$$
$$= 2\pi^2 (450)^2 (0.004)^2 \sqrt{(225)(15 \times 10^{-3})} = 117 \text{ W}$$

20.7. The linear density of a nonuniform wire under constant tension changes gradually with length so that an incident wave is transmitted without reflection. The wave speed and the wave shape will change as the wave travels. Suppose that a sinusoidal wave, initially of amplitude A_1, travels far enough so that the linear density becomes one-half its original value. What is the new amplitude in terms of the initial amplitude?

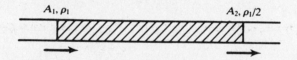

Fig. 20-2

Let us apply conservation of energy to the section of wire between station *1*, where the wave amplitude is A_1 and the linear density is ρ_1, and station *2*, where the wave amplitude is A_2 and the linear density is $\rho_1/2$. The rate at which energy enters at station *1* is, from Problem 20.6,

$$P_{\text{avg},1} = 2\pi^2 f^2 A_1^2 \sqrt{F\rho_1}$$

and the rate at which energy leaves at station *2* is

$$P_{\text{avg},2} = 2\pi^2 f^2 A_2^2 \sqrt{F(\rho_1/2)}$$

The net rate of energy creation in the section must be zero. Thus,

$$P_{\text{avg},1} = P_{\text{avg},2}$$
$$2\pi^2 f^2 A_1^2 \sqrt{F\rho_1} = 2\pi^2 f^2 A_2^2 \sqrt{F(\rho_1/2)}$$
$$A_2 = 2^{1/4} A_1 = 1.19 A_1$$

20.8. Sources separated by 20 m vibrate according to the equations

$$y_1' = 0.06 \sin \pi t \quad \text{(m)} \qquad y_2' = 0.02 \sin \pi t \quad \text{(m)}$$

They send out waves of speed 3 m/s. What is the equation of motion of a particle 12 m from the first source and 8 m from the second?

Refer to Fig. 20-3. Let source *1* send out waves in the $+X$-direction, so that

$$y_1 = A_1 \sin 2\pi f_1 \left(t - \frac{x_1}{v} \right)$$

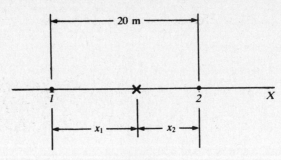

Fig. 20-3

and let source *2* send out waves in the $-X$-direction, so that

$$y_2 = A_2 \sin 2\pi f_2\left(t + \frac{x_2}{v}\right)$$

Here, $v = 3$ m/s; and equating y_1 at $x_1 = 0$ to y_1', and y_2 at $x_2 = 0$ to y_2', we obtain

$$A_1 = 0.06 \text{ m} \qquad f_1 = f_2 = \frac{1}{2} \text{ Hz} \qquad A_2 = 0.02 \text{ m}$$

The resultant disturbance at the point $x_1 = 12$ m, $x_2 = -8$ m is then

$$y = y_1 + y_2 = 0.06 \sin \pi\left(t - \frac{12}{3}\right) + 0.02 \sin \pi\left(t - \frac{8}{3}\right)$$

$$= 0.06 \sin \pi t + 0.02 \sin\left(\pi t - \frac{2\pi}{3}\right)$$

$$= 0.06 \sin \pi t + 0.02\left(\sin \pi t \cos \frac{2\pi}{3} - \cos \pi t \sin \frac{2\pi}{3}\right)$$

$$= 0.06 \sin \pi t + 0.02\left[(\sin \pi t)\left(-\frac{1}{2}\right) - (\cos \pi t)\left(\frac{\sqrt{3}}{2}\right)\right]$$

$$= 0.05 \sin \pi t - 0.0173 \cos \pi t$$

20.9. When two sinusoidal waves that are close together in frequency are superposed, the resultant disturbance exhibits *beats*: the amplitude at a given location oscillates periodically between a maximum and a minimum value. Find the beat frequency.

One can, with no loss in generality, choose $x = 0$ as the fixed observation point and write the component waves there as

$$y_1 = A_1 \cos \omega_1 t \qquad y_2 = A_2 \cos (\omega_2 t + \phi)$$

where ω_2 is slightly larger than ω_1 and ϕ is a phase constant. The resultant wave is then given by

$$y = y_1 + y_2 = A_1 \cos \omega_1 t + A_2 \cos (\omega_2 t + \phi)$$

Let

$$\alpha \equiv \frac{\omega_1 + \omega_2}{2} t + \frac{\phi}{2} \qquad \beta \equiv \frac{\omega_2 - \omega_1}{2} t + \frac{\phi}{2}$$

Then

$$y = A_1 \cos (\alpha - \beta) + A_2 \cos (\alpha + \beta)$$

$$= A_1 (\cos \alpha \cos \beta + \sin \alpha \sin \beta) + A_2 (\cos \alpha \cos \beta - \sin \alpha \sin \beta)$$

$$= (\cos \alpha)[(A_1 + A_2) \cos \beta] + (\sin \alpha)[(A_1 - A_2) \sin \beta]$$

To make the multipliers of $\cos \alpha$ and $\sin \alpha$ in the last line less than unity in absolute value, multiply and divide by

$$B \equiv \{[(A_1 + A_2) \cos \beta]^2 + [(A_1 - A_2) \sin \beta]^2\}^{1/2}$$

$$= \{A_1^2 + A_2^2 + 2A_1 A_2 \cos 2\beta\}^{1/2}$$

$$y = B\left[(\cos \alpha)\frac{(A_1 + A_2) \cos \beta}{B} + (\sin \alpha)\frac{(A_1 - A_2) \sin \beta}{B}\right]$$

Finally, define angle σ by

$$\frac{(A_1 + A_2)\cos\beta}{B} = \cos\sigma \qquad \frac{(A_1 - A_2)\sin\beta}{B} = \sin\sigma$$

that is,
$$\sigma = \arctan\left(\frac{A_1 - A_2}{A_1 + A_2}\tan\beta\right)$$

The displacement y can now be written as

$$y = B\cos(\alpha - \sigma)$$

This is a sinusoidal disturbance whose amplitude, B, is itself a periodic function of time, varying as

$$\cos 2\beta = \cos[(\omega_2 - \omega_1)t + \phi]$$

Thus the beat frequency is $(\omega_2 - \omega_1)/2\pi$. The maximum and minimum values of B are $A_1 + A_2$ and $|A_1 - A_2|$, respectively representing constructive and destructive interference between the two component waves.

20.10. An oscillator is to have its frequency adjusted to 10^3 Hz, to within 1%, by comparison with a standard oscillator having a frequency exactly 10^3 Hz. What is the maximum permissible beat frequency between the two oscillators?

$$f_{beat} = |f - f_{standard}| \le (0.01)(10^3) = 10 \text{ Hz}$$

20.11. When fixed at both ends, a wire under tension of 900 N and having a linear density 10^{-2} kg/m is resonant at the frequency of 420 Hz. The next higher frequency at which it resonates is 490 Hz. What is the wire's length?

We must first find which integral multiple of the fundamental frequency 420 Hz is:

$$\frac{420}{490} = \frac{6}{7}$$

Thus, 420 Hz is the sixth harmonic. Since, for a standing wave in a wire fixed at both ends,

$$f_6 = 6f_1 = 6\frac{\sqrt{F/\rho}}{2L}$$

we have
$$L = \frac{3}{f_6}\sqrt{\frac{F}{\rho}} = \frac{3}{420}\sqrt{\frac{900}{10^{-2}}} = 2.1 \text{ m}$$

20.12. Two wires, having different densities, are joined at $x = 0$ (see Fig. 20-4). An incident wave,

$$y_i = A_i\sin(\omega t - k_1 x)$$

traveling to the right in the wire $x \le 0$, is partly reflected and partly transmitted at $x = 0$. Find the reflected and transmitted amplitudes in terms of the incident amplitude.

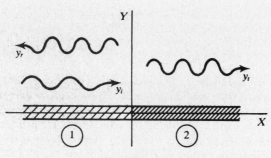

Fig. 20-4

The reflected and transmitted waves will have the forms

$$y_r = A_r \sin(\omega t + k_1 x) \qquad y_t = A_t \sin(\omega t - k_2 x)$$

where A_r may possibly be negative (corresponding to a 180° phase change upon reflection). The boundary conditions at $x = 0$ are that the displacement y and the slope $\partial y/\partial x$ be continuous. Thus:

$$y_i|_{x=0} + y_r|_{x=0} = y_t|_{x=0}$$

$$A_i + A_r = A_t \qquad (1)$$

and

$$\left.\frac{\partial y_i}{\partial x}\right|_{x=0} + \left.\frac{\partial y_r}{\partial x}\right|_{x=0} = \left.\frac{\partial y_t}{\partial x}\right|_{x=0}$$

$$-k_1 A_i + k_1 A_r = -k_2 A_t \qquad (2)$$

Solving (1) and (2) simultaneously,

$$A_r = \frac{k_1 - k_2}{k_1 + k_2} A_i \qquad A_t = \frac{2k_1}{k_1 + k_2} A_i$$

It is seen that A_r is indeed negative when $k_2 > k_1$, which will be the case if wire 2 is denser than wire 1.

20.13. Transverse waves may move in two-dimensional space along the surface of a flexible membrane under tension. Find the wave speed.

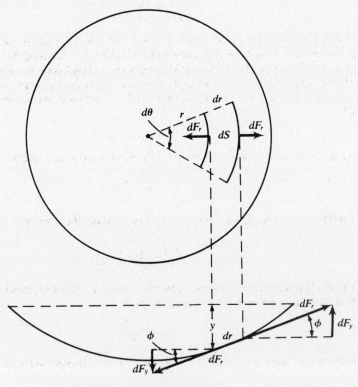

Fig. 20-5

Suppose that the boundary of the membrane is a very large circle, as shown in Fig. 20-5. Set up the equation of motion for an element of area, $dS = r\,dr\,d\theta$, of the membrane vibrating with circular symmetry; the vertical displacement of dS is $y(r, t)$. The membrane will be assumed to be thin, uniform, and perfectly elastic. Let σ be the area density of the membrane and let F be the tension to which the edge of the membrane is stretched. This tension is distributed uniformly throughout the membrane; so, the material on opposite sides of a line segment of length ds will tend to be pulled apart with a force $F\,ds$.

The radial force acting across the inner boundary of dS is therefore

$$dF_r = Fr\,d\theta$$

The vertical component of this force is $dF_y = (dF_r)\sin\phi$, where ϕ is the angle between a radial line tangent to the surface at dS and the plane of the undistorted membrane. For small displacements,

$$\sin\phi \approx \tan\phi = \frac{\partial y}{\partial r}$$

and

$$(dF_y)_r = F\left(r\,\frac{\partial y}{\partial r}\right)_r d\theta$$

Similarly, the vertical component of the force acting across the outer boundary of dS is

$$(dF_y)_{r+dr} = F\left(r\,\frac{\partial y}{\partial r}\right)_{r+dr} d\theta$$

and so the net vertical force acting upon dS is

$$(dF_y)_{r+dr} - (dF_y)_r = F\,d\theta\left[\frac{\partial}{\partial r}\left(r\,\frac{\partial y}{\partial r}\right)dr\right]$$

The mass of element dS is $\sigma r\,dr\,d\theta$; hence, Newton's second law gives

$$F\,d\theta\left[\frac{\partial}{\partial r}\left(r\,\frac{\partial y}{\partial r}\right)dr\right] = (\sigma r\,dr\,d\theta)\frac{\partial^2 y}{\partial t^2}$$

or

$$\frac{1}{r}\frac{\partial}{\partial r}\left(r\,\frac{\partial y}{\partial r}\right) = \frac{\sigma}{F}\frac{\partial^2 y}{\partial t^2} \tag{1}$$

Equation (1) is, in fact, the wave equation for circular waves, with the polar coordinate r playing a role analogous to that of the Cartesian coordinate x in (20.2). The analogy can be brought out by an energy argument. Imagine that at $t = 0$ the undisturbed membrane is struck impulsively at its center. At a later instant of time we expect that *most* of the wave's energy will be passing through the circular wavefront of radius r, at a rate proportional to the square of the amplitude of the wave [see (20.9)]. Thus

$$\text{energy} \propto rA^2$$

so that, by conservation of energy, $A \propto 1/\sqrt{r}$. Then, the change of variable

$$y = \frac{u}{\sqrt{r}}$$

ought to simplify (1). Indeed, when the substitution is made, the result is

$$\frac{\partial^2 u}{\partial r^2} + \frac{u}{4r^2} = \frac{\sigma}{F}\frac{\partial^2 u}{\partial t^2} \tag{2}$$

When r is large, the term $u/4r^2$ may be neglected, and it is seen by comparison with (20.2) that the disturbance u acts like a linear wave traveling at speed

$$v = \sqrt{\frac{F}{\sigma}}$$

Unlike a linear wave, however, the full circular wave, $u/\sqrt{r}$, *changes shape* as it travels in the uniform medium.

Supplementary Problems

20.14. A transverse wave on a stretched string is a pulse that at $t = 0$ is described by

$$y = \frac{12.0}{x^2 + 4.0} \quad \text{(m)}$$

The wave travels in the positive X-direction at a constant speed of 6.0 m/s and maintains its waveform. (*a*) What is the function $y(x, t)$ that describes this wave? (*b*) At $t = 2.0$ s, what is the velocity of the particle located at $x = 8.0$ m?
Ans. (*a*) $y(x, t) = \dfrac{12.0}{(x - 6.0\,t)^2 + 4.0}$ (m); (*b*) $v_y = -1.4$ m/s

20.15. A man has his boat anchored on a lake. He notices that his fishing-line float makes 20 oscillations in 20 s, and that it takes 5 s for the trough of a wave to move the 15 m length of his boat. Find the number of waves along the length of his boat at any time. *Ans.* five

20.16. A wave on a string is described by $y = 0.04 \cos{(12.0\,x + 24.0\,t)}$ (m). Find (*a*) the wave speed, (*b*) the wavelength, (*c*) the period. *Ans.* (*a*) 2.0 m/s; (*b*) 1.05 m; (*c*) 0.52 s

20.17. Derive (*20.9*) from (*20.4*).

20.18. A sinusoidal wave is moving along a horizontal rope of linear density 0.5 kg/m that is held under a tension of 10 N. (*a*) What is the speed of the wave? (*b*) How much power is transmitted by the wave if the wavelength is 6 m and the amplitude is 0.2 m? *Ans.* (*a*) 4.5 m/s; (*b*) 1.00 W

20.19. For a wave on a string, show that the energy per unit length (i.e. the mechanical energy of the particles composing a unit length of string) is given by

$$E_\ell = \frac{P_{\text{avg}}}{v}$$

where P_{avg} is the average power transmitted by the wave and v is the wave speed. (*Hint*: Suppose the wave to be incident on an undisturbed section of the string.)

20.20. A wire of density ρ (kg/m³) is stretched so that its initial length L_0 is increased to $L_0 + \Delta L$. Find the speed with which a transverse wave would move along the stretched wire, in terms of ΔL, L_0, ρ, and Y (Young's modulus).
Ans. $v = \sqrt{\dfrac{Y\,\Delta L}{\rho L_0}}$

20.21. In Problem 20.8, what is the phase difference between the disturbance y and either source?
Ans. -0.333 rad $= -19.1°$

20.22. For the superposition principle to be valid it is necessary that the wave speed be large compared to the transverse speed of a particle in the medium. Show that for a sinusoidal wave this implies that the amplitude must be small compared to the wavelength.

Chapter 21

Sound Waves

Sound is a longitudinal wave in a medium in which the particles of the medium oscillate along the same direction as that in which energy and linear momentum are transmitted by the wave. The results of Chapter 20 apply equally to transverse and longitudinal waves. Thus, sound waves obey the superposition principle (provided that the amplitudes are sufficiently small), exhibit beats, standing waves, etc.

21.1 SOUND SPEED

The longitudinal displacement χ of the medium and the excess pressure p_e are related by

$$p_e = -B \frac{\partial \chi}{\partial x}$$

where B is the bulk modulus of the medium (Problem 14.5). Both p_e and χ satisfy the wave equation, (20.2). The wave speed is given by

$$v = \sqrt{\frac{B}{\rho_0}}$$

where ρ_0 is the density of the medium when it is in equilibrium [see Problem 21.1(b)]. In particular, if the medium is a gas,

$$v = \sqrt{\frac{B_{\text{adiabatic}}}{\rho_0}} = \sqrt{\frac{\gamma p_0}{\rho_0}} = \sqrt{\frac{\gamma k T}{m}}$$

where p_0 is the undisturbed gas pressure, γ is the specific heat ratio, ρ_0 is the undisturbed density of the gas, k is Boltzmann's constant, and m is the mass of a molecule of the gas. Note that if γ is constant, the speed of sound in a given gas varies as the square root of the absolute temperature.

21.2 INTENSITY AND LOUDNESS OF SOUND WAVES

The *intensity* I of a sinusoidal sound wave is the average (over one period) rate of transfer of energy per unit area transverse to the direction of propagation. It is given by the two-dimensional form of (20.9):

$$I = \frac{1}{\text{area}} \left(\frac{\partial U}{\partial t} \right)_{\text{avg}} = \frac{1}{2} \omega^2 A^2 \rho v$$

and has the units W/m^2.

In a three-dimensional medium, the intensity of sound waves radiated from a point source decreases inversely as the square of the distance from the source, if there is no absorption by the medium.

Loudness or *intensity level* is defined by

$$\text{intensity level} = 10 \log_{10} \left(\frac{I}{10^{-12}} \right) \ (\text{dB})$$

in which I is in W/m^2. The loudness unit, the *decibel* (dB), is a dimensionless unit, like the radian.

21.3 THE DOPPLER EFFECT

Suppose that a source, emitting sound waves of frequency f_s, and an observer move along the same straight line. Then the observer will hear sound of frequency

$$f_o = f_s \frac{v + v_o}{v - v_s}$$

where v_o and v_s are the velocities of the observer and source relative to the transmitting medium and v is the sound speed in this medium. Velocity v_o is positive if the observer is heading *toward* the source; v_s is positive if the source is heading *toward* the observer. See Problem 21.10 for an extension to the case of noncollinear motion of source and observer.

Transverse waves (light) also exhibit a Doppler effect, but in the case of light the effect depends only on the relative velocity between source and observer, there being no transmitting medium.

Solved Problems

21.1. Figure 21-1 shows at one instant of time a longitudinal pulse traveling at 4.5 m/s to the right in a long slender rod of density 10^4 kg/m^3. (*a*) What speed is imparted to a particle by the pulse? (*b*) What change in pressure is associated with this pulse? (*c*) By what fraction is the density of the solid changed by the pulse?

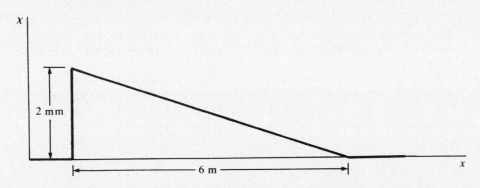

Fig. 21-1

(*a*) Consider the particle at the leading edge of the pulse shown in Fig. 21-1. It takes

$$\frac{6 \text{ m}}{4.5 \text{ m/s}} = 1.333 \text{ s}$$

for the pulse to pass this particle, during which time the particle is displaced longitudinally 2 mm. The average speed of the particle over this time interval is thus

$$v_p = \frac{2 \times 10^{-3}}{1.333} = 1.5 \times 10^{-3} \text{ m/s}$$

Indeed, because the pulse profile is linear, the particle moves at speed v_p at each instant of the time interval.

(*b*) Consider the particles making up the part of the rod which lies to the right of the leading edge of the pulse in Fig. 21-1. In a small time interval Δt, the pulse will penetrate a distance $v \Delta t$ to the right, so that all of the particles in a volume $Sv \Delta t$ (S = cross-sectional area of rod) are set into motion at speed v_p. The mass of these particles is approximately $\rho_0(Sv \Delta t)$, ρ_0 being the undeformed density. Thus, this collection of particles gains momentum $\rho_0(Sv\,\Delta t)v_p$ in time Δt. There must then act on it a force

$$F = \frac{\rho_0(Sv\,\Delta t)v_p}{\Delta t} = \rho_0 S v v_p$$

exerted to the right by the leftward-lying material. In short, there exists locally in the rod a

$$p_e = \frac{F}{S} = \rho_0 v v_p = 10^4(4.5)(1.5\times 10^{-3}) = 67.5 \text{ Pa}$$

over and above any pressure present at equilibrium.

The relation $p_e = \rho_0 v v_p$, which holds for all sound waves, when combined with

$$v_p = -v\frac{\partial\chi}{\partial x}$$

[see part (a)] yields

$$p_e = -\rho_0 v^2 \frac{\partial\chi}{\partial x}$$

Comparison of this last equation with

$$p_e = -B\frac{\partial\chi}{\partial x}$$

gives the formula for the wave speed:

$$v = \sqrt{\frac{B}{\rho_0}}$$

(c) When the pulse passes through the small volume bounded by cross sections of the rod at x and $x + \Delta x$, the particles composing the cross section at x are sent to $x + \chi$, those composing the cross section at $x + \Delta x$ to

$$(x + \Delta x) + \left(\chi + \frac{\partial\chi}{\partial x}\Delta x\right)$$

and particles in between these two cross sections are sent to intermediate locations. Thus, a mass Δm which initially occupied a volume of $S\,\Delta x$ now occupies a volume of $S[\Delta x + (\partial\chi/\partial x)\,\Delta x]$.

$$\rho_0 = \frac{\Delta m}{S\,\Delta x} \qquad \rho_0 + \Delta\rho = \frac{\Delta m}{S\left(\Delta x + \dfrac{\partial\chi}{\partial x}\Delta x\right)}$$

whence

$$\frac{\Delta\rho}{\rho_0} = \frac{-\dfrac{\partial\chi}{\partial x}}{1 + \dfrac{\partial\chi}{\partial x}} = \frac{v_p}{v - v_p}$$

where we have used $v_p/v = -\partial\chi/\partial x$ from (b). Substituting numerical values,

$$\frac{\Delta\rho}{\rho_0} = \frac{1.5\times 10^{-3}}{4.5 - (1.5\times 10^{-3})} \approx \frac{1}{3000} = 0.033\%$$

Another Method

$$\rho_0 = \frac{m}{V} \qquad \rho_0 + \Delta\rho = \frac{m}{V + \Delta V}$$

whence

$$\frac{\Delta\rho}{\rho_0} = \frac{-\Delta V}{V + \Delta V} = \frac{p_e\,\Delta V}{-p_e V - p_e\,\Delta V} = \frac{p_e}{-\dfrac{p_e}{\Delta V/V} - p_e} = \frac{p_e}{B - p_e} = \frac{p_e/B}{1 - (p_e/B)} = \frac{-\dfrac{\partial\chi}{\partial x}}{1 + \dfrac{\partial\chi}{\partial x}}$$

where the definition of the bulk modulus and the relation

$$p_e = -B\frac{\partial\chi}{\partial x}$$

have been used.

21.2. Longitudinal waves travel through water at a speed of 1450 m/s.   What is the bulk modulus of water?   The density of water is 10^3 kg/m^3.

$$B = \rho_0 v^2 = 10^3 (1450)^2 = 2.1 \times 10^9 \text{ Pa}$$

or 2100 MPa.

21.3. What is the amplitude of the pressure wave in air under standard conditions corresponding to a sinusoidal displacement wave of amplitude $\chi_0 = 1\ \mu$m and frequency is 2.0 kHz?   Standard pressure is $p_0 = 101$ kPa, $\gamma_{\text{air}} = 1.40$, and the speed of sound in air at S.T.P. is $v = 331$ m/s.

For a displacement wave of the form $\chi = \chi_0 \cos(kx + \omega t)$, the pressure wave has the form

$$p_e = -B_{\text{adiabatic}} \frac{\partial \chi}{\partial x} = -\gamma p_0 \frac{\partial \chi}{\partial x} = \gamma p_0 k \chi_0 \sin(kx + \omega t)$$

the amplitude being

$$\gamma p_0 k \chi_0 = (1.40)(101 \times 10^3)\left[\frac{2\pi(2.0 \times 10^3)}{331}\right](1 \times 10^{-6}) = 5.3 \text{ Pa}$$

21.4. A loudspeaker with a diaphragm of 0.1 m radius is to generate acoustic radiation of 1 kHz with a power of 40 W.   What is the minimum oscillation amplitude of the diaphragm?   At the temperature under consideration, the density of air is 1.29 kg/m^3 and the speed of sound is 344 m/s.

The intensity of the sound is to be

$$I = \frac{\text{power}}{\text{area}} = \frac{40}{\pi (0.1)^2} \text{ W/m}^2$$

Then, from $I = \frac{1}{2}\omega^2 A^2 \rho v$,

$$A = \frac{1}{\omega}\sqrt{\frac{2I}{\rho v}} = \frac{1}{2\pi(1 \times 10^3)}\sqrt{\frac{2(40)/\pi(0.1)^2}{(1.29)(344)}} = 3.8 \times 10^{-4} \text{ m}$$

or 0.38 mm.

21.5. What is the noise level, in dB, of (a) ordinary conversation ($I = 10^{-6}$ W/m^2)?   (b) the threshold of pain ($I = 1$ W/m^2)?

(a)
$$10 \log \frac{10^{-6}}{10^{-12}} = 10 \log 10^6 = 10(6) = 60 \text{ dB}$$

(b)
$$10 \log \frac{1}{10^{-12}} = 10 \log 10^{12} = 10(12) = 120 \text{ dB}$$

21.6. A pipe initially closed at one end has an oscillation frequency of 210 Hz.   When both ends of the pipe are open, the pipe oscillates at 840 Hz (not the fundamental frequency).   What is the smallest pipe length (L) that will satisfy the conditions?   The speed of sound is 330 m/s.

Figure 21-2 indicates the type of boundary conditions obeyed by the standing pressure wave in the two cases.   In (a),

$$L = \left(i + \frac{1}{2}\right)\frac{\lambda}{2} \qquad \text{or} \qquad v = \frac{4fL}{2i + 1}$$

where $i = 0, 1, 2, \ldots$.   In (b),

$$L = j\frac{\lambda'}{2} \qquad \text{or} \qquad v = \frac{2f'L}{j}$$

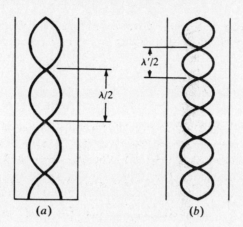

Fig. 21-2

where $j = 1, 2, 3, \ldots$. Since v is the same in both cases,

$$\frac{4fL}{2i+1} = \frac{2f'L}{j} \qquad \text{or} \qquad \frac{2i+1}{j} = \frac{2f}{f'} = \frac{2(210)}{840} = \frac{1}{2}$$

The smallest i and j satisfying this equation are $i = 0$, $j = 2$; and either of these gives

$$L_{\min} = \frac{330}{840} = 0.39 \text{ m}$$

21.7. What is the minimum power output of an isotropic point source of sound that will produce an intensity level of -40 dB at the ear of a listener 10 km distant from the source?

The intensity I is given by

$$10 \log \frac{I}{10^{-12}} = -40 \qquad \text{or} \qquad I = 10^{-16} \text{ W/m}^2$$

and this intensity must be maintained over the surface of a sphere of radius 10 km. Hence,

$$\text{power} = \text{intensity} \times \text{area} = 10^{-16} 4\pi (10^4)^2 = 1.3 \times 10^{-7} \text{ W}$$

or 0.13 μW.

21.8. An object traveling in a straight line through a medium at speed $V > v$, where v is the sound speed in the medium, creates a shock wave. Show that the angle θ of the shock wave is given by

$$\sin \theta = \frac{1}{n}$$

where $n = V/v$ is the *Mach number*.

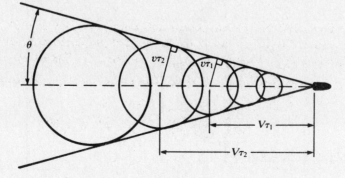

Fig. 21-3

Figure 21-3 shows, in longitudinal cross section, the object at one instant of time, together with the spherical wavefronts emitted by the object . . . , τ_1, τ_2, . . . seconds previously. It is clear from the geometry that the wavefronts are all tangent internally to a right circular cone having apex angle 2θ, where

$$\sin\theta = \frac{v}{V} = \frac{1}{n}$$

This conical envelope constitutes the shock wave.

21.9. Two trains, each moving at 17 m/s, are receding from each other. One of the locomotive whistles sounds at a frequency $f_s = 200$ Hz. What is the frequency as heard by the brakeman in the caboose of the other train? Assume the speed of sound to be $v = 330$ m/s.

Using the sign convention given in Section 21.3,

$$f_o = f_s \frac{v + v_o}{v - v_s} = (200)\frac{330 + (-17)}{330 - (-17)} = 180 \text{ Hz}$$

21.10. A train has just completed a long U-curve in a track, as shown in Fig. 21-4(a). The caboose is just coming into the curve, which is a semicircle. The engineer blows the horn, which has a frequency $f_s = 200$ Hz. The speed of the train is 29 m/s and the speed of sound is $v = 344$ m/s. What frequency is heard by (a) a workman down the track where the train is headed? (b) the engineer, who is just behind the horn? (c) the conductor in the caboose? (d) a passenger in the middle of the train?

(a)
$$f_o = f_s \frac{v + v_o}{v - v_s} = (200)\frac{344 + 0}{344 - 29} = 218 \text{ Hz}$$

(b)
$$f_o = f_s \frac{v + v_o}{v - v_s} = f_s \frac{v + v_{\text{train}}}{v - (-v_{\text{train}})} = f_s = 200 \text{ Hz}$$

(c) The motions of both the source and the observer are perpendicular to the line joining them. Hence $v_s = v_o = 0$ in the Doppler equation, and $f_o = f_s = 200$ Hz.

(d) The Doppler equation holds for noncollinear motion if v_s and v_o are taken to be the components along the line joining source and observer of the source velocity and the observer velocity, respectively. In this case we have [see Fig. 21-4(b)]

$$v_s = v_{\text{train}} \cos 135° = -\frac{29}{\sqrt{2}} \text{ m/s} \qquad v_o = v_{\text{train}} \cos 45° = \frac{29}{\sqrt{2}} \text{ m/s}$$

and so
$$f_o = f_s \frac{v + v_o}{v - v_s} = (200)\frac{29 + (29/\sqrt{2})}{29 - (-29/\sqrt{2})} = 200 \text{ Hz}$$

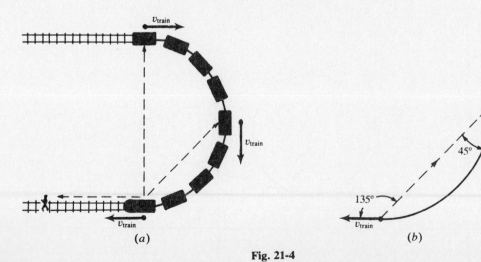

v_{train}

v_{train}

v_{train}

v_{train}

v_{train}

v_{train}

45°

135°

(a)   (b)

Fig. 21-4

Supplementary Problems

21.11. The distance to a submarine is computed by measuring the time interval between the transmission and reception of a sound pulse at a submerged generator. The bulk modulus of water is 2100 MPa and its density is 10^3 kg/m^3. What is the scale factor for converting measured time into distance? *Ans.* $d_{km} = 0.725 \, t_s$

21.12. The speed of sound in molecular hydrogen gas at 0 °C is 1.3 km/s. Are intermolecular vibrations significant at this temperature? (*Hint*: If the energy of the gas derived solely from molecular translation and rotation, then γ would equal 7/5.) *Ans.* yes

21.13. The (average) intensity of a sound wave is [see (20.4)]

$$I = -B\left(\frac{\partial \chi}{\partial x} \frac{\partial \chi}{\partial t}\right)_{avg}$$

From this and the relation between excess pressure and displacement, derive an expression for the intensity of the wave $\chi = f(x - vt)$ in a medium of rest density ρ_0.

Ans. $I = \dfrac{(p_e^2)_{avg}}{\sqrt{\rho_0 B}}$

21.14. The fundamental frequency of a certain violin string of length 0.8 m is 450 Hz. Find (*a*) the speed of waves along the string, (*b*) the wavelength associated with the fundamental frequency, (*c*) the wavelength of the third harmonic. *Ans.* (*a*) 720 m/s; (*b*) 1.6 m; (*c*) 0.533 m

21.15. A train approaches a station platform at a speed of 20 m/s, blowing its whistle at a frequency of 200 Hz, as measured by the engineer. The speed of sound in air is 330 m/s. (*a*) What is the wavelength in the still air in front of the train? (*b*) What is the frequency that will be measured by a man standing on the platform? (*c*) What frequency will be measured by the driver of a car moving at 40 m/s relative to the highway, approaching the station from a direction opposite to that of the train? *Ans.* (*a*) 1.55 m; (*b*) 213 Hz; (*c*) 239 Hz

21.16. A girl who holds a tuning fork oscillating at 440 Hz runs directly toward a wall at a speed of 3 m/s. What beat frequency will she hear between the direct and reflected waves? The speed of sound is 330 m/s. (*Hint*: Replace the wall by a mirror image of the tuning fork.) *Ans.* 8.1 Hz

Chapter 22

Electric Charge and Coulomb's Law

22.1 ELECTRIC CHARGE

In order to express electromagnetic quantities, a new fundamental dimension and corresponding base unit (Section 1.1) must be added to those used for mechanical quantities. In the SI, one chooses *electric current*, measured in *amperes* (A). Since an electric current (in a wire, for example) corresponds to the transport of a certain amount of electric charge through a fixed cross section in a certain time interval, the derived unit for electric charge is the $A \cdot s$ or *coulomb* ($1\,C = 1\,A \cdot s$). One coulomb turns out to be rather a large amount of charge; hence the microcoulomb (μC) and the nanocoulomb (nC) are frequently employed.

To the best of present knowledge, every observed charge q, large or small, positive or negative, is, in magnitude, an integral multiple of

$$e = 1.6022 \times 10^{-19}\ C$$

where $-e$ is the charge on a single electron.

EXAMPLE 22.1. (*a*) One coulomb contains $(1.6022 \times 10^{-19})^{-1}$, or about 6 billion billion electronic charges. (The exact number need not be integral, as the coulomb is a *defined* amount of charge.) (*b*) The charge on the nucleus of a copper atom (atomic number 29) is $(29)(1.6022 \times 10^{-19}) = 4.646 \times 10^{-18}\ C$. (*c*) The charge passing any cross section of a wire in two minutes during which there is a current of 15 A is

$$q = (15)(2 \times 60) = 1800\ C$$

or $1.1234 \times 10^{22} e$.

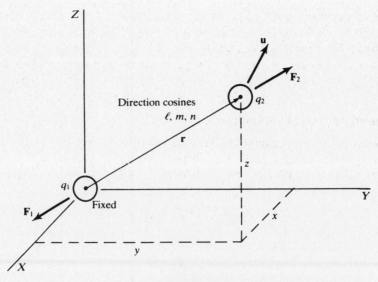

Fig. 22-1

22.2 FORCE BETWEEN POINT CHARGES

Figure 22-1 shows two point charges, one of which, q_1, is fixed at the origin of an inertial frame X, Y, Z. According to *Coulomb's law*, charge q_2 experiences an electrical force given by

$$\mathbf{F}_2 = \frac{bq_1q_2}{r^2}\,\mathbf{e} = \frac{bq_1q_2}{r^3}\,\mathbf{r} \qquad (22.1)$$

where $\mathbf{r} = r\mathbf{e} = r(\ell\mathbf{i} + m\mathbf{j} + n\mathbf{k})$ (see Section 1.6) is the vector displacement from q_1 to q_2 and b is a positive constant. Note the analogy between Coulomb's law and Newton's law of gravitation (Section 13.2). Coulomb's law remains valid if q_1 and q_2 represent two, nonoverlapping, spherically symmetric, charge distributions, where r is the distance between the two centers.

By experiment it is found that

$$b = 8.98742 \times 10^9 \text{ N} \cdot \text{m}^2/\text{C}^2 \approx 9 \times 10^9 \text{ N} \cdot \text{m}^2/\text{C}^2$$

For mathematical reasons, it is convenient to replace b by another constant, ϵ_0, called the *permittivity of empty space* and defined by

$$\epsilon_0 = \frac{1}{4\pi b} = 8.85432 \times 10^{-12} \text{ C}^2/\text{N} \cdot \text{m}^2 = 8.85432 \times 10^{-12} \text{ F/m}$$

(see Problem 25.10 for the farad, F, which is the unit of electrical capacitance). In terms of ϵ_0, Coulomb's law is written

$$\mathbf{F}_2 = \frac{q_1q_2}{4\pi\epsilon_0 r^2}\,\mathbf{e} \qquad (22.2)$$

When q_1 and q_2 are immersed in an isotropic dielectric medium such as oil, ϵ_0 in (22.2) must be replaced by $K\epsilon_0$, where K (a dimensionless quantity) is the *dielectric constant* of the medium.

On the Validity of Coulomb's Law

All laws of "classical electrodynamics" must meet the requirements of special relativity (Chapter 38). Accordingly, conditions for the validity of (22.1) or (22.2) stated briefly are:

Charge q_1 *must be at rest* relative to an inertial frame of reference (it need not, of course, be located at the origin), while q_2 may have any velocity $\mathbf{u}$, even approaching the speed of light, c.

Likewise, if we wish to compute $\mathbf{F}_1$, the force on q_1 due to q_2, by applying (22.1) or (22.2) with the subscripts interchanged, then q_2 must be at rest and q_1 can have any velocity $\mathbf{v}$.

However, if $\mathbf{u}$ and $\mathbf{v}$ are small such that $u^2/c^2 \ll 1$, $v^2/c^2 \ll 1$, Coulomb's law holds very nearly for both forces, so that $\mathbf{F}_2$ and $\mathbf{F}_1$ may be regarded as equal in magnitude and opposite in direction. (It must be remembered that $\mathbf{F}_1$ and $\mathbf{F}_2$, computed as mentioned above, are only electrostatic forces and do not account for magnetic forces which exist when both charges are in motion).

Total Force Exerted by Several Point Charges

Forces between point charges act *independently*. Suppose that charges $q_1, q_2, q_3, \ldots$ are fixed in an inertial frame. Then the force $\mathbf{F}_{(1)}$ on a charge q due to q_1 is computed by Coulomb's law as if $q_2, q_3, \ldots$ did not exist; etc. Finally, the total force on q is just the vector sum of the individual forces:

$$\mathbf{F}_{\text{total}} = \mathbf{F}_{(1)} + \mathbf{F}_{(2)} + \mathbf{F}_{(3)} + \cdots \qquad (22.3)$$

Solved Problems

22.1. In Fig. 22-1, let $q_1 = 200 \ \mu C$, $q_2 = 30 \ \mu C$, $x = 20$ cm, $y = 25$ cm, $z = 30$ cm. Find the magnitude and direction of $\mathbf{F}_2$ (a) in free space, (b) in a medium of dielectric constant 2.5.

(a)
$$r = (20^2 + 25^2 + 30^2)^{1/2} = 43.875 \text{ cm} = 0.43875 \text{ m}$$

and the magnitude of $\mathbf{F}_2$ is

$$F_2 = \frac{(9 \times 10^9)(200 \times 10^{-6})(30 \times 10^{-6})}{(0.43875)^2} = 280.52 \text{ N}$$

The direction cosines of $\mathbf{F}_2$ are

$$\ell = \frac{x}{r} = \frac{20}{43.875} = 0.4558 \qquad m = 0.5698 \qquad n = 0.6838$$

(b)
$$F_2 = \frac{280.52}{2.5} = 112.21 \text{ N}$$

The direction of $\mathbf{F}_2$ is the same as in (a).

22.2. In Fig. 22-1, q_1 is an α-particle (charge $+2e$), q_2 is an electron, and $r = 1 \ \text{Å} = 10^{-10}$ m; ℓ, m, n are as found in Problem 22.1. If the particles are released from rest, find the initial acceleration of each particle.

The force on the electron is

$$\mathbf{F}_2 = \frac{(9 \times 10^9)(3.2 \times 10^{-19})(-1.6 \times 10^{-19})}{(10^{-10})^2} \mathbf{e} = -4.61 \times 10^{-8} \mathbf{e} \quad \text{N}$$

where $\mathbf{e} = \ell \mathbf{i} + m \mathbf{j} + n \mathbf{k}$. The mass of an electron is $m_e = 9.11 \times 10^{-31}$ kg, so that, by Newton's second law, the acceleration of the electron is

$$\mathbf{a}_2 = \frac{1}{m_e} \mathbf{F}_2 = \frac{-4.61 \times 10^{-8} \mathbf{e}}{9.11 \times 10^{-31}} = -5.06 \times 10^{22} \mathbf{e} \quad \text{m/s}^2$$

The force on the α-particle is $-\mathbf{F}_2$ and its mass is 6.65×10^{-27} kg; hence

$$\mathbf{a}_1 = \frac{+4.61 \times 10^{-8} \mathbf{e}}{6.65 \times 10^{-27}} = 6.93 \times 10^{18} \mathbf{e} \quad \text{m/s}^2$$

22.3. Two metal spheres are located 100 km apart (about 60 miles) in space. One sphere carries as a surface charge all the electrons, and the other all the positive nuclei, in one cubic centimeter of copper. What is the force of attraction between the spheres? Copper has atomic weight 63.54, atomic number 29, density 8.96×10^3 kg/m^3; Avogadro's number is 6.022×10^{23} atoms/mol.

The number of atoms in one cubic centimeter is

$$\left(\frac{8.96}{63.54} \right)(6.022 \times 10^{23}) = 8.49 \times 10^{22}$$

The total negative charge of electrons is then

$$(8.49 \times 10^{22})(29)(1.6 \times 10^{-19}) = 3.94 \times 10^5 \text{ C}$$

which, of course, is also the total positive charge of the nuclei. Hence, the force of attraction is

$$F = (9 \times 10^9) \frac{(3.94 \times 10^5)^2}{(10^5)^2} = 1.4 \times 10^{11} \text{ N}$$

or about 16 million tons!

22.4. In Fig. 22-2, $q_1 = 300$ μC, $q_2 = 400$ μC, $q_3 = 500$ μC, $r_{12} = 9$ m, $r_{13} = 12$ m. Compute the magnitude and direction of $\mathbf{F}_3$, the total force on q_3. Charges are uniformly distributed over spheres located in empty space.

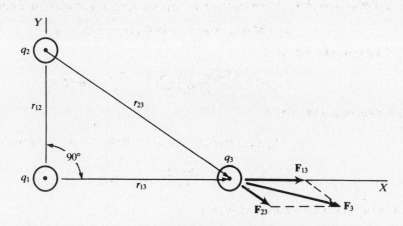

Fig. 22-2

The force on q_3 by q_1 is

$$\mathbf{F}_{13} = (9 \times 10^9) \frac{(300)(500)(10^{-12})}{(12)^2} \mathbf{i} = 9.375\mathbf{i} \quad \text{N}$$

As seen from the diagram,

$$\mathbf{r}_{23} = 12\mathbf{i} - 9\mathbf{j} \qquad \text{and} \qquad r_{23} = \sqrt{(12)^2 + (9)^2} = 15 \text{ m}$$

whence

$$\mathbf{F}_{23} = (9 \times 10^9) \frac{(400)(500)(10^{-12})}{(15)^3} (12\mathbf{i} - 9\mathbf{j}) = 6.4\mathbf{i} - 4.8\mathbf{j} \quad \text{N}$$

and

$$\mathbf{F}_3 = \mathbf{F}_{13} + \mathbf{F}_{23} = 15.775\mathbf{i} - 4.8\mathbf{j} \quad \text{N}$$

From this, $F_3 = [(15.775)^2 + (4.8)^2]^{1/2} = 16.49$ N, and the direction of $\mathbf{F}_3$ is given by

$$\ell = \frac{15.775}{16.49} = 0.9567 \qquad m = \frac{-4.8}{16.49} = -0.2911 \qquad n = 0$$

22.5. In Fig. 22-3, q_1 is located on the X axis at a distance x_1 from the origin, q_2 at z_2 on Z, and q at a point $P(x, y, z)$ in space. Given that $q_1 = 40$ μC, $q_2 = 50$ μC, $q = 8$ μC, $x_1 = 0.80$ m, $z_2 = 0.75$ m, $x = 0.40$ m, $y = 0.50$ m, $z = 0.60$ m, find $\mathbf{F}$, the total force on q.

$$\mathbf{r}_1 = (x - x_1)\mathbf{i} + (y - 0)\mathbf{j} + (z - 0)\mathbf{k} = -0.40\mathbf{i} + 0.50\mathbf{j} + 0.60\mathbf{k}$$

and $r_1 = 0.8775$ m. Hence the force on q due to q_1 is

$$\mathbf{F}_{(1)} = (9 \times 10^9) \frac{(40)(8)(10^{-12})}{(0.8775)^3} (-0.40\mathbf{i} + 0.50\mathbf{j} + 0.60\mathbf{k}) = -1.7047\mathbf{i} + 2.1310\mathbf{j} + 2.5574\mathbf{k}$$

Likewise,

$$\mathbf{r}_2 = 0.40\mathbf{i} + 0.50\mathbf{j} - 0.15\mathbf{k} \qquad r_2 = 0.6576$$

and the force on q due to q_2 is

$$\mathbf{F}_{(2)} = (9 \times 10^9) \frac{(50)(8)(10^{-12})}{(0.6576)^3} (0.40\mathbf{i} + 0.50\mathbf{j} - 0.15\mathbf{k}) = 5.063\mathbf{i} + 6.3284\mathbf{j} - 1.8985\mathbf{k}$$

The total force is then

$$\mathbf{F} = \mathbf{F}_{(1)} + \mathbf{F}_{(2)} = 3.358\mathbf{i} + 8.459\mathbf{j} + 0.659\mathbf{k} \quad \text{N}$$

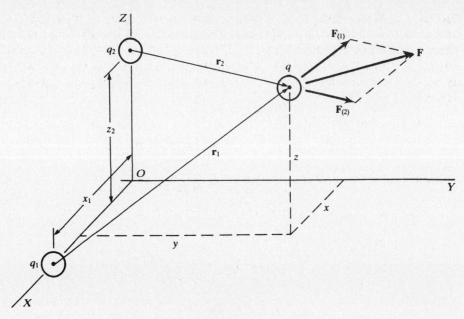

Fig. 22-3

22.6. A charge q_1 is fixed at y and another equal charge at $-y$, as in Fig. 22-4.   Charge q can be moved along X.   (*a*) For what value of x is the total force **F** on q of maximum magnitude?   (*b*) Discuss the variation of the force in the vicinity of the origin.

(*a*)   By symmetry, **F** is directed along X; its (signed) magnitude is

$$F = \frac{2bq_1^2}{r^2} \cos \theta = 2bq_1^2 \frac{x}{(x^2+y^2)^{3/2}}$$

Differentiating F with respect to x and setting the result equal to zero, we get

$$(x^2+y^2)^{3/2} - 3x^2(x^2+y^2)^{1/2} = 0$$

from which $x = \pm y/\sqrt{2}$.   These two points are obviously maxima, not minima.

(*b*)   For $x = 0$, $F = 0$, which means that q is in equilibrium at the origin.   However, the equilibrium is unstable, since for $x \neq 0$, **F** is directed away from the origin.   This is a special case of the following general result: A movable charge acted upon by any distribution of fixed charges is never in *stable* equilibrium.   (Problems 22.7 and 22.10 below show that stable equilibrium can be attained if other forces, e.g. tension in a string, act in addition to the electrical forces.)

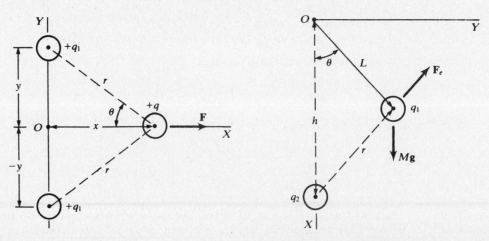

Fig. 22-4   Fig. 22-5

22.7. Figure 22-5 shows a sphere of mass M, carrying a uniformly distributed charge $+q_1$, suspended as a pendulum by a nonconducting string of length L. Another sphere, having uniform charge $+q_2$, is fixed on the X axis at a distance $h > L$ vertically below the origin. For $q_1 = 60\ \mu C$, $q_2 = 40\ \mu C$, $L = 1.5$ m, $h = 1.8$ m, $M = 0.8$ kg, and $g = 9.8$ m/s^2, find the value of θ at which the pendulum can remain at rest. (Motion of the pendulum is confined to the vertical XY plane).

The vector from the origin to q_1 is

$$\mathbf{L} = (L \cos \theta)\mathbf{i} + (L \sin \theta)\mathbf{j}$$

and the vector from q_2 to q_1 is

$$\mathbf{r} = \mathbf{L} - h\mathbf{i} = (L \cos \theta - h)\mathbf{i} + (L \sin \theta)\mathbf{j}$$

Hence the total force on q_1, excluding the tension in the string, is given by

$$\mathbf{F} = \mathbf{F}_e + M\mathbf{g} = \frac{bq_1q_2}{r^3}\mathbf{r} + Mg\mathbf{i}$$

$$= \frac{bq_1q_2[(L \cos \theta - h)\mathbf{i} + (L \sin \theta)\mathbf{j}]}{[(L \cos \theta - h)^2 + (L \sin \theta)^2]^{3/2}} + Mg\mathbf{i}$$

$$= \left[\frac{bq_1q_2(L \cos \theta - h)}{(L^2 + h^2 - 2Lh \cos \theta)^{3/2}} + Mg\right]\mathbf{i} + \left[\frac{bq_1q_2L \sin \theta}{(L^2 + h^2 - 2Lh \cos \theta)^{3/2}}\right]\mathbf{j}$$

The resultant torque on the pendulum about O is (see Section 10.1)

$$\boldsymbol{\tau} = \mathbf{L} \times \mathbf{F} = (L_xF_y - L_yF_x)\mathbf{k}$$

For equilibrium,

$$\tau = L_xF_y - L_yF_x = \frac{bq_1q_2Lh \sin \theta}{(L^2 + h^2 - 2Lh \cos \theta)^{3/2}} - MgL \sin \theta = 0 \qquad (1)$$

This equation has two solutions. The first of these, $\sin \theta = 0$, corresponds to an equilibrium position at $\theta = 0$ or at $\theta = 180°$. We discard these possibilities, which could, however, be realized for certain ranges of the physical parameters (see Problems 22.10 and 22.14). The other solution is

$$\cos \theta = \frac{L^2 + h^2 - (bq_1q_2h/Mg)^{2/3}}{2Lh} \qquad (2)$$

Substituting the numerical values,

$$\cos \theta = \frac{5.49 - 2.90}{5.4} = 0.48 \qquad \text{or} \qquad \theta = 61.3°$$

22.8. A point charge $-q_1$ is fixed at the origin of X, Y. Charge $+q_2$ is uniformly distributed over a sphere of mass m which revolves with angular velocity ω in a circular orbit of radius r about the origin. Neglecting gravity and any other forces, find an expression for r.

The force of attraction on q_2,

$$F = \frac{1}{4\pi\epsilon_0}\left(\frac{q_1q_2}{r^2}\right)$$

is clearly directed toward the origin. Equating this to the centripetal force, $mr\omega^2$, gives

$$r = \left(\frac{q_1q_2}{4\pi\epsilon_0 m\omega^2}\right)^{1/3}$$

22.9. The very long nonconducting cord ab shown along the X axis in Fig. 22-6 is uniformly painted with a charge of $+\sigma$ coulombs per meter. Find the magnitude and direction of the force on q_1, placed as shown.

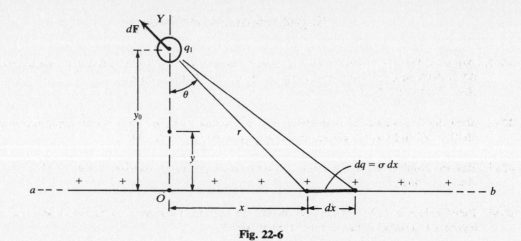

Fig. 22-6

It is clear from symmetry that the resultant force, $\mathbf{F}$, will have only a Y component.   We have

$$dF_y = \frac{q_1(\sigma\,dx)}{4\pi\epsilon_0 r^2}\cos\theta = \frac{q_1\sigma y_0\,dx}{4\pi\epsilon_0(x^2 + y_0^2)^{3/2}}$$

Hence

$$F_y = \frac{q_1\sigma y_0}{4\pi\epsilon_0}\int_{-\infty}^{\infty}\frac{dx}{(x^2 + y_0^2)^{3/2}} = \frac{q_1\sigma}{4\pi\epsilon_0 y_0}\left[\frac{x}{(x^2 + y_0^2)^{1/2}}\right]_{-\infty}^{\infty} = \frac{q_1\sigma}{2\pi\epsilon_0 y_0}$$

22.10. Referring to Problem 22.7 and Fig. 22-5, suppose that $q_1 = 0.60\ \mu\text{C}$ and $q_2 = 0.40\ \mu\text{C}$, all other data remaining the same.   (*a*) What is now the rest position of the pendulum?   (*b*) If the pendulum bob is displaced from the rest position through a small angle θ_0 and released from rest, what is the subsequent motion?

(*a*)   With both charges reduced by a factor of 100, solution (*2*) of Problem 22.7 no longer applies; indeed,

$$\frac{L^2 + h^2 - (bq_1q_2h/Mg)^{2/3}}{2Lh} = \frac{5.49 - (2.91))(10^{-8/3})}{5.4} > 1$$

Now the gravitational force predominates over the electrical force, and the rest position is $\theta = 0$.

(*b*)   For $|\theta| \le |\theta_0|$, we may approximate $\sin\theta$ and $\cos\theta$ in (*1*) of Problem 22.7 by θ and 1, respectively.   This gives

$$\tau = \left[\frac{bq_1q_2Lh}{(h - L)^3} - MgL\right]\theta = -11.54\,\theta$$

It is seen that, for the given values of the physical parameters, the torque is a linear *restoring* torque; therefore (Chapter 14) the motion of the pendulum will be angular SHM, with equation

$$\tau = I\ddot{\theta} \qquad\text{or}\qquad -11.54\,\theta = 1.80\,\ddot{\theta}$$

since the moment of inertia about O is $I = ML^2 = 1.80\ \text{kg}\cdot\text{m}^2$.   The period of the motion is then

$$T = 2\pi\sqrt{\frac{1.80}{11.54}} = 2.48\ \text{s}$$

as compared to

$$T_0 = 2\pi\sqrt{\frac{L}{g}} = 2\pi\sqrt{\frac{1.5}{9.8}} = 2.46\ \text{s}$$

in the absence of the (weak) electrical forces.

Supplementary Problems

22.11. Assuming that the electron of a hydrogen atom revolves about the proton in a circular orbit of radius 0.5 Å (0.05 nm), find the force of attraction and the electron's speed.
Ans. 9.24×10^{-8} N; 2.25×10^{6} m/s

22.12. Show that the number of electrons associated with the atoms in 5 cm^3 of water (molecular weight = 18) is 16.73×10^{23} and that the corresponding negative charge is 2.68×10^{5} C.

22.13. Rework Problem 22.7 if the entire system is submerged in oil (dielectric constant = 2.5).
Ans. $\theta = 43.5°$

22.14. Refer to Problem 22.7.   What conditions must the physical parameters satisfy in order that $\theta = 180°$ be a point of (*a*) equilibrium, (*b*) stable equilibrium?
Ans. (*a*) $Mg < \dfrac{bq_1q_2}{(L+h)^2}$;   (*b*) $Mg < \dfrac{bq_1q_2h}{(L+h)^3}$

22.15. In Problem 22.9, q_1 is negative and is distributed uniformly over a small sphere having mass m.   If the sphere is released from rest at the position shown in Fig. 22-6, how long does it take to reach the line charge?   *Hint*:   Apply the work-energy principle and make use of the definite integral

$$\int_0^1 \frac{du}{\sqrt{\ln(1/u)}} = \sqrt{\pi}$$

Ans. $t = y_0\pi \sqrt{\dfrac{\epsilon_0 m}{|q_1|\,\sigma}}$

22.16. In Fig. 22-7, two spheres, each of mass M and carrying a uniformly distributed charge $+q$, are suspended by silk threads of length r from point P.   Another charge, Q, is fixed at a distance $h > r$ directly below P.   Let $q = 40\ \mu$C, $r = 0.8$ m, $M = 0.5$ kg, $h = 1.5$ m, and $g = 9.8$ m/s^2.   The pendulums are confined to the vertical XY plane.   Find the equilibrium angle θ when (*a*) $Q = 0$, (*b*) $Q = -43.13\ \mu$C.
Ans. (*a*) $\approx 58°$;   (*b*) $31°$

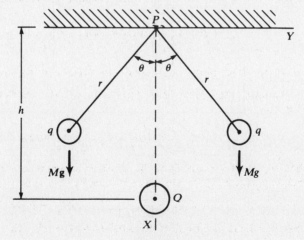

Fig. 22-7

22.17. Verify that $r = |m|$ in Problem 22.8 and that $t = |s|$ in Problem 22.15.

Chapter 23

The Electric Field
Established by Charges at Rest

23.1 GENERAL DEFINITION OF E

Consider a set of charges at rest relative to an inertial frame X, Y, Z, as in Fig. 23-1, where $+q_1$ and $-q_2$ are distributed over bodies B_1 and B_2, and $-q_3$ and $+q_4$ are point charges. A *test charge* $+q'$, a positive point charge of small magnitude, is located at point $P(x, y, z)$, and an observer (also at rest with respect to X, Y, Z) measures a net force $\mathbf{F}$ on the test charge. Then the *electric field* $\mathbf{E}$ at $P(x, y, z)$ set up by the given set of charges is defined by

$$\mathbf{E} = \frac{1}{q'}\mathbf{F} \quad (\text{N/C}) \tag{23.1}$$

It is seen that $\mathbf{E}$ is a vector in the direction of $\mathbf{F}$ and represents the force (in newtons) on a unit (1 C) charge at the point considered. Equivalently, the force on a point charge q located at a point at which the electric field is $\mathbf{E}$ (regardless of what arrangement of charges may have established $\mathbf{E}$) is given by

$$\mathbf{F} = q\mathbf{E} \quad (\text{N}) \tag{23.2}$$

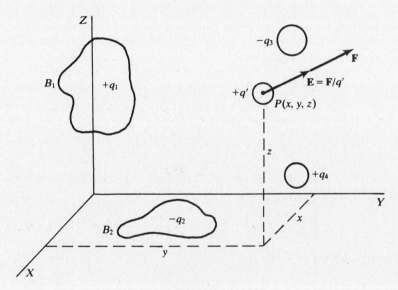

Fig. 23-1

When the set of stationary charges reduces to a single point charge q_1, (22.2) and (23.1) give

$$\mathbf{E} = \frac{q_1}{4\pi\epsilon_0 r^2}\mathbf{e} = \frac{q_1}{4\pi\epsilon_0 r^3}\mathbf{r} \tag{23.3}$$

wherein ϵ_0 is to be replaced by $K\epsilon_0$ when an isotropic dielectric medium is assumed. Equation (23.3) is valid even if the test charge is in motion; see Section 22.2.

23.2 SUPERPOSITION PRINCIPLE FOR E

From (22.3) and (23.1),

$$\mathbf{E}_{\text{total}} = \mathbf{E}_{(1)} + \mathbf{E}_{(2)} + \mathbf{E}_{(3)} + \cdots$$

that is, the electric field at any point due to a distribution of charges (or charge elements) $q_1, q_2, q_3, \ldots$ is found by adding (or integrating) the fields independently established at that point by the individual charges.

Solved Problems

23.1. In the inertial frame of Fig. 23-2, $q = 60 \ \mu C$. Find the magnitude and direction of **E** at the point $x = 50$ cm, $y = 60$ cm, $z = 80$ cm. Assume empty space.

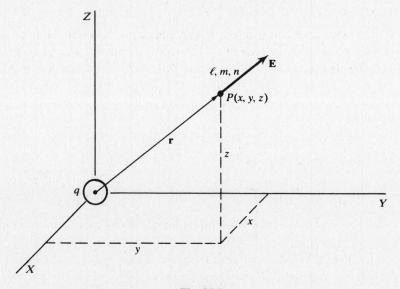

Fig. 23-2

From (23.3), the magnitude of **E** is

$$E = \frac{q}{4\pi\epsilon_0 r^2} = (9 \times 10^9) \frac{60 \times 10^{-6}}{(0.50)^2 + (0.60)^2 + (0.80)^2} = 4.32 \times 10^5 \ \text{N/C}$$

and the direction cosines of **E** are

$$\ell = \frac{x}{r} = \frac{0.50}{\sqrt{1.25}} = 0.4472 \qquad m = 0.5367 \qquad n = 0.7155$$

23.2. Referring to Problem 22.5, find **E** at the point occupied by q.

The total force on $q = 8 \times 10^{-6}$ C was found to be

$$\mathbf{F} = 3.36\mathbf{i} + 8.46\mathbf{j} + 0.66\mathbf{k} \quad \text{N}$$

so that

$$\mathbf{E} = \frac{1}{q}\mathbf{F} = 0.42\mathbf{i} + 1.06\mathbf{j} + 0.082\mathbf{k} \quad \text{MN/C}$$

23.3. At point $P(x, y, z)$, Fig. 23-3, electric fields

$$\mathbf{E}_1 = -7\mathbf{i} - 6\mathbf{j} + 8\mathbf{k} \qquad \mathbf{E}_2 = 6\mathbf{i} + 3\mathbf{i} + 7\mathbf{k} \qquad \mathbf{E}_3 = 3\mathbf{i} - 4\mathbf{j} - 6\mathbf{k}$$

are established by individual charges not shown.   Find (a) the magnitude and direction of $\mathbf{E}_{total}$ at P; (b) the force on a point charge, $q = 5 \times 10^{-9}$, placed at P.

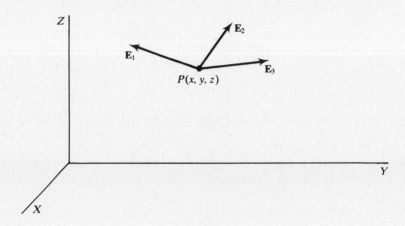

Fig. 23-3

(a)   As seen from the figure,

$$\mathbf{E}_{total} = \mathbf{E}_1 + \mathbf{E}_2 + \mathbf{E}_3 = 2\mathbf{i} - 7\mathbf{j} + 9\mathbf{k}$$

In magnitude,

$$E_{total} = (2^2 + 7^2 + 9^2)^{1/2} = 11.576$$

and the direction of $\mathbf{E}_{total}$ is given by

$$\ell = \frac{2}{11.576} = 0.1728 \qquad m = 0.6047 \qquad n = 0.7775$$

(b) $$\mathbf{F} = q\mathbf{E}_{total} = (10\mathbf{i} - 35\mathbf{j} + 45\mathbf{k}) \times 10^{-9}$$

23.4. A static electric field $\mathbf{E}$ gives rise to a conservative force (Section 9.1):   When a charge of 1 C moves through a *potential difference* of $+1$ *volt* (V), it loses exactly 1 J of kinetic energy.   Show that the units of $\mathbf{E}$ may be stated as V/m.

Since $1\text{ V} = 1\text{ J/C}$ and $1\text{ J} = 1\text{ N} \cdot \text{m}$,

$$\left|\frac{N}{C}\right| = \left|\frac{J/m}{C}\right| = \left|\frac{J/C}{m}\right| = \left|\frac{V}{m}\right|$$

23.5. Referring to Problem 22.9 and Fig. 22-6, find an expression for $\mathbf{E}$ due to the charged cord, at the location of q_1.

Since, in general, $\mathbf{E} = \mathbf{F}/q$, the electric field too is purely in the radial (Y) direction, and is given by

$$E_y = \frac{\sigma}{2\pi\epsilon_0 y_0}$$

23.6. Two very long nonconducting cords are stretched along the X and Y axes, as shown in Fig. 23-4.   They are uniformly charged, with charge densities σ_x and σ_y, respectively.   Find the magnitude and direction of the electric field at P; let $\sigma_x = 5\ \mu\text{C/m}$, $\sigma_y = 8\ \mu\text{C/m}$, $x = 3\text{ m}$, $y = 4\text{ m}$.

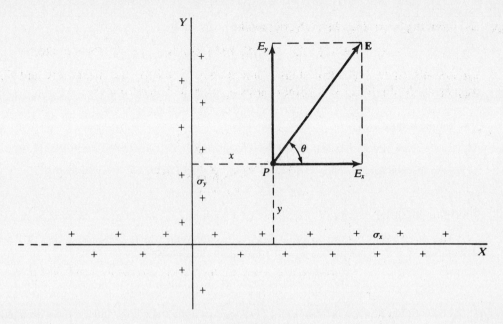

Fig. 23-4

Applying the result of Problem 23.5,

$$E_x = \frac{\sigma_y}{2\pi\epsilon_0 x} = \frac{2\sigma_y}{x}(9\times10^9) = 4.8\times10^4 \text{ N/C}$$

$$E_y = \frac{2\sigma_x}{y}(9\times10^9) = 2.25\times10^4 \text{ N/C}$$

(or $E_x = 48$ kV/m, $E_y = 22.5$ kV/m; see Problem 23.4).   Thus

$$E = (E_x^2 + E_y^2)^{1/2} = 5.30\times10^4 \text{ N/C}$$

and

$$\sin\theta = \frac{E_y}{E} = 0.4245 \qquad \theta = 25.12°$$

23.7. Two small spheres, carrying uniformly distributed charges $+q$ and $-q$ and connected by a thin nonconducting rod of length l, constitute an *electric dipole*.   Figure 23-5 shows such a dipole in a uniform electric field **E**; the axes are chosen such that **E** is along X and the dipole lies in the XY plane.   Given that $q = 4\ \mu C$, $E = 5\times10^5$ N/C, and $l = 20$ cm, find the *dipole moment* **P** and the torque τ tending to change the angle θ.

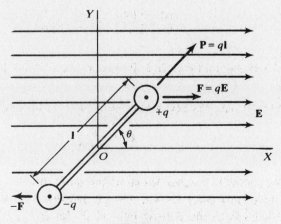

Fig. 23-5

The field exerts a torque on the electric dipole, given by

$$\tau = \frac{1}{2} \times q\mathbf{E} + \left(-\frac{1}{2}\right) \times (-q\mathbf{E}) = q\mathbf{l} \times \mathbf{E}$$

The vector $q\mathbf{l}$ is defined as the dipole moment $\mathbf{P}$:

$$\mathbf{P} = (4 \times 10^{-6})(0.20)(\mathbf{i} \cos \theta + \mathbf{j} \sin \theta) \quad C \cdot m$$

The torque is then

$$\tau = \mathbf{P} \times \mathbf{E} = (0.80 \times 10^{-6})(\mathbf{i} \cos \theta + \mathbf{j} \sin \theta) \times (5 \times 10^{5})\mathbf{i} = (0.40 \sin \theta)(-\mathbf{k}) \quad N \cdot m$$

It is seen that the torque tends to decrease θ, i.e. to align $\mathbf{P}$ with the external field $\mathbf{E}$.

23.8. Find the potential energy of the dipole of Problem 23.7 when it is oriented as in Fig. 23-5.

The work done by the external field in rotating the dipole (always in the XY plane) from orientation $\theta_0 = \theta$ to orientation $\theta_0 = 0$ (the equilibrium position) is

$$W = \int_{\theta}^{0} \tau \, d\theta_0 = \int_{\theta}^{0} (-0.40 \sin \theta_0) \, d\theta_0 = 0.40 \, (1 - \cos \theta)$$

If the equilibrium position is taken as the zero level of potential energy, then the potential energy is just equal to the above work:

$$U_\theta = 0.40 \, (1 - \cos \theta) \quad J$$

Observe that, in this case, $-\mathbf{P} \cdot \mathbf{E} = -0.40 \cos \theta$. Thus, to within an additive constant (which may be made to vanish by changing the reference level),

$$U_\theta = -\mathbf{P} \cdot \mathbf{E}$$

This result holds generally.

23.9. Two small spheres, Fig. 23-6, attached to a thin nonconducting rod, carry uniformly distributed charges $-q_1$ and $+q_2$. A third sphere, carrying charge $+q$, is fixed at the origin. Find the net force on the dumbbell, if $q_1 = 40 \, \mu C$, $q_2 = 50 \, \mu C$, $q = 60 \, \mu C$, $x_1 = 2$ m, $y_1 = 2.5$ m, $x_2 = 3.5$ m, $y_2 = 4$ m.

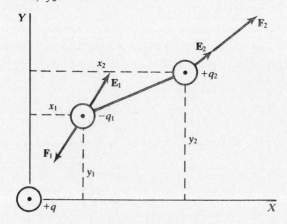

Fig. 23-6

Only the electric field due to $+q$ need be considered (why?). Its values at (x_1, y_1) and (x_2, y_2) are

$$\mathbf{E}_1 = \frac{q}{4\pi\epsilon_0} \frac{x_1\mathbf{i} + y_1\mathbf{j}}{(x_1^2 + y_1^2)^{3/2}} = (3.291\mathbf{i} + 4.114\mathbf{j}) \times 10^4 \quad N/C$$

$$\mathbf{E}_2 = \frac{q}{4\pi\epsilon_0} \frac{x_2\mathbf{i} + y_2\mathbf{j}}{(x_2^2 + y_2^2)^{3/2}} = (1.259\mathbf{i} + 1.438\mathbf{j}) \times 10^4 \quad N/C$$

whence the forces on the dumbbell are

$$\mathbf{F}_1 = -q_1\mathbf{E}_1 = -1.316\mathbf{i} - 1.646\mathbf{j} \quad N$$

$$\mathbf{F}_2 = +q_2\mathbf{E}_2 = 0.629\mathbf{i} + 0.719\mathbf{j} \quad N$$

and the net force is

$$\mathbf{F}_1 + \mathbf{F}_2 = -0.687\mathbf{i} - 0.927\mathbf{j} \quad N$$

23.10. In Fig. 23-7, charge q is uniformly distributed over a sphere which is fixed on X at $x = a$. An equal charge is fixed at $x = -a$. A small sphere, of mass m_1 and bearing charge q_1, is free to slide along X. (*a*) Find an exact expression for the total force on m_1 in the position shown. (*b*) Assuming that m_1 is released from rest at a point $x_0 \ll a$, find the subsequent motion.

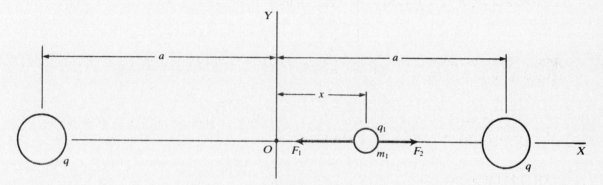

Fig. 23-7

(*a*)
$$F_1 = q_1 E_1 = -\frac{q_1 q}{4\pi\epsilon_0(a-x)^2} \qquad F_2 = q_1 E_2 = +\frac{q_1 q}{4\pi\epsilon_0(a+x)^2}$$

and the total force is

$$F = F_1 + F_2 = -\frac{q_1 q}{\pi\epsilon_0}\frac{ax}{(a^2-x^2)^2}$$

(*b*) Assuming that $|x|$ remains very small compared with a, x^2 can be neglected in the expression for F, which becomes a linear restoring force:

$$F = -\left(\frac{q_1 q}{\pi\epsilon_0 a^3}\right)x$$

Thus (Section 14.2) the motion is simple harmonic, $x = x_0 \cos \omega t$, of frequency

$$\omega = \sqrt{\frac{q_1 q}{\pi\epsilon_0 a^3 m_1}} \tag{1}$$

Supplementary Problems

23.11. Refer to Problem 23.10. (*a*) Find the electric field at any point x between the two fixed charges. (*b*) If $q = 30$ μC, $q_1 = 5$ μC, $a = 0.8$ m, and $m_1 = 0.9$ kg, find the period of oscillation of m_1.
Ans. (*a*) $E = -qax/\pi\epsilon_0(a^2 - x^2)^2$; (*b*) 1.835 s

23.12. Make a dimensional analysis to verify that (*1*) of Problem 23.10 gives ω in s^{-1}.

23.13. Find the force on the 6 μC charge in Fig. 23-8.   *Ans.* $\mathbf{F} = 5.555\mathbf{i} + 4.444\mathbf{j} + 7.333\mathbf{k}$ N

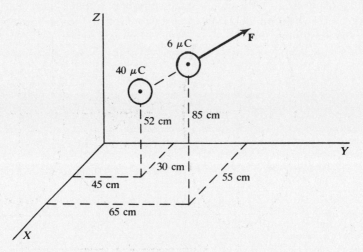

Fig. 23-8

23.14. Figure 23-9 shows an electric dipole, of dipole moment $\mathbf{P}$, in a uniform electric field $\mathbf{E}$. The direction cosines of $\mathbf{P}$ are

$$\ell = 0.462 \qquad m = -0.580 \qquad n = 0.671$$

and the rectangular components of $\mathbf{E}$ are

$$E_x = 40 \text{ kV/m} \qquad E_y = 45 \text{ kV/m} \qquad E_z = 50 \text{ kV/m}$$

Using the appropriate vector relations, find (*a*) the torque on the dipole, (*b*) the potential energy of the dipole relative to the orientation perpendicular to $\mathbf{E}$.
Ans. (*a*) $\boldsymbol{\tau} = -0.710\mathbf{i} + 0.045\mathbf{j} + 0.528\mathbf{k}$ N·m; (*b*) $U = -0.311$ J

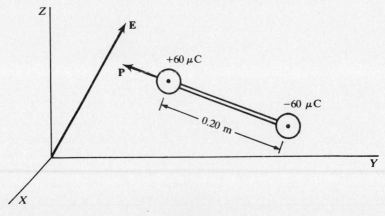

Fig. 23-9

23.15. Two uniformly charged circular rings, C_1 and C_2, are parallel and coaxial, as shown in Fig. 23-10.   At an arbitrary point P between the rings on the X axis, (a) determine the electric field due to C_1; (b) by analogy, find the field due to C_2; (c) find the resultant field.   (d) If $q_1 = 8$ μC, $r_1 = 0.5$ m, $q_2 = 4$ μC, $r_2 = 0.3$ m, and $d = 1.5$ m, show that the resultant field vanishes on the X axis at $x = 0.85$ m.   (e) Does the resultant field vanish anywhere else on the X axis?

Ans.   (a) $\mathbf{E}_1 = \dfrac{q_1 x}{4\pi\epsilon_0(r_1^2 + x^2)^{3/2}} \mathbf{i}$;   (b) $\mathbf{E}_2 = \dfrac{q_2(d - x)}{4\pi\epsilon_0[r_2^2 + (d - x)^2]^{3/2}}(-\mathbf{i})$;   (c) $\mathbf{E} = \mathbf{E}_1 + \mathbf{E}_2$;   (e) no

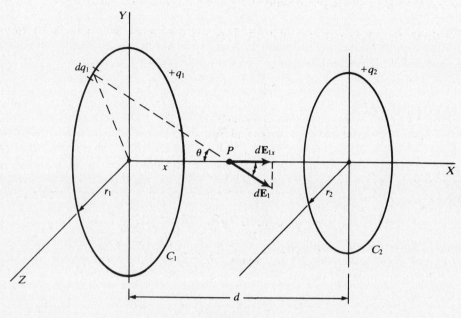

Fig. 23-10

23.16. Refer to Problem 22.11.   What is the magnitude of the electric field in which the electron moves?   *Ans.*   0.577 TV/m

Chapter 24

Electric Flux and Gauss's Law

In Section 13.5 we defined the *flux* of the gravitational field **g** through a closed surface and saw that this flux is proportional to the mass enclosed by the surface. An exactly analogous result holds for the electric field **E** (which like **g** is an inverse-square field.)

24.1 ELECTRIC FLUX

In Fig. 24-1, Q represents the algebraic sum of charges (positive and negative) distributed throughout the region of free space shown. The dashed line indicates a surface S of arbitrary shape completely enclosing the charges Q. **E** is the electric field due to Q at some point P of S, and $d\mathbf{S} = \mathbf{n}\,dS$ is the vector element of area at P. The *electric flux* through the elementary area dS is defined as

$$d\psi = \mathbf{E} \cdot d\mathbf{S} \quad (\mathrm{N} \cdot \mathrm{m}^2/\mathrm{C}) \tag{24.1}$$

In terms of "lines of force" (indicated by curved arrows in Fig. 24-1), which are lines to which **E** is tangent at each point, $d\psi$ can be interpreted as the number of lines cutting dS, provided the lines are supposed drawn with normal density E.

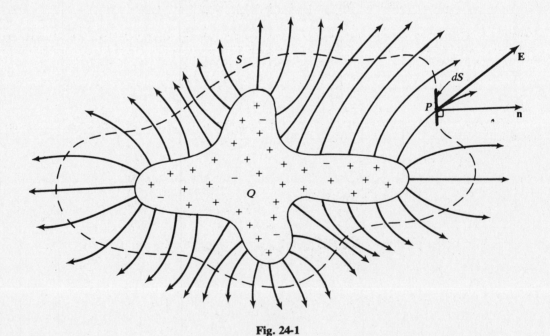

Fig. 24-1

The total flux through S in the outward direction is given by integration of (*24.1*) as

$$\psi = \int_S \mathbf{E} \cdot d\mathbf{S} \quad (\mathrm{N} \cdot \mathrm{m}^2/\mathrm{C}) \tag{24.2}$$

212

24.2 GAUSS'S LAW

In terms of the quantities defined above, Gauss's law reads:

$$\psi = \frac{Q}{\epsilon_0} \qquad (24.3)$$

That is, the total electric flux out of an arbitrary closed surface in free space is proportional to the net electric charge within the surface; the constant of proportionality is $1/\epsilon_0 = 4\pi b$. Gauss's law is valid even for enclosed charges in motion.

Solved Problems

24.1. Derive Gauss's law from Coulomb's law.

By superposition, it suffices to consider the case of a single point charge, q, at an arbitrary location inside a closed surface S. At a point P of S (see Fig. 24-2), the field due to q is given by Coulomb's law as

$$\mathbf{E} = \frac{q}{4\pi\epsilon_0 r^2}\mathbf{e}$$

The flux through dS is then

$$d\psi = \mathbf{E} \cdot d\mathbf{S} = \frac{q}{4\pi\epsilon_0 r^2}\mathbf{e} \cdot d\mathbf{S} = \frac{q}{4\pi\epsilon_0}\frac{dS\cos\theta}{r^2} = \frac{q}{4\pi\epsilon_0}d\Omega$$

where $d\Omega$ is the infinitesimal solid angle subtended by dS at the location of q. At that location, the total solid angle subtended by the entire closed surface S is 4π steradians. Thus,

$$\psi = \int_S d\psi = \frac{q}{4\pi\epsilon_0}\int_S d\Omega = \frac{q}{4\pi\epsilon_0}(4\pi) = \frac{q}{\epsilon_0}$$

which is Gauss's law.

Conversely, Coulomb's law can be derived from Gauss's law (Problem 24.8); hence the two laws are equivalent.

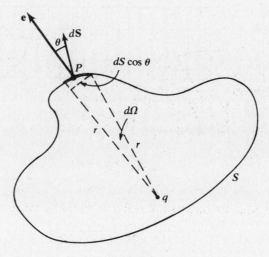

Fig. 24-2

24.2. Use Gauss's law to verify that a charge Q, uniformly distributed over the surface of a sphere, is equivalent externally to a point charge Q at the center of the sphere.

In Fig. 24-3, the sphere, of radius a, is surrounded by a concentric spherical "Gaussian surface," of radius $r > a$. By symmetry, $\mathbf{E}$ has constant magnitude E on the Gaussian surface and is everywhere normal to the surface. Hence

$$\frac{Q}{\epsilon_0} = \psi = \int_S \mathbf{E} \cdot d\mathbf{S} = \int_S E\, dS = E(4\pi r^2) \qquad \text{or} \qquad E = \frac{Q}{4\pi\epsilon_0 r^2}$$

which is the same field as would be produced by a point charge Q located at the center of the sphere.

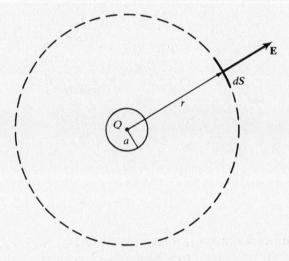

Fig. 24-3

24.3. The plates of a capacitor, Fig. 24-4, carry charges $+Q$ and $-Q$ as shown. Each plate has area $A = 600 \text{ cm}^2$. Between the plates, the field is constant at $E = 300 \text{ kV/m}$ (see Problem 23.4), and the field is assumed zero outside the plates (i.e. no fringing). Evaluate Q.

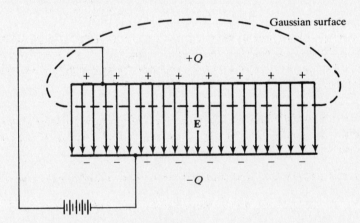

Fig. 24-4

Employing the Gaussian surface shown in Fig. 24-4, and assuming $\mathbf{E}$ normal to the inner surface of the plate, we have

$$\frac{Q}{\epsilon_0} = \int_S \mathbf{E} \cdot d\mathbf{S} = EA$$

or $\qquad Q = \epsilon_0 EA = (8.854 \times 10^{-12})(300 \times 10^3)(600 \times 10^{-4}) = 1.59 \times 10^{-7} \text{ C} = 0.159 \ \mu\text{C}$

24.4. In Fig. 24-5, a small sphere carrying charge $+Q$ is located at the center of a spherical cavity in a large metal sphere. Find, by Gauss's law, the field E at points P_1 in the space between small sphere and cavity wall, at points P_2 in the metal of the large sphere, and at points P_3 outside the large sphere.

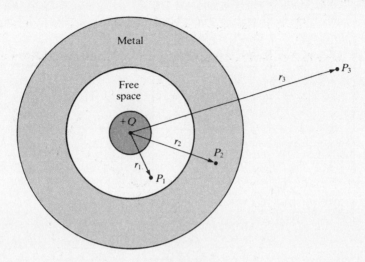

Fig. 24-5

On the inside surface of the cavity there is a uniformly distributed induced charge $-Q$. On the outside surface of the large sphere there is an induced charge $+Q$.

For a spherical Gaussian surface of radius r_1 enclosing $+Q$,

$$\psi = \frac{Q}{\epsilon_0} = \int \mathbf{E}_1 \cdot d\mathbf{S} = 4\pi r_1^2 E_1 \qquad \text{or} \qquad E_1 = \frac{Q}{4\pi\epsilon_0 r_1^2}$$

In the same way,

$$E_2 = 0 \qquad E_3 = \frac{Q}{4\pi\epsilon_0 r_3^2}$$

Actually, we might proceed the other way around and infer the induced charges from the fields, via Gauss's law. Thus, for a perfectly conducting metal, $E_2 = 0$ everywhere on the surface of a Gaussian sphere of radius r_2. There must, then, be zero net charge within the Gaussian sphere, which implies a charge $-Q$ on the cavity wall (it can't be anywhere else).

24.5. A long, cylindrical, nonconducting solid, Fig. 24-6, has a uniform charge density $\sigma = 5\ \mu C/m^3$ throughout. Compute E_1 at P_1 and E_2 at P_2, given $r_1 = 150\ mm$, $r_2 = 300\ mm$, and $R = 200\ mm$.

Considering a cylindrical Gaussian surface of radius r_1 and length ℓ, we write

$$\frac{1}{\epsilon_0} \times (\text{charge inside cylinder}) = E_1 \times (\text{area of cylinder})$$

$$\frac{1}{\epsilon_0}(\pi r_1^2 \ell \sigma) = E_1(2\pi r_1 \ell)$$

$$E_1 = \frac{r_1 \sigma}{2\epsilon_0} = \frac{(150 \times 10^{-3})(5 \times 10^{-6})}{2(8.854 \times 10^{-12})} = 42.4\ kV/m$$

In like manner,

$$E_2 = \frac{R^2 \sigma}{2r_2 \epsilon_0} = 37.6\ kV/m$$

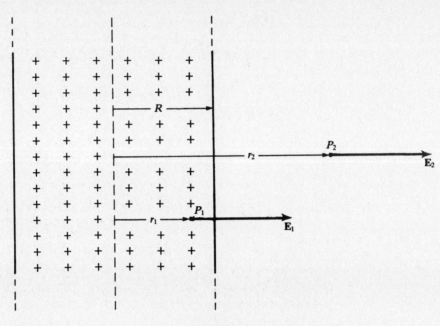

Fig. 24-6

24.6. Make a unit-dimensional check of the expression for E_2 found in Problem 24.5.

From Section 22.2, $\epsilon_0 = |F/m| = |C^2/N \cdot m^2|$, and so

$$E_2 = \frac{|m^2| \, |C/m^3|}{|m| \, |C^2/N \cdot m^2|} = \left| \frac{N}{C} \right|$$

or $|V/m|$, as required.

24.7. A metal can, suspended by a silk thread, carries a positive charge (not uniformly distributed). At a certain point very near the surface of the can, the electric field has a value $E = 600$ kV/m. Evaluate the surface charge density σ near that point.

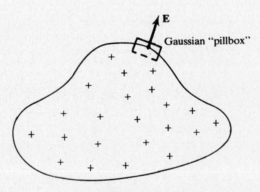

Fig. 24-7

Apply Gauss's law much as in Problem 24.4, this time choosing as Gaussian surface a very small, thin "pillbox" sunk halfway into the conducting surface (Fig. 24-7). Since **E** must be perpendicular to the surface, the flux through one face of the pillbox is $E \, \Delta A$, and the flux through the other two faces is zero. Hence,

$$E \, \Delta A = \frac{\sigma \, \Delta A}{\epsilon_0} \qquad \text{or} \qquad \sigma = \epsilon_0 E = (8.854 \times 10^{-12})(600 \times 10^3) = 5.313 \ \mu C/m^2$$

Supplementary Problems

24.8. Apply Gauss's law to an appropriate surface and so obtain Coulomb's law.

24.9. Compute the total flux ψ from the nucleus of a copper atom (atomic number = 29).
Ans. 5.25×10^{-7} N · m^2/C

24.10. A small cube, with a net charge of $+30$ μC, is suspended from silk threads inside a closed metal container of ellipsoidal shape, which is also suspended from silk threads. Determine the total charge on the outside surface of the container.   *Ans.* $+30$ μC

24.11. The largest dimension of the system of Problem 24.10 is 0.04 m.   What is the approximate value of E at a distance of 5 m from the cube?   *Ans.*   10.8 kV/m

24.12. The net charge Q in Fig. 24-1 is 63 μC.   Find the total flux threading the dashed surface.
Ans. 7.115×10^6 N · m^2/C

24.13. The total flux into the surface of a charged potato is 4×10^3 N · m^2/C.   What charge does the potato carry, measured in units of the electronic charge?   *Ans.* $-2.21 \times 10^{11} e$

24.14. In Fig. 24-5, the large sphere has an initial charge $-Q_2$.   Then the small sphere having charge $+Q_1$ is introduced.   Determine the magnitude of the electric field at P_3.   *Ans.* $E_3 = (Q_1 - Q_2)/4\pi\epsilon_0 r_3^2$

24.15. Show that electric flux $\psi = \left| \dfrac{\text{Nm}^2}{\text{C}} \right| = \left| \dfrac{\text{J}}{e_0 \text{V}} \right| = \left| \dfrac{\text{N}}{\epsilon_0 \text{E}} \right| .$

Chapter 25

Electric Potential

25.1 ELECTRIC POTENTIAL ENERGY

The electric force $\mathbf{F}$ exerted on a test charge q' by some stationary distribution of charges is a conservative force. Therefore, the test charge possesses electric potential energy U as given by (9.2) or (9.3). With the reference point A taken at infinity, (9.3) gives for the absolute potential energy of the test charge at the location $B(x, y, z)$:

$$U(x, y, z) = -\int_{\infty}^{(x, y, z)} \mathbf{F} \cdot d\mathbf{s} = -\int_{\infty}^{(x, y, z)} F_x\, dx + F_y\, dy + F_z\, dz \quad \text{(J)} \qquad (25.1)$$

In words: The absolute electric potential energy of a test charge at a given location is the work that the external field would do in moving the charge from that location to infinity, along any path.

Conversely, if the electric potential energy is known as a function of position, then the component of the electric force along an arbitrary direction $d\mathbf{s}$ can be calculated via (9.1) as

$$F_s = -\frac{dU}{ds}$$

In particular, the components of $\mathbf{F}$ along the X, Y, and Z axes are given by

$$F_x = -\frac{\partial U}{\partial x} \qquad F_y = -\frac{\partial U}{\partial y} \qquad F_z = -\frac{\partial U}{\partial z}$$

so that, in vector form,

$$\mathbf{F} = -\left(\frac{\partial U}{\partial x}\mathbf{i} + \frac{\partial U}{\partial y}\mathbf{j} + \frac{\partial U}{\partial z}\mathbf{k}\right) \quad \text{(N)} \qquad (25.2)$$

If the only force acting on a point charge is the electric force $\mathbf{F}$, conservation of energy takes the form

$$\Delta K + \Delta U = 0 \qquad (25.3)$$

where, as usual, K is the kinetic energy of the point charge.

25.2 ELECTRIC POTENTIAL OR VOLTAGE

The electric potential energy per unit test charge is called the *electric potential*, ϕ, or the *voltage*, V. (We shall here use the former term; in circuit applications, the latter is employed.) Since $\phi = U/q'$ and $\mathbf{E} = \mathbf{F}/q'$, ϕ and $\mathbf{E}$ enjoy the same relation as U and $\mathbf{F}$:

$$\phi(x, y, z) = -\int_{\infty}^{(x, y, z)} \mathbf{E} \cdot d\mathbf{s} = -\int_{\infty}^{(x, y, z)} E_x\, dx + E_y\, dy + E_z\, dz \quad \text{(V)} \qquad (25.4)$$

$$\mathbf{E} = -\left(\frac{\partial \phi}{\partial x}\mathbf{i} + \frac{\partial \phi}{\partial y}\mathbf{j} + \frac{\partial \phi}{\partial z}\mathbf{k}\right) \quad \text{(V/m)} \qquad (25.5)$$

Notice that the unit of electric potential is the *volt* (V), where $1\,\text{V} = 1\,\text{J/C}$. Also, as previously noted, $1\,\text{V/m} = 1\,\text{N/C}$.

EXAMPLE 25.1 (The Potential Near a Point Charge).   If the source of the field is a point charge q located at the origin (Fig. 25-1), then, by (23.3), the field at $P'(x', y', z')$ is given by

$$\mathbf{E}' = \frac{q}{4\pi\epsilon_0 r'^3}\mathbf{r}'$$

We choose as the path of integration in (25.4) the line from ∞ to $P(x, y, z)$ that, extended, meets q.   Along this path,

$$ds = -dr' \qquad \text{and} \qquad E_s = -E' = -\frac{q}{4\pi\epsilon_0 r'^2}$$

Consequently,

$$\phi = -\int_\infty^P E_s\, ds = -\int_\infty^P \frac{q\, dr'}{4\pi\epsilon_0 r'^2} = -\frac{1}{4\pi\epsilon_0}\int_\infty^r \frac{dr'}{r'^2} = \frac{q}{4\pi\epsilon_0 r}$$

As a check, let us verify that $E_z = -\partial\phi/\partial z$.   Since

$$\mathbf{E} = \frac{q}{4\pi\epsilon_0 r^3}(x\mathbf{i} + y\mathbf{j} + z\mathbf{k})$$

$E_z = (q/4\pi\epsilon_0 r^3)z$.   On the other hand,

$$-\frac{\partial\phi}{\partial z} = -\frac{\partial}{\partial z}\left[\frac{q}{4\pi\epsilon_0(x^2 + y^2 + z^2)^{1/2}}\right] = \frac{qz}{4\pi\epsilon_0(x^2 + y^2 + z^2)^{3/2}} = \frac{qz}{4\pi\epsilon_0 r^3}$$

which agrees.

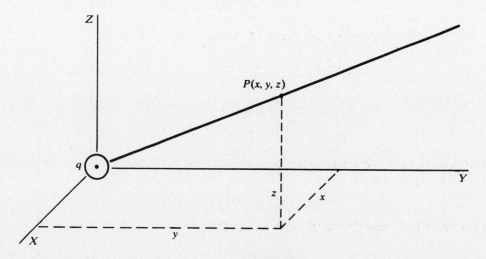

Fig. 25-1

EXAMPLE 25.2 (Equipotentials).   A surface (or, in two dimensions, a curve) on which the potential is constant is called an *equipotential surface (curve)*.   Thus, in Cartesian coordinates, the equipotential surfaces are given by the equation

$$\phi(x, y, z) = c$$

with one surface for each value of the constant c.   In terms of the equipotentials, the relation between ϕ and $\mathbf{E}$ may be geometrically expressed as follows: *the field lines and the equipotentials are everywhere perpendicular.*   Indeed, for a direction $d\mathbf{s}$ lying in an equipotential surface,

$$d\phi = 0 = \mathbf{E} \cdot d\mathbf{s}$$

and so $\mathbf{E}$ must be perpendicular to $d\mathbf{s}$.

25.3 SUPERPOSITION PRINCIPLE FOR ϕ

It follows at once from (23.4) that

$$\phi_{\text{total}} = \phi_{(1)} + \phi_{(2)} + \phi_{(3)} + \cdots \tag{25.7}$$

It is usually much simpler to perform the *scalar* addition in (25.7), and then to differentiate ϕ_{total}, than it is to carry out the *vector* addition in (23.4). (See Problem 25.6.) In particular, if charge is distributed throughout a region of space with density τ (C/m³), as indicated in Fig. 25-2, the charge element $dq = \tau\, dv$ gives rise to the potential

$$d\phi = \frac{\tau\, dv}{4\pi\epsilon_0 r}$$

at the fixed point P, according to (25.6). The total potential at P is then given by (25.7) as

$$\phi(P) = \frac{1}{4\pi\epsilon_0} \int_V \frac{\tau\, dv}{r} \tag{25.8}$$

It should be noted that (25.8), as it stands, cannot be applied when the charge distribution extends to infinity.

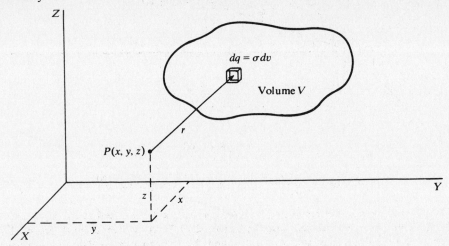

Fig. 25-2

25.4 THE ELECTRON VOLT

When an electron (charge, $-e$) "falls" through a potential difference of 1 V—i.e., when it moves from a given location to a location where the electric potential is 1 V *higher*—the electron loses

$$e(1\ \text{V}) = 1.602 \times 10^{-19}\ \text{J}$$

of electric potential energy and gains [see (25.3)] the same amount of kinetic energy. This quantity of energy is referred to as one *electron volt* (eV). In general, any energy expressed in joules may be converted to electron volts by dividing by 1.602×10^{-19} J/eV.

Solved Problems

25.1. If, in Fig. 25-1, $q = 40\ \mu\text{C}$, $x = 0.5$ m, $y = 0.8$ m, $z = 0.6$ m, find the potential at point P.

Here $r = (0.5^2 + 0.8^2 + 0.6^2)^{1/2} = 1.118$ m. Hence, from (25.6),

$$\phi = (9 \times 10^9)\left(\frac{40 \times 10^{-6}}{1.118}\right) = 3.22 \times 10^5\ \text{V} = 322\ \text{kV}$$

25.2. In Problem 25.1, a charge $q_1 = 9\ \mu C$ is placed at P.   Compute its electric potential energy.

$$U = q_1\phi = (9 \times 10^{-6})(3.22 \times 10^5) = 2.9\ \text{J}$$

25.3. In Fig. 25-3, a sphere carrying a uniformly distributed charge $Q = 40\ \mu C$ is located at the origin; $r_1 = 0.5$ m and $r_2 = 1.2$ m.   (a) Find the potential at P_1 and at P_2 due to Q.   What is the potential at P_1 with respect to P_2?   (b) A small sphere carrying charge $q = 8\ \mu C$ is placed at P_1.   What is its potential energy (relative to ∞)?   (c) Suppose that the small sphere moves freely from P_1 through P_2.   In the trip from P_1 to P_2, what is its change in kinetic energy?

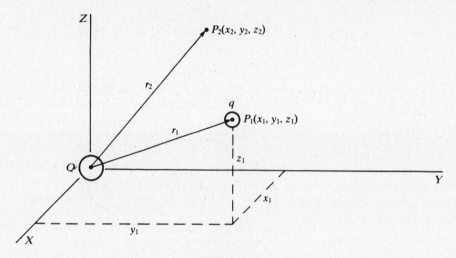

Fig. 25-3

(a)
$$\phi_1 = (9 \times 10^9)\left(\frac{40 \times 10^{-6}}{0.5}\right) = 720\ \text{kV} \qquad \phi_2 = (9 \times 10^9)\left(\frac{40 \times 10^{-6}}{1.2}\right) = 300\ \text{kV}$$

The potential at P_1 with respect to P_2 is the work done by the field in moving a unit test charge from P_1 to P_2.   It is thus the absolute potential at P_1 minus the absolute potential at P_2.   We denote this relative potential as ϕ_{12} or V_{12}, and also refer to it as *the difference in potential between P_1 and P_2* (note the order).

$$\phi_{12} = \phi_1 - \phi_2 = 720 - 300 = 420\ \text{kV}$$

(b)
$$U_1 = q\phi_1 = (8 \times 10^{-6})(720 \times 10^3) = 5.76\ \text{J}$$

(c)
$$K_2 - K_1 = U_1 - U_2 = q(\phi_1 - \phi_2) = q\phi_{12} = (8 \times 10^{-6})(420 \times 10^3) = 3.36\ \text{J}$$

25.4. The difference in potential V_{AB} between wires A and B, Fig. 25-4, as measured by a voltmeter is 6000 V.   A small sphere having mass 0.150 kg and carrying a charge $q = +500\ \mu C$ is released from rest at a point very near wire A and allowed to move to wire B.   (a) Does V_{AB} depend on s, the distance between the wires?   (b) What work W can the sphere do?   (c) With what speed u will the sphere arrive at B?   (d) What is the average field E_{avg} between A and B?

(a)   No.

(b)
$$W = U_A - U_B = qV_{AB} = (500 \times 10^{-6})(6000) = 3\ \text{J}$$

(c)
$$W = \frac{1}{2}mu^2 \qquad \text{or} \qquad u = \sqrt{\frac{2W}{m}} = \sqrt{\frac{2(3)}{0.150}} = 6.324\ \text{m/s}$$

(d)
$$E_{avg} = \frac{V_{AB}}{s} = \frac{6000}{0.20} = 30\ \text{kV/m}$$

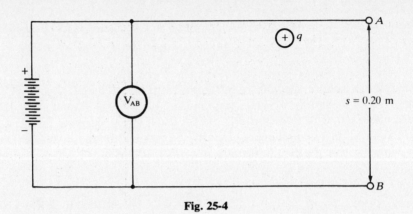

Fig. 25-4

25.5. Electrons (charge-to-mass ratio, $e/m_e = 1.759 \times 10^{11}$ C/kg) from the hot filament f, Fig. 25-5, are attracted to the anode P. The difference in potential between f and P, as read by the voltmeter, is $V = 600$ V.   (a) With what speed u do electrons arrive at P, assuming they start from rest at f?   (b) If the distance from f to P is 3 centimeters, what is the average electric field in which the electrons move?   (c) If electrons now arrive at P with a speed of 4×10^7 m/s, what is the new applied voltage?   (d) With how many electron volts of energy does each electron arrive at P?   (e) How much energy has been lost in the anode after 6×10^{23} electrons (about one mole) have flowed?   (Neglecting a slight amount of X-radiation, this energy will appear as heat.)

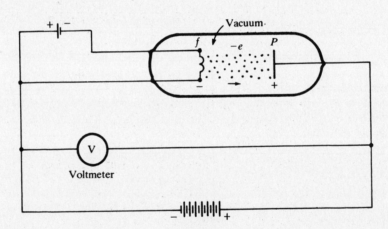

Fig. 25-5

(a)
$$eV = \frac{1}{2} m_e u^2$$

$$u = \sqrt{2(e/m_e)V} = \sqrt{2(1.759 \times 10^{11})(600)} = 1.453 \times 10^7 \text{ m/s}$$

(b)
$$E_{\text{avg}} = \frac{600 \text{ V}}{0.30 \text{ m}} = 20 \text{ kV/m}$$

(c)   From (a),
$$V = \frac{1}{2} \frac{u^2}{e/m_e} = \frac{1}{2} \frac{(4 \times 10^7)^2}{1.759 \times 10^{11}} = 4550 \text{ V}$$

(d)   600 eV (each electron falls through 600 V).

(e)
$$\mathcal{E} = (6 \times 10^{23} \text{ electrons})(600 \text{ eV/electron})(1.602 \times 10^{-19} \text{ J/eV}) = 57.6 \text{ MJ}$$

25.6. Referring to Fig. 25-6, (a) write an expression for the potential $\phi(x, y)$ at a general point $P(x, y)$ in the XY plane. (b) Find expressions for E_x and E_y at P.

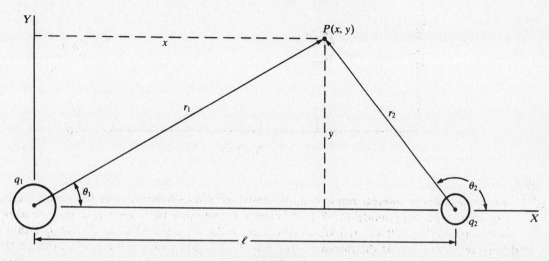

Fig. 25-6

(a) Since $r_1 = (x^2 + y^2)^{1/2}$ and $r_2 = [(\ell - x)^2 + y^2]^{1/2}$,

$$\phi(x, y) = \frac{1}{4\pi\epsilon_0}\left(\frac{q_1}{r_1} + \frac{q_2}{r_2}\right) = \frac{1}{4\pi\epsilon_0}\left[\frac{q_1}{(x^2 + y^2)^{1/2}} + \frac{q_2}{(x^2 + y^2 + \ell^2 - 2x\ell)^{1/2}}\right]$$

(b)

$$E_x = -\frac{\partial\phi}{\partial x} = \frac{1}{4\pi\epsilon_0}\left[\frac{q_1 x}{(x^2 + y^2)^{3/2}} + \frac{q_2(x - \ell)}{(x^2 + y^2 + \ell^2 - 2x\ell)^{3/2}}\right]$$

$$= \frac{1}{4\pi\epsilon_0}\left[\frac{q_1 \cos\theta_1}{r_1^2} + \frac{q_2 \cos\theta_2}{r_2^2}\right]$$

$$E_y = -\frac{\partial\phi}{\partial y} = \frac{1}{4\pi\epsilon_0}\left[\frac{q_1 y}{(x^2 + y^2)^{3/2}} + \frac{q_2 y}{(x^2 + y^2 + \ell^2 - 2x\ell)^{3/2}}\right]$$

$$= \frac{1}{4\pi\epsilon_0}\left[\frac{q_1 \sin\theta_1}{r_1^2} + \frac{q_2 \sin\theta_2}{r_2^2}\right]$$

25.7. (a) Show that the potential $\phi(x, y, z)$ due to a point charge q located at (x_0, y_0, z_0) satisfies *Laplace's equation*,

$$\frac{\partial^2\phi}{\partial x^2} + \frac{\partial^2\phi}{\partial y^2} + \frac{\partial^2\phi}{\partial z^2} = 0$$

(b) Generalize the result of (a).

(a) The distance r from (x_0, y_0, z_0) to (x, y, z) is given by

$$r = [(x - x_0)^2 + (y - y_0)^2 + (z - z_0)^2]^{1/2}$$

and so $$\phi(x, y, z) = \frac{q}{4\pi\epsilon_0 r} = C[(x - x_0)^2 + (y - y_0)^2 + (z - z_0)^2]^{-1/2}$$

From this,

$$\frac{\partial\phi}{\partial x} = -C[(x - x_0)^2 + (y - y_0)^2 + (z - z_0)^2]^{-3/2}(x - x_0)$$

$$\frac{\partial^2\phi}{\partial x^2} = 3C[(x - x_0)^2 + (y - y_0)^2 + (z - z_0)^2]^{-5/2}(x - x_0)^2 - C[(x - x_0)^2 + (y - y_0)^2 + (z - z_0)^2]^{-3/2}$$

$$= \frac{C}{r^5}[3(x - x_0)^2 - r^2]$$

Similarly,

$$\frac{\partial^2 \phi}{\partial y^2} = \frac{C}{r^5}[3(y - y_0)^2 - r^2] \qquad \frac{\partial^2 \phi}{\partial z^2} = \frac{C}{r^5}[3(z - z_0)^2 - r^2]$$

Adding the three second derivatives,

$$\frac{\partial^2 \phi}{\partial x^2} + \frac{\partial^2 \phi}{\partial y^2} + \frac{\partial^2 \phi}{\partial z^2} = \frac{C}{r^5}\{3[(x - x_0)^2 + (y - y_0)^2 + (z - z_0)^2] - 3r^2\} = \frac{C}{r^5}(3r^2 - 3r^2)$$

The last expression vanishes everywhere except at $r = 0$, yielding Laplace's equation.

(b) Because Laplace's equation is a linear equation, it follows from the superposition principle, (25.7), that the potential arising from *any* distribution of charges satisfies Laplace's equation. In other words, we obtain Laplace's equation upon differentiating under the integral sign in (25.8).

25.8. In Problem 22.6(b), we cited an important result which may be rephrased as follows: *The electric potential cannot assume a minimum (or a maximum) value inside a charge-free region.* Give an informal proof of this, using Gauss's law.

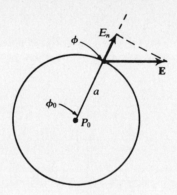

Fig. 25-7

Suppose, on the contrary, that the potential took on a local minimum value, ϕ_0, at the interior point P_0. Then (Fig. 25-7) we could enclose P_0 in a Gaussian sphere of so small a radius, a, that (i) the sphere lies entirely within the charge-free region, and (ii) $\phi \geq \phi_0$ at every point of the spherical surface. Gauss's law, applied to this sphere, would give

$$0 = \int_S \mathbf{E} \cdot d\mathbf{S} = \int_S E_n \, dS \tag{1}$$

where E_n is the normal (radial) component of the field at the surface of the sphere. But, by definition of the electric potential,

$$E_n = -\frac{d\phi}{dn} \tag{2}$$

i.e., the derivative of the potential in the radial direction. Now, by decreasing a if necessary, we can preserve conditions (i) and (ii) above and at the same time allow the derivative in (2) to be approximated by a difference-quotient to any required degree of accuracy. Thus,

$$E_n = -\frac{d\phi}{dn} \approx -\frac{\phi - \phi_0}{a} \tag{3}$$

and (1) becomes

$$0 = \int_S (\phi - \phi_0) \, dS \tag{4}$$

But (4) is impossible: $\phi - \phi_0$ is nonnegative at each point of S and so its integral over S must be positive. This contradiction establishes the desired result.

25.9. Find the potential due to a thin, uniformly charged rod of length $2a$.

Take the rod along the Z axis, as in Fig. 25-8. By the rotational symmetry of the problem, it is enough to determine the potential at an arbitrary point of the YZ plane. At the point (y, z), the potential due to the element of charge $dq = \lambda \, ds$ is

$$d\phi = \frac{dq}{4\pi\epsilon_0 r} = \frac{\lambda \, ds}{4\pi\epsilon_0 [y^2 + (z-s)^2]^{1/2}}$$

Hence,
$$\phi(y, z) = \frac{1}{4\pi\epsilon_0} \int_{-a}^{a} \frac{\lambda \, ds}{[y^2 + (z-s)^2]^{1/2}} = \frac{\lambda}{4\pi\epsilon_0} \int_{-a}^{a} \frac{ds}{[s^2 - 2zs + y^2 + z^2]^{1/2}}$$
$$= \frac{\lambda}{4\pi\epsilon_0} \ln \frac{[y^2 + (z-a)^2]^{1/2} - (z-a)}{[y^2 + (z+a)^2]^{1/2} - (z+a)}$$

The potential at point (x, y, z) is obtained by replacing y^2 by $x^2 + y^2$ in the above expression.

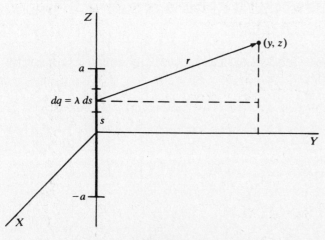

Fig. 25-8

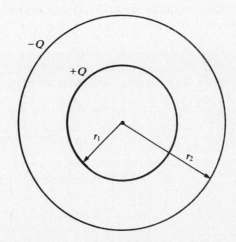

Fig. 25-9

25.10. Consider a system of two separated conducting bodies (equipotentials), body 1 carrying a charge $+Q$ and body 2 a charge $-Q$. Then the *capacitance* of the system is defined as

$$C = \frac{Q}{\phi_{12}} \quad \text{(F)} \tag{25.9}$$

i.e., the ratio of the magnitude of the charge to the magnitude of the difference in potential between the bodies. The SI unit of capacitance is the *farad* (F), where $1 \text{ F} = 1 \text{ C/V}$. The capacitance depends only on the geometry of the system. Calculate the capacitance of the two concentric spherical shells shown in Fig. 25-9.

We know (see Problem 24.4) that the electric field between the shells is as if produced by a point charge $+Q$ located at the center of the system. Thus,

$$\phi_{12} = \phi_1 - \phi_2 = \frac{Q}{4\pi\epsilon_0 r_1} - \frac{Q}{4\pi\epsilon_0 r_2}$$

$$C = \frac{Q}{\phi_{12}} = \frac{4\pi\epsilon_0}{(1/r_1) - (1/r_2)}$$

From this result it is seen that ϵ_0 may be expressed in F/m; indeed, these are the official SI units for permittivity. As $r_2 \to \infty$, we obtain as the *capacitance of an isolated sphere*:

$$C_\infty = 4\pi\epsilon_0 r_1$$

Supplementary Problems

25.11. A lead pellet (mass, 2 grams) fired from an air rifle has a speed of 150 ft/sec. Through what difference in potential would this pellet have to fall to acquire the same speed, assuming it carries a charge of 1 μC? *Ans.* 2.09 MV

25.12. (*a*) A charge of 3 microcoulombs falls through a difference in potential of 50 volts; (*b*) an α-particle (charge, $+2e$) falls through a difference in potential of 1 MV; (*c*) a bullet (mass, 4 grams) has a speed of 0.3048 km/s. Express the energy in each of the above cases in electron volts.
Ans. (*a*) 936 TeV; (*b*) 2 MeV; (*c*) 1.16×10^{21} eV

25.13. Refer to Fig. 25-4. An electric motor (assumed to be 100% efficient) is connected between A and B and a charge of 10^4 C is allowed to flow through from the battery. What external work can the motor do, if $V_{AB} = 50$ V?   *Ans.* 500 kJ

25.14. It can be shown that the period T of oscillation of electrons in a plasma is given by

$$T = \frac{1}{2\pi} \left(\frac{\epsilon_0 m_e}{ne^2} \right)^{1/2}$$

where m_e is the electronic mass; e, the magnitude of the electronic charge; and n, the number of electrons per unit volume. Verify that, unit-dimensionally, $T = |s|$.

25.15. The electric field **E** set up by a certain distribution of charges is two-dimensional, with X component

$$E_x = A[(x - \ell)(x^2 + y^2 + \ell^2 - 2x\ell)^{-3/2} - x(x^2 + y^2)^{-3/2}]$$

where A and ℓ are constants. Determine E_y.

Ans. $\phi = -\int_{\infty}^{x} E_x \, dx = A[(x^2 + y^2 + \ell^2 - 2x\ell)^{-1/2} - (x^2 + y^2)^{-1/2}]$

$$E_y = -\frac{\partial \phi}{\partial y} = A[y(x^2 + y^2 + \ell^2 - 2x\ell)^{-3/2} - y(x^2 + y^2)^{-3/2}]$$

25.16. A circular disk carries a surface charge σ (C/m²). Show that the electric field at any point on the axis of the disk depends only on σ and the angle α subtended there by the disk.

Ans. $E = \frac{\sigma}{2\epsilon_0} \left(1 - \cos \frac{\alpha}{2} \right)$

25.17. In Problem 25.9, set $\lambda = q/2a$ and show that as $a \to 0$ the potential due to a point charge q located at the origin is obtained.
Ans. Using L'Hospital's rule,

$$\phi(y, z) \to \frac{q}{4\pi\epsilon_0 r}$$

where $r = (y^2 + z^2)^{1/2}$.

Chapter 26

Electric Current, Resistance, and Power

26.1 CURRENT AND CURRENT DENSITY

The rate of flow of electric charge across a given area (within a conductor) is defined as the *electric current I* through that area. Thus,

$$I = \frac{dq}{dt} \quad \text{(A)} \tag{26.1}$$

See Section 22.1. The *electric current density* **J** at a point (within a conductor) is a vector whose direction is the direction of flow of charge at that point and whose magnitude is the current through a *unit area perpendicular to the flow direction* at that point. Thus, the current through an element of area dS, arbitrarily oriented with respect to the flow direction, is given by (see Fig. 26-1)

$$dI = \mathbf{J} \cdot d\mathbf{S} = J \, dA \tag{26.2}$$

where $dA = dS \cos \theta$ is the projection of dS perpendicular to the flow direction. The total current through a surface S (e.g. a cross section of the conductor) is then

$$I = \int_S \mathbf{J} \cdot d\mathbf{S} = \int_S J \, dA \tag{26.3}$$

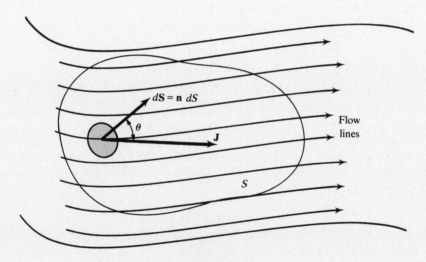

Fig. 26-1

EXAMPLE 26.1. If charge flows uniformly in a wire of constant cross-sectional area A, (26.3) gives

$$I = JA \quad \text{or} \quad J = \frac{I}{A} \tag{26.4}$$

as the relation between the constant current I in the wire and the constant current density J. From this (or directly from the definition of **J**) we see that the units of current density are A/m^2. (One must be careful not to confuse A for amperes with A for area, or J for joules with J for current density.)

227

26.2 OHM'S LAW; RESISTANCE

Empirically it is found that the current I in a "resistor" (long wire, conducting bar, etc.) is very nearly proportional to the difference, v, in electric potential between the ends of the resistor. This proportionality is expressed by *Ohm's law*:

$$v = IR \qquad (26.5)$$

where the proportionality factor, R, is called the *resistance*. The unit of resistance, the *ohm* (Ω), is defined by (26.5) as $1\,\Omega = 1\,\text{V/A}$. (It should be noted here that the definition of the volt used in Chapter 25—namely, $1\,\text{V} = 1\,\text{J/C}$—is not the *primary* definition in the SI; see Section 26.5.)

The resistance of a conductor of length L and uniform cross-sectional area A is given by

$$R = \rho \frac{L}{A} \qquad (26.6)$$

where ρ, the *resistivity*, is a property of the material and is temperature-dependent (see Section 26.3). The units of ρ are seen to be $\Omega \cdot \text{m}$. Very often one specifies, instead of ρ, the *conductivity*

$$\sigma = \frac{1}{\rho} \qquad (26.7)$$

The SI unit of conductivity is the *siemens per meter* (S/m).

The average electric field in the conductor just mentioned is $E = v/L$, or $v = EL$. Substituting this, along with (26.4) and (26.6), in (26.5), we obtain

$$E = \rho J \qquad (26.8)$$

as the expression of Ohm's law in terms of "field" quantities. More generally, the vector relation

$$\mathbf{E} = \rho \mathbf{J} \qquad \text{or} \qquad \mathbf{J} = \sigma \mathbf{E}$$

holds at each point of the conductor.

26.3 TEMPERATURE COEFFICIENT OF RESISTANCE

For many purposes, the thermal variation of resistance or resistivity is suitably approximated by the following linear relation (here written for resistivity):

$$\rho = \rho_0(1 + \alpha\theta) \qquad (26.9)$$

In (26.9), θ is the temperature in °C (see Section 17.1 for the Celsius scale) and α, the *temperature coefficient of resistance*, has the units $°\text{C}^{-1}$ or K^{-1}. Table 26-1 gives some typical values of α and of ρ at room temperature.

Table 26-1

Material	$\rho_{20°C}$, $\Omega \cdot \text{m}$	α, $°\text{C}^{-1}$
Silver	1.6×10^{-8}	3.8×10^{-3}
Copper	1.72×10^{-8}	3.9×10^{-3}
Iron	10×10^{-8}	5.2×10^{-3}
Nichrome	108×10^{-8}	4×10^{-4}
Carbon	3.5×10^{-5}	-5×10^{-4} (negative)
Fused Quartz	7.5×10^{17}	

26.4 ELECTRICAL ENERGY SOURCES

The difference in potential between the terminals of any source of electrical energy such as a battery or mechanically driven generator, *when delivering no current*, is a measure of the "electromotive force" of the source, which will be indicated as v_e. This is often called the *open-circuit voltage* of a battery or generator.

An actual source of electrical energy is usually attributed an *internal resistance*, r; as, for example, the resistance between plates of a cell or the armature resistance of a generator. If current I is passing through the source, its *terminal voltage*, v_t, is given by

$$v_t = v_e \pm Ir \qquad (26.10)$$

The negative sign is used in (26.10) when the source is delivering current and the positive sign when I has the opposite direction (charging a storage cell).

26.5 ELECTRICAL POWER

Like mechanical power (Section 8.4), electrical power is defined as the time rate of doing work (on charged bodies, by electric fields), or, equivalently, as the time rate of change of electric energy $\mathscr{E}$:

$$P = \frac{d\mathscr{E}}{dt} \quad \text{(W)}$$

In particular, if an amount of charge dq moves through a difference in potential v, the change in electric energy is $d\mathscr{E} = (dq)v$, and so

$$P = \frac{d\mathscr{E}}{dt} = \left(\frac{dq}{dt}\right)v = Iv \quad \text{(W)} \qquad (26.11)$$

From (26.11) (1 A)(1 V) = 1 W or 1 V = 1 W/A; this is actually the way the volt is defined in the SI.

The power output of an energy source having terminal voltage v_t and supplying current I is given by (26.11) and (26.10) as

$$P = Iv_t = I(v_e - Ir) \qquad (26.12)$$

and the power absorbed by a pure resistance R when the potential difference across it is v_{ab} is given by (26.11) and (26.5) as

$$P = Iv_{ab} = I^2R = \frac{v_{ab}^2}{R} \qquad (26.13)$$

This latter power is dissipated as heat.

Solved Problems

26.1. (a) A steady current of 5 A flows in a wire. How many electrons, on the average, pass through a cross section of the wire per minute? (b) In a two-element radio tube (see Fig. 25-5), 6×10^{14} electrons leave the filament and arrive on the plate per second. What is the reading of a microammeter in the plate circuit?

(a) $$\frac{(5 \text{ C/s})(60 \text{ s/min})}{1.602 \times 10^{-19} \text{ C/electron}} = 1.873 \times 10^{21} \text{ electrons/min}$$

(b) $$(6 \times 10^{14} \text{ electrons/s})(1.602 \times 10^{-19} \text{ C/electron}) = 9.612 \times 10^{-5} \text{ A} = 96.12 \ \mu\text{A}$$

26.2. A long iron wire carries a current of 6.32 A. An accurate voltmeter connected to its terminals reads 48.25 V. (a) What is the resistance of the wire? (b) How many calories of heat are generated per minute in it? (1 cal = 4.184 J.)

(a)
$$R = \frac{48.25}{6.32} = 7.63 \ \Omega$$

(b) The power loss is $P = (6.32)(48.24) = 304.94 \ \text{W} = 304.94 \ \text{J/s}$. Converting to calories per minute,

$$P = \frac{304.94}{4.184} \times 60 = 4372 \ \text{cal/min}$$

26.3. As shown in Fig. 26-2, a metal rod of radius r_1 is concentric with a metal cylindrical shell of radius r_2 and length L. The space between rod and cylinder is tightly packed with a high-resistance material of resistivity ρ. A battery having a terminal voltage v_t is connected as shown. Neglecting resistances of rod and cylinder, derive expressions for (a) the total current I, (b) the current density J and the electric field E at any point P between rod and cylinder, (c) the resistance R between rod and cylinder.

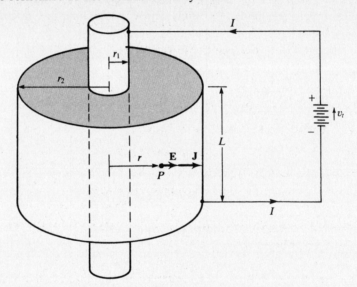

Fig. 26-2

(a) Assuming radial flow of charge between rod and cylinder, we have at P:

$$J = \frac{I}{2\pi rL} \qquad \text{and} \qquad E = \rho J = \frac{\rho I}{2\pi rL}$$

with both **J** and **E** in the direction of **r**. Then, by definition of the potential,

$$dv = -\mathbf{E} \cdot d\mathbf{s} = -E \, dr = -\frac{\rho I}{2\pi L} \frac{dr}{r}$$

and so, noting the polarity of v_t,

$$-v_t = \int_{r_1}^{r_2} dv = -\frac{\rho I}{2\pi L} \int_{r_1}^{r_2} \frac{dr}{r} = -\frac{\rho I}{2\pi L} \ln \frac{r_2}{r_1}$$

Solving for I,
$$I = \frac{2\pi L v_t}{\rho \ln (r_2/r_1)}$$

(b) From (a), $J = \dfrac{I}{2\pi rL} = \dfrac{v_t}{\rho r \ln (r_2/r_1)}$ and $E = \rho J = \dfrac{v_t}{r \ln (r_2/r_1)}$

(c) From Ohm's law,
$$R = \frac{v_t}{I} = \frac{\rho \ln (r_2/r_1)}{2\pi L}$$

26.4. A metal wire of diameter 2 mm and of length 300 m has a resistance of 1.6424 Ω at 20 °C, and 2.415 Ω at 150 °C. Find the values of α, R_0, ρ_0 and $\rho_{20°C}$. Identify the metal.

$$R_{150°C} = 2.415 = R_0(1 + \alpha 150) \qquad R_{20°C} = 1.6424 = R_0(1 + \alpha 20)$$

Solving these relations simultaneously, $\alpha = 3.9 \times 10^{-3}$ °C^{-1} and $R_0 = 1.5236$ Ω.

From $R_0 = \rho_0 L/A$,

$$1.5236 = \frac{\rho_0(300)}{\pi(2 \times 10^{-3})^2/4} \qquad \text{or} \qquad \rho_0 = 1.596 \times 10^{-8} \ \Omega \cdot \text{m}$$

Then, $\qquad \rho_{20°C} = \rho_0(1 + \alpha 20) = (1.596 \times 10^{-8})[1 + (3.9 \times 10^{-3})(20)] = 1.720 \times 10^{-8} \ \Omega \cdot \text{m}$

Table 26-1 indicates that the metal is copper.

26.5. At a certain point P in a slab of copper through which current is flowing, $J = 5$ MA/m^2; the temperature of the slab is 50 °C. Compute the magnitude of **E** at P.

From the Table 26-1, $\rho_{20°C} = 1.72 \times 10^{-8}$ $\Omega \cdot$ m and $\alpha = 3.9 \times 10^{-3}$ °C^{-1}. Applying (26.9),

$$\frac{\rho_{50°C}}{\rho_{20°C}} = \frac{1 + \alpha 50}{1 + \alpha 20} \qquad \text{whence} \qquad \rho_{50°C} = 1.907 \times 10^{-8} \ \Omega \cdot \text{m}$$

Thus $\qquad\qquad E = \rho_{50°C} J = (1.907 \times 10^{-8})(5 \times 10^6) = 0.0954$ V/m

26.6. A bar of copper having mass 1.5 kg is to be drawn into a wire having resistance 250 Ω at 20 °C. Determine the length L and diameter d of the wire.

The density of copper is 8.9×10^3 kg/m^3, and so

$$(8.9 \times 10^3)LA = 1.5$$

On the other hand, Table 26-1 gives for $R_{20°C}$:

$$(1.72 \times 10^{-8})\frac{L}{A} = 250$$

Solving the above two equations simultaneously, $L = 1.565$ km and $A = 0.1077$ mm^2. But $A = \pi d^2/4$, from which $d = 0.37$ mm.

26.7. An iron wire, of length 2000 m, and a copper wire, of length 3000 m, are connected in parallel across a source having a terminal voltage of 200 V. The diameter of the copper wire is 1 mm; the temperature of the wires is 100 °C. If each wire carries the same current, find the current, the diameter of the iron wire, and the electric field strength in each.

As found in Problem 26.4, $\rho_0 = 1.596 \times 10^{-8}$ $\Omega \cdot$ m for copper. Likewise, $\rho_0 = 9.06 \times 10^{-8}$ $\Omega \cdot$ m for iron. Hence, at 100 °C,

$$R_{\text{Cu}} = \frac{(1.596 \times 10^{-8})(3000)}{\pi(10^{-6})/4}[1 + (3.9 \times 10^{-3})(100)] = 84.74 \ \Omega$$

Thus $\qquad\qquad I_{\text{Cu}} = \frac{200}{84.74} = 2.36$ A

Since the current is assumed to be the same in each wire,

$$R_{\text{Fe}} = R_{\text{Cu}} = 84.74 = \frac{(9.06 \times 10^{-8})(2000)}{\pi d^2/4}[1 + (5.2 \times 10^{-3})(100)]$$

from which the diameter of the iron wire is $d = 2.034$ mm. The electric fields are

$$E_{\text{Cu}} = \frac{200 \text{ V}}{3000 \text{ m}} = \frac{1}{15} \text{ V/m} \qquad E_{\text{Fe}} = \frac{200 \text{ V}}{2000 \text{ m}} = \frac{1}{10} \text{ V/m}$$

These values also follow from $E = \rho I/A$.

26.8. A carbon rod, of length 5 m, is in series with a copper wire having a resistance $R_1 = 10 \ \Omega$ at 0 °C. What must be the resistance R_2 of the carbon rod at 0 °C so that the resistance of the combination does not appreciably change with temperature?

The total resistance, as a function of temperature θ, is given by

$$R_{\text{tot}} = R_1(1 + \alpha_1\theta) + R_2(1 + \alpha_2\theta) = (R_1 + R_2) + (\alpha_1 R_1 + \alpha_2 R_2)\theta$$

This will be independent of θ if

$$\alpha_1 R_1 + \alpha_2 R_2 = 0 \qquad \text{or} \qquad R_2 = \frac{\alpha_1 R_1}{-\alpha_2} = \frac{(3.9 \times 10^{-3})(10)}{5 \times 10^{-4}} = 78 \ \Omega$$

26.9. In Fig. 26-3, the battery has electromotive force $v_e = 100$ V and internal resistance $r = 0.5 \ \Omega$; for the resistor, $R = 19.5 \ \Omega$. With switch S closed, find (a) I; (b) v_{ab}, the "voltage drop" across R; (c) the "voltage lost" in the battery; (d) power taken by R; (e) heat generated in R and in the battery.

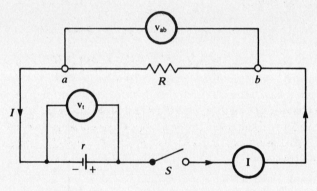

Fig. 26-3

(a) From $v_t = v_e - rI$ and $v_t = v_{ab} = RI$,

$$v_e - rI = RI \qquad \text{or} \qquad I = \frac{v_e}{r + R} = \frac{100}{20} = 5 \text{ A}$$

(b) $v_{ab} = (19.5)(5) = 97.5$ V

(c)   "voltage lost" = $rI = (0.5)(5) = 2.5$ V

(d) $P = I^2 R = (5^2)(19.5) = 487.5$ W

(e)   heat in $R = P = 487.5$ W

                              heat in battery = $I^2 r = 12.5$ W

Note that the total heat is $I^2 R + I^2 r = I[I(R + r)] = Iv_e$; this relation effectively defines the internal resistance r.

26.10. In the simple parallel circuit depicted in Fig. 26-4, the electromotive force of the battery is $v_e = 80$ V; internal resistance is $r = 0.4 \ \Omega$; resistances are $R_1 = 5 \ \Omega$, $R_2 = 10 \ \Omega$, $R_3 = 20 \ \Omega$. Find (a) R_{tot}, the resistance equivalent to R_1, R_2, and R_3; (b) the power delivered by the battery to the parallel circuit.

(a) The equivalent resistance must account for the actual power loss; thus,

$$P = \frac{v_1^2}{R_{\text{tot}}} = \frac{v_1^2}{R_1} + \frac{v_1^2}{R_2} + \frac{v_1^2}{R_3} \qquad \text{or} \qquad \frac{1}{R_{\text{tot}}} = \frac{1}{R_1} + \frac{1}{R_2} + \frac{1}{R_3} = \frac{1}{5} + \frac{1}{10} + \frac{1}{20} = \frac{7}{20} \text{ S}$$

and $R_{\text{tot}} = 2.8571 \ \Omega$. The SI unit of reciprocal resistance is the *siemens* (S), replacing the older "mho."

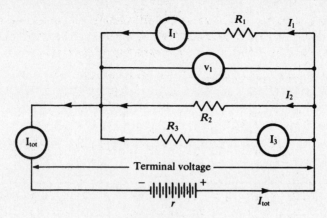

Fig. 26-4

(b)   As in Problem 26.9(a),

$$I_{tot} = \frac{v_e}{r + R_{tot}}$$

and so

$$P = I_{tot}^2 R_{tot} = \frac{v_e^2 R_{tot}}{(r + R_{tot})^2} = \frac{(80)^2(2.8571)}{(3.2571)^2} = 1723.5 \text{ W}$$

26.11. In the simple series circuit depicted in Fig. 26-5, the electromotive force of the battery is $v_e = 200$ V; internal resistance is $r = 0.6\,\Omega$; resistances are $R_1 = 8\,\Omega$, $R_2 = 15\,\Omega$, $R_3 = 20\,\Omega$. Find (a) R_{tot}, the resistance equivalent to R_1, R_2, and R_3; (b) the power delivered by the battery to the series circuit.

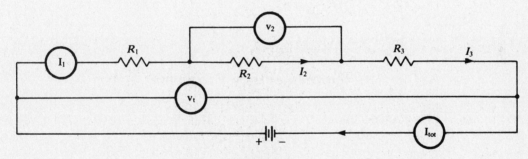

Fig. 26-5

(a)   Since $I_1 = I_2 = I_3 = I_{tot}$, the power loss is given by

$$I_{tot}^2 R_1 + I_{tot}^2 R_2 + I_{tot}^2 R_3 = I_{tot}^2 R_{tot}$$

or

$$R_{tot} = R_1 + R_2 + R_3 = 43\,\Omega$$

(b)   As in Problem 26.10(b),

$$P = \frac{v_e^2 R_{tot}}{(r + R_{tot})^2} = \frac{(200)^2(43)}{(43.6)^2} = 904.78 \text{ W}$$

26.12. Figure 26-6 shows a series-parallel circuit for which $R_1 = 10\,\Omega$, $R_2 = 30\,\Omega$, $R_3 = 20\,\Omega$, $R_4 = 25\,\Omega$, $R_5 = 35\,\Omega$, $R_6 = 40\,\Omega$; the total applied voltage is $v_{tot} = 250$ V.   Find: (a) R_{23} and R_{456}, the equivalent resistances of the two parallel groups; (b) the total resistance R_{tot} as measured between A and B; (c) total current I_{tot}; (d) power delivered by the source.

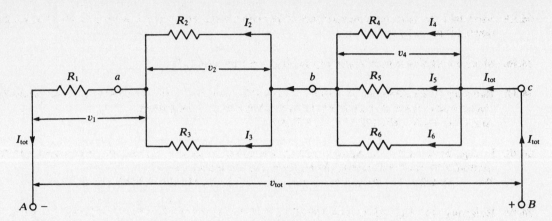

Fig. 26-6

(a) From the result of Problem 26.10(a),

$$\frac{1}{R_{23}} = \frac{1}{R_2} + \frac{1}{R_3} \qquad \text{or} \qquad R_{23} = 12 \ \Omega$$

$$\frac{1}{R_{456}} = \frac{1}{R_4} + \frac{1}{R_5} + \frac{1}{R_6} \qquad \text{or} \qquad R_{456} = 10.687 \ \Omega$$

(b) From the result of Problem 26.11(a),

$$R_{\text{tot}} = R_1 + R_{23} + R_{456} = 32.687 \ \Omega$$

(c)

$$I_{\text{tot}} = \frac{v_{\text{tot}}}{R_{\text{tot}}} = 7.648 \ \text{A}$$

(d)

$$P = \frac{v_{\text{tot}}^2}{R_{\text{tot}}} = 1912 \ \text{W}$$

Supplementary Problems

26.13. A certain wire, of uniform diameter 2 mm and length 2.5 km, has resistance 96 Ω at 40 °C and 83.6 Ω at 10 °C.   Of what metal is the wire?   *Ans.*   iron

26.14. In the circuit of Fig. 26-7, $v_t = 216$ V and, at 0 °C, $R_C = 30 \ \Omega$, $R_{Cu} = 40 \ \Omega$, and $R_{Fe} = 60 \ \Omega$.   Find the power delivered by the battery (a) at 0 °C, (b) at 100 °C.   *Ans.*   (a) 864 W;   (b) 740 W

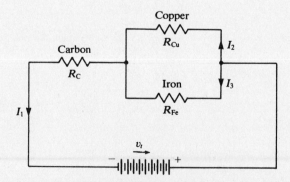

Fig. 26-7

26.15. Show that the heat generated per second per cubic meter at any point in a conductor in which current is flowing is given by ρJ^2 (W/m^3).

26.16. Make a unit-dimensional check of the results of Problem 26.3.

26.17. Referring to Fig. 26-3, let $R = 10\ \Omega$, $v_{ab} = 150$ V, and $r = 1.5\ \Omega$. Determine (a) the current I, (b) the electromotive force of the battery, (c) the voltage drop in the battery.
Ans.   (a) 15 A;   (b) 172.5 V;   (c) 22.5 V

26.18. In Fig. 26-4, let $I_1 = 10$ A, $R_2 = 12.5\ \Omega$, $r = 0.5\ \Omega$, $v_1 = 120$ V, $I_{tot} = 25$ A. Determine (a) the electromotive force of the battery, (b) I_2, (c) I_3, (d) R_3, (e) R_{tot}, (f) R_1.
Ans.   (a) 132.5 V;   (b) 9.6 A;   (c) 5.4 A;   (d) 22.222 Ω;   (e) 4.8 Ω;   (f) 12 Ω

26.19. Referring to Fig. 26-8, $v_{e1} = 160$ V, $v_{e2} = 40$ V; $R_1 = 9\ \Omega$, $R_2 = 8\ \Omega$; $r_1 = 1.6\ \Omega$, $r_2 = 1.4\ \Omega$. Determine (a) I, (b) v_{ab}, (c) v_{bc}, (d) v_{dg}, (e) v_{bf}, (f) power converted to heat, (g) power expended by first battery, (h) power absorbed by second battery. Show that your answers to (f), (g), and (h) are consistent.
Ans.   (a) 6 A;   (b) 150.4 V;   (c) 48.4 V;   (d) 102 V;   (e) 96.4 V;   (f) 720 W;   (g) 960 W;
(h) 240 W

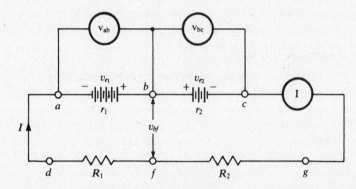

Fig. 26-8

26.20. Show that $R = \left|\dfrac{\text{Nms}}{\text{C}^2}\right|$; $\rho = |\Omega\text{m}| = \left|\dfrac{\text{Nm}^2\text{s}}{\text{C}^2}\right|$; $\dfrac{v}{r\ln(r_2/r_1)} = \left|\dfrac{\text{N}}{\text{C}}\right|$, see Problem 26.3.

Chapter 27

Kirchhoff's Laws for Resistive Circuits

Consider a network composed of resistors and sources (batteries), such as the one shown in Fig. 27-5. Given the values of a certain number of the electromotive forces v_e, currents I, and resistances R, the values of the remaining quantities can be found by an application of *Kirchhoff's laws*.

27.1 PRELIMINARY STEPS

On a carefully drawn diagram of the network, indicate by heavy arrows the directions of the *given* currents. Arbitrarily choose a direction for each unknown current and indicate same with an arrow. Every current, known or unknown, must be shown by an arrow.

Indicate the direction of each known electromotive force with a light or open arrow pointing from the − to the + terminal. (This indicates the direction in the external circuit in which positive charge carriers would move under the influence of the electromotive force.) If there are unknown electromotive forces, assign an *arbitrary* direction to each and indicate same with a voltage arrow.

Label the resistances $R_1, R_2, \ldots$, and if the sources have internal resistance, show these as $r_1, r_2, \ldots$ (see Fig. 27-2).

As a matter of practical importance it is well to enter all *given values* of voltage, current, and resistance on the diagrams.

27.2. KIRCHHOFF'S JUNCTION LAW

The algebraic sum of all currents *into any junction* (as, for example, a, b, c, g in Fig. 27-5) is zero. That is, at any junction,

$$\sum I = 0 \qquad (27.1)$$

In (27.1) a current *out of* the junction is counted as negative.

27.3 KIRCHHOFF'S LOOP LAW

The algebraic sum of all electromotive forces *around any closed circuit* (*loop*) is equal to the algebraic sum of the voltage drops (the RI and the rI) around the loop:

$$\sum v_e \text{ (electromotive force)} = \sum RI_R + \sum rI_r \qquad (27.2)$$

where I_R denotes the current in the corresponding resistance R and I_r denotes the current passing through a source (battery or generator) having internal resistance r. In (27.2) the loop is continuously traversed in one direction or the other. An electromotive force v_e is counted as positive if its voltage arrow points in the direction of traversal; a voltage drop RI_R or rI_r is counted as positive if the current arrow for I_R or I_r points in the direction of traversal.

A useful extension of (27.2) is as follows: The difference in potential, v_{12}, between any two points P_1 and P_2 in the network is given by

$$v_{12} = \sum (RI_R + rI_r) - \sum v_e \qquad\qquad (27.3)$$

where the summations are taken along any continuously directed path from P_1 to P_2 in the network.   Apply the same rules for the algebraic signs of v_e, RI_R, and rI_r as given above.   If v_{12} comes out positive, this means that P_1 is at a positive potential relative to P_2.

27.4   APPLICATION OF THE TWO LAWS

If the network contains n unknown quantities (resistances, currents, electromotive forces), we need to select n *independent* equations from among all possible junction equations (27.1) and loop equations (27.2).   To this end, we observe that any set of distinct junctions, *except the set of all junctions*, gives rise to junction equations which are independent of one another and of any loop equations.   Because (27.1) is trivial for a junction at which only two branches meet, we shall, in writing down equations, consider as "junctions" only those points at which three or more branches meet.

As regards the loop equations, an independent set of them may be obtained by picking the loops one at a time, in such manner that each new equation contains a term not contained in any previous equation.   A simpler procedure may be followed when the network diagram can be drawn in a plane (the usual case).   Then, the boundaries of the *finite* areas into which the diagram decomposes the plane give rise to independent loop equations; one writes as many of these as necessary.

Solved Problems

27.1. Referring to Fig. 27-1, find currents I_1, I_2, and I_3.   Directions shown for the currents are arbitrary.

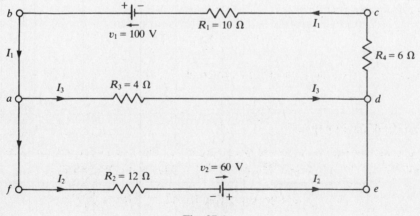

Fig. 27-1

Junction equations.   At junction a,

$$I_1 - I_2 - I_3 = 0 \qquad\qquad (1)$$

[Another junction equation can be written at d, but it is the negative of (1).]

Loop equations. Traversing the loop *adcba* of the planar network, we obtain

$$100 = 4I_3 + 6I_1 + 10I_1 \tag{2}$$

and traversing *afeda*,

$$60 = 12I_2 + 4(-I_3) \tag{3}$$

(It is assumed that the batteries have no internal resistance.) Note that the loop *afedcba* would not give an independent equation, but rather the sum of (2) and (3).

The independent equations (*1*), (*2*), (*3*) can now be solved simultaneously (applying usual methods) for I_1, I_2, I_3, yielding

$$I_1 = 6.053 \text{ A} \qquad I_2 = 5.263 \text{ A} \qquad I_3 = 0.7895 \text{ A}$$

Note that each value of I is positive; thus the arbitrarily chosen directions indicated in Fig. 27-1 happen to be correct.

27.2. Referring again to Fig. 27-1, now let $v_1 = 100$ V, $I_1 = 4$ A, $I_2 = 6$ A, $R_2 = 12 \, \Omega$, $R_3 = 4 \, \Omega$, $R_4 = 6 \, \Omega$. Find R_1, I_3, v_2, and the voltage v_{ef} between points *e* and *f*.

Arrows indicating directions of the currents may be left as shown. Proceeding as in Problem 27.1, we obtain the junction equation

$$4 - 6 - I_3 = 0 \tag{1}$$

and the loop equations

$$100 = 4I_3 + (6)(4) + R_1(4) \tag{2}$$

$$v_2 = (12)(6) + 4(-I_3) \tag{3}$$

A simultaneous solution of (*1*), (*2*), (*3*) gives

$$R_1 = 21 \, \Omega \qquad v_2 = 80 \text{ V} \qquad I_3 = -2 \text{ A}$$

Here the minus sign indicates that the direction of I_3 is opposite to that shown in Fig. 27-1.

Applying relation (*27.3*) along the direct branch from *e* to *f*,

$$v_{ef} = (-6)(12) - (-80) = 8 \text{ V}$$

i.e. point *e* is 8 volts above point *f*. The same result is obtained by traversing the path *edcbaf* or *edaf*.

27.3. In Fig. 27-2, find all currents and the difference in potential, v_{12}, between points p_1 and p_2.

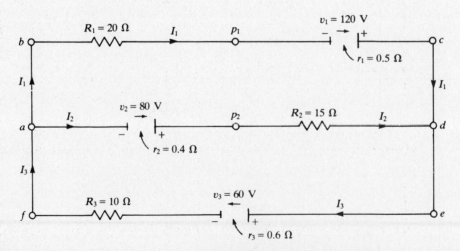

Fig. 27-2

Electromotive forces have the directions shown and current arrows indicate assumed directions of I's. The junction equation at a is

$$I_3 - I_1 - I_2 = 0 \qquad (1)$$

The loop equation for *abcda* is

$$120 - 80 = 20I_1 + 0.5I_1 - 15I_2 - 0.4I_2 \qquad (2)$$

and for path *adefa*, $v_2 + v_3 = R_2I_2 + r_2I_2 + R_3I_3 + r_3I_3$, or

$$80 + 60 = 15I_2 + 0.4I_2 + 10I_3 + 0.6I_3 \qquad (3)$$

Solving (1), (2), (3), we get

$$I_1 = 4.590 \text{ A} \qquad I_2 = 3.513 \text{ A} \qquad I_3 = 8.103 \text{ A}$$

Applying (27.3) along the path p_1cdp_2,

$$v_{12} = (0.5)(4.590) + (15)(-3.513) - 120 = -170.4 \text{ V}$$

Thus p_1 is at a negative potential relative to p_2.

27.4. Find the total power delivered by the sources (the batteries) in Problem 27.3 and compare this with the total power converted into heat in all resistances.

The total power delivered is

$$P_1 = v_1I_1 + v_2I_2 + v_3I_3 = 1318.0 \text{ W}$$

The power absorbed by the resistances is

$$P_2 = (R_1 + r_1)I_1^2 + (R_2 + r_2)I_2^2 + (R_3 + r_3)I_3^2 = 1318 \text{ W}$$

As required by energy conservation, $P_1 = P_2$.

27.5. The network shown in Fig. 27-3 is exactly the same as the one in Fig. 27-1, except for the two capacitors C_1 and C_2. A short while after the capacitors have been connected into the network there is no current to the capacitors. Then I_1, I_2, I_3 are as found in Problem 27-1, and C_1 and C_2 have constant charges q_1 and q_2, respectively. If $C_1 = 10 \ \mu\text{F}$ and $C_2 = 15 \ \mu\text{F}$, find q_1 and q_2.

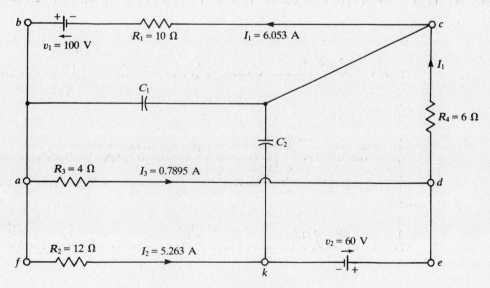

Fig. 27-3

From (25.9), Problem 25.10, $q_1 = C_1|v_{ca}|$, $q_2 = C_2|v_{ck}|$. Now, by (27.3),

$$v_{ca} = R_1I_1 - v_1 = -39.47 \text{ V}$$

(The minus sign merely means that point c is negative relative to a.) Thus, $q_1 = (10)(39.47) = 394.7 \ \mu C$. Likewise,

$$v_{ck} = -R_4I_1 - (-v_2) = 23.68 \text{ V}$$

and $q_2 = (15)(23.68) = 355.2 \ \mu C$.

27.6. In the network of Fig. 27-4, the batteries have negligible internal resistances. Assuming directions of voltages and currents as indicated, evaluate the unspecified currents, voltages, and resistances.

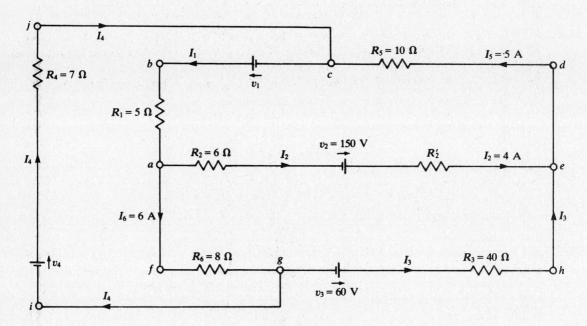

Fig. 27-4

Six independent equations are required. Choosing three of the four junctions—say, a, e, and g—we obtain

$$I_1 - I_6 - I_2 = 0 \tag{1}$$

$$I_2 + I_3 - I_5 = 0 \tag{2}$$

$$I_6 - I_3 - I_4 = 0 \tag{3}$$

Also choosing all three elementary loops $aedcba$, $aehgfa$, and $jcbafgij$, we have

$$v_1 + v_2 = R_1I_1 + (R_2 + R_2')I_2 + R_5I_5 \tag{4}$$

$$v_2 - v_3 = (R_2 + R_2')I_2 - R_3I_3 - R_6I_6 \tag{5}$$

$$v_1 + v_4 = R_1I_1 + R_4I_4 + R_6I_6 \tag{6}$$

Inserting the data and solving equations (1) through (6), we find:

$$I_1 = 10 \text{ A} \qquad I_3 = 0.1 \text{ A} \qquad I_4 = 5 \text{ A} \qquad v_1 = 92 \text{ V} \qquad v_4 = 41 \text{ V} \qquad R_2' = 29.5 \ \Omega$$

Supplementary Problems

27.7. Refer to Fig. 27-4. (*a*) What would be the reading of a high-resistance voltmeter connected between points *b* and *h*? (*b*) How much energy would be stored in (the electric field of) a 25-μF capacitor connected between points *b* and *h*? (*Hint*: $d\mathscr{E} = q\,dv = Cv\,dv$.)
Ans. (*a*) $v_{bh} = +42$ V; (*b*) $\mathscr{E} = 0.02205$ J

27.8. What is the maximum number of independent equations derivable from Kirchhoff's laws for the network of Fig. 27-5? Ans. 3 junction and 3 loop

27.9. In Fig. 27-5, let $I_1 = 5$ A, $I_4 = 8$ A, $I_5 = -7$ A; $R_1 = 10\ \Omega$, $R_2 = 6\ \Omega$, $R_3 = 7\ \Omega$, $R_4 = 8\ \Omega$, $R_6 = 4\ \Omega$, $R_7 = 3\ \Omega$; $v_1 = 100$ V, $v_2 = 20$ V, $v_3 = 40$ V, $v_5 = -30$ V, $v_7 = 70$ V. Calculate the remaining quantities.
Ans. $I_2 = 3$ A, $I_3 = 10$ A, $I_6 = 2$ A; $R_5 = 11.143\ \Omega$; $v_4 = 92$ V, $v_6 = 162$ V

27.10. In Fig. 27-5, let $v_1 = 100$ V, $v_2 = 60$ V, $v_3 = 40$ V, $v_4 = 80$ V, $v_5 = -70$ V, $v_6 = 120$ V, $v_7 = -35$ V; $R_1 = 10\ \Omega$, $R_2 = 8\ \Omega$, $R_3 = 6\ \Omega$, $R_4 = 9\ \Omega$, $R_5 = 4\ \Omega$, $R_6 = 5\ \Omega$, $R_7 = 2\ \Omega$. Determine the currents and the difference in potential between points *e* and *a*.
Ans. $I_1 = 5.151$ A, $I_2 = 1.312$ A, $I_3 = 18.559$ A, $I_4 = 6.462$ A, $I_5 = -17.246$ A, $I_6 = 12.095$ A;
$v_{ea} = 70.332$ V

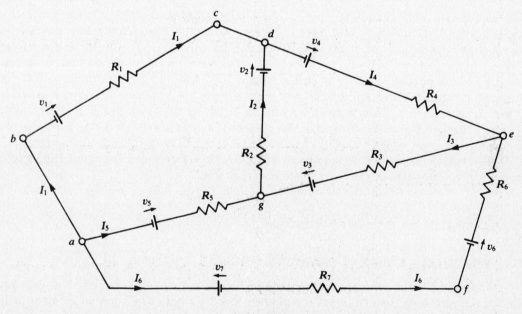

Fig. 27-5

Chapter 28

Magnetic Forces on Moving Charges

28.1 THE MAGNETIC FIELD

A region of space is said to be the site of a *magnetic field* if a test charge, moving in the region, experiences a force *by virtue of its motion with respect to an inertial frame*. This force may be described in terms of a field vector **B**, called the *magnetic induction* or the *magnetic flux density*, or simply, the *magnetic field*. If **u** is the velocity of the charge, then the force is given by

$$\mathbf{F} = q(\mathbf{u} \times \mathbf{B}) \tag{28.1}$$

or $F = quB \sin \theta$, where θ is the smaller angle between the vectors **u** and **B**. All three vectors are measured in the same inertial frame. The unit of **B** or B is the *tesla* (T). (Do not confuse T for tesla with the SI symbol for multiplication by 10^{12}.) By (28.1),

$$1\,\text{T} = 1\,\frac{\text{N} \cdot \text{s}}{\text{C} \cdot \text{m}} = 1\,\frac{\text{N}}{\text{A} \cdot \text{m}} = 1\,\frac{\text{V} \cdot \text{s}}{\text{m}^2} \tag{28.2}$$

It is seen from (28.1) that the magnetic force differs significantly from the electric force $q\mathbf{E}$: (i) At each point of the magnetic field there is a certain direction of motion (namely, the direction of **B**) for which the magnetic force is zero. (ii) The magnetic force is perpendicular to the velocity of the moving charge and thus performs no work; hence there is no scalar potential function for the force, and the force is nonconservative, as well as noncentral. Despite these differences, there is a deep connection between the magnetic and the electric field. A magnetic field itself arises from the motion of charges (i.e., from a changing electric field), as discussed in Chapter 29. Indeed, the Special Theory of Relativity shows that the magnetic and electric fields are two aspects of a single *electromagnetic field*; which "face" the observer sees depends on his coordinate system.

When both a magnetic field **B** and an electric field **E** are present in a region, the force on a test charge is the vector sum of the electric and magnetic forces:

$$\mathbf{F}_{\text{tot}} = q[\mathbf{E} + (\mathbf{u} \times \mathbf{B})] \tag{28.3}$$

This relation is known as the *Lorentz equation*.

28.2 FORCE ON A CURRENT-CARRYING WIRE

Consider a wire of arbitrary shape, as ab in Fig. 28-1, carrying current I in a magnetic field **B** which may vary from point to point. It follows directly from (28.1) that an element of length ds experiences a force $d\mathbf{F}$ given by

$$d\mathbf{F} = I(d\mathbf{s} \times \mathbf{B}) \tag{28.4}$$

Here the vector $d\mathbf{s}$ is of magnitude ds and is in the direction of motion of positive charge (conventional current). Actually, (28.4) gives the force on the charge-carriers within the length ds. However, this force is converted, by collisions, into a force on the wire as a whole—a force which, moreover, is capable of doing work on the wire.

The net force on the wire ab is obtained by integrating (28.4) from a to b. In the special case of a straight wire of length L in a uniform field **B**, the integration gives

$$\mathbf{F} = I(\mathbf{L} \times \mathbf{B}) \tag{28.5}$$

or $F = ILB \sin \theta$; in component form,

$$F_x = I(B_z L_y - B_y L_z) \qquad F_y = I(B_x L_z - B_z L_x) \qquad F_z = I(B_y L_x - B_x L_y)$$

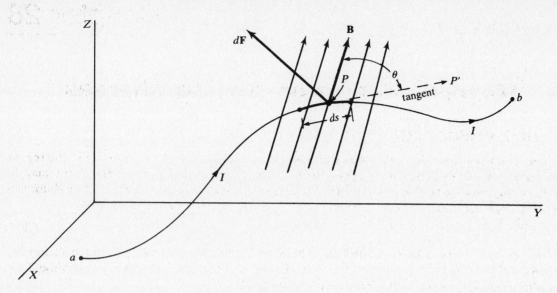

Fig. 28-1

EXAMPLE 28.1 (Torque on a Planar Coil). Consider a planar coil of arbitrary shape, consisting of n closely packed turns each carrying current I. The coil is located in a *uniform* magnetic field **B** which (presently) makes an angle θ with the normal to the plane of the coil. We choose this plane as the XY plane (Fig. 28-2).

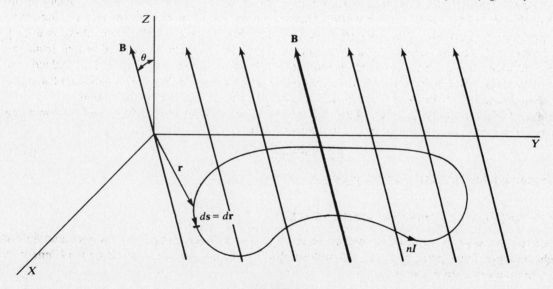

Fig. 28-2

Let us calculate the net force on the coil in this orientation. Using the determinant form of the cross product (see Section 1.5), we have from (28.4):

$$d\mathbf{F} = nI \begin{vmatrix} \mathbf{i} & \mathbf{j} & \mathbf{k} \\ dx & dy & 0 \\ B_x & B_y & B_z \end{vmatrix} = \mathbf{i}nIB_z\,dy - \mathbf{j}nIB_z\,dx + \mathbf{k}nI(B_y\,dx - B_x\,dy)$$

When this is integrated around the perimeter of the coil, the result is zero, since x and y return to their starting values (remember: B_x, B_y, B_z are constants). Thus, *the coil experiences no net force.* It does, however, experience a couple, and therefore a resultant torque τ. By Problem 10.6, we may choose any convenient point—say, the origin—about which to calculate the torque. We have

$$d\boldsymbol{\tau} = \mathbf{r} \times d\mathbf{F}$$

Using the above expression for $d\mathbf{F}$, this becomes

$$d\boldsymbol{\tau} = \begin{vmatrix} \mathbf{i} & \mathbf{j} & \mathbf{k} \\ x & y & 0 \\ nIB_z\,dy & -nIB_z\,dx & nI(B_y\,dx - B_x\,dy) \end{vmatrix}$$

$$= \mathbf{i}(nIyB_y\,dx - nIyB_x\,dy) - \mathbf{j}(nIxB_y\,dx - nIxB_x\,dy) + \mathbf{k}(-nIxB_z\,dx - nIyB_z\,dy)$$

When this is integrated around the perimeter of the coil, all terms in $x\,dx = d(x^2/2)$ or $y\,dy = d(y^2/2)$ vanish, since $x^2/2$ and $y^2/2$ return to their starting values. We are left with:

$$\boldsymbol{\tau} = \mathbf{i}nIB_y \oint y\,dx + \mathbf{j}nIB_x \oint x\,dy$$

But we know that the area of the coil is given by

$$A = \oint x\,dy = -\oint y\,dx$$

Thus, finally,

$$\boldsymbol{\tau} = nIA(-\mathbf{i}B_y + \mathbf{j}B_x) = nIA(\mathbf{k} \times \mathbf{B}) \tag{28.6}$$

Note that the torque has no Z component; it therefore tends to rotate the coil in such a way as to make the plane of the coil perpendicular to $\mathbf{B}$. If we let $\mathbf{A}$ represent a vector area having magnitude A and the direction given by the right-hand rule (with fingers curled in the direction of I, the extended thumb points in the direction of $\mathbf{A}$), then (28.6) can be put in a form that is independent of any particular coordinate system:

$$\boldsymbol{\tau} = nI(\mathbf{A} \times \mathbf{B}) \tag{28.7}$$

or $\tau = nIAB \sin\theta$, where θ is the angle between $\mathbf{A}$ and $\mathbf{B}$.

28.3　MAGNETIC FLUX

Just as electric flux ψ is associated with the electric field $\mathbf{E}$ (see Section 24.1), so *magnetic flux* Φ is associated with the magnetic field $\mathbf{B}$. Thus, the magnetic flux through an elementary area dS is defined as

$$d\Phi = \mathbf{B} \cdot d\mathbf{S} \quad \text{(Wb)} \tag{28.8}$$

and through an open or closed surface S,

$$\Phi = \int_S \mathbf{B} \cdot d\mathbf{S} \quad \text{(Wb)} \tag{28.9}$$

Note that the SI unit of magnetic flux is the *weber* (Wb), where

$$1\,\text{Wb} = 1\,\text{T} \cdot \text{m}^2 = 1\,\text{V} \cdot \text{s} = 1\,\text{J/A} \tag{28.10}$$

Not uncommonly, $\mathbf{B}$ (the "magnetic flux density") is given in Wb/m^2.

EXAMPLE 28.2 The flux through the coil of Fig. 28-2 is

$$\Phi = \int B_z\,dS = B_zA = BA\cos\theta = \Phi_0\cos\theta$$

where Φ_0 is the flux at $\theta = 0$, the equilibrium position of the coil. The torque equation may now be rewritten as

$$\tau = nI\Phi_0 \sin\theta$$

which is the basic equation of electric machines.

One can interpret (28.8) or (28.9) in terms of "lines of force" or "lines of flux," exactly as in the electrical case. However, it can be shown that for magnetic fields Gauss's law becomes

$$\Phi_{\text{closed surface}} = 0 \tag{28.11}$$

which means that magnetic lines of force do not terminate on some sort of "magnetic charge" but close on themselves.

Solved Problems

28.1. In Fig. 28-3, charge q is uniformly distributed over a small sphere having a mass of 5 grams. The sphere is moving normal to a uniform magnetic field. (a) Given $q = 3 \mu C$, $u = 4$ km/s, and $F = 2.4$ mN, find B. (b) Given $B = 0.5$ T, $q = 40 \mu C$, and $u = 60$ km/s, find F and the acceleration, a, of the sphere. (c) In the above the sphere is replaced by an electron having speed $u = 5 \times 10^6$ m/s. (The electronic charge and mass are $-e = -1.602 \times 10^{-19}$ C and $m = 9.11 \times 10^{-31}$ kg.) For $B = 0.1$ T, find F and acceleration.

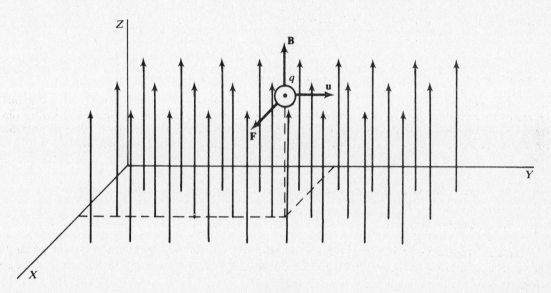

Fig. 28-3

(a)
$$2.4 \times 10^{-3} = (3 \times 10^{-6})(4 \times 10^3)B \qquad \text{or} \qquad B = 0.2 \text{ T}$$

(b)
$$F = F_x = (40 \times 10^{-6})(60 \times 10^3)(0.5) = 1.2 \text{ N} \qquad \text{and} \qquad a = a_x = \frac{1.2}{0.005} = 240 \text{ m/s}^2$$

(c)
$$F_x = (-1.602 \times 10^{-19})(5 \times 10^6)(0.1) = -8.01 \times 10^{-14} \text{ N}$$

$$a_x = \frac{-8.01 \times 10^{-14}}{9.11 \times 10^{-31}} = -8.79 \times 10^{16} \text{ m/s}^2$$

28.2. In Fig. 28-4, let $q = 40 \mu C$, $u = 5$ km/s, and $B = 2$ T. The direction of **B** is given by its direction cosines $\beta_1 = 0.5$, $\beta_2 = 0.6$, $\beta_3 = 0.6245$; the direction cosines of **u** are $\alpha_1 = 0.3$, $\alpha_2 = 0.4$, $\alpha_3 = 0.866$. Find (a) F_x, F_y, F_z, the rectangular components of **F**; (b) the magnitude and direction of **F**; (c) the angle θ.

(a) By (28.1), $F_x = q(u_y B_z - u_z B_y)$. But

$$u_y = u\alpha_2 = 2 \times 10^3 \qquad B_y = B\beta_2 = 1.2$$

$$u_z = u\alpha_3 = 4.33 \times 10^3 \qquad B_z = B\beta_3 = 1.249$$

Hence,
$$F_x = (4 \times 10^{-2})(2 \times 1.249 - 4.33 \times 1.2) = -0.108 \text{ N}$$

Likewise, $F_y = 0.09826$ N, $F_z = -0.008$ N.

(b)
$$F = (F_x^2 + F_y^2 + F_z^2)^{1/2} = 0.14623 \text{ N}$$

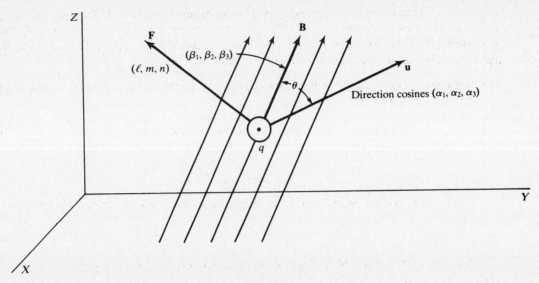

Fig. 28-4

The direction cosines of $\mathbf{F}$ are then:

$$\ell = \frac{F_x}{F} = -0.7386 \qquad m = \frac{F_y}{F} = 0.6720 \qquad n = \frac{F_z}{F} = -0.0547$$

(c) From $F = quB \sin \theta$,

$$\sin \theta = \frac{F}{quB} = 0.3656 \qquad \text{or} \qquad \theta = 21.44°$$

Alternatively, apply $\cos \theta = \alpha_1\beta_1 + \alpha_2\beta_2 + \alpha_3\beta_3$.

28.3. In Fig. 28-5, electrons from the heated filament f are accelerated to plate P_1 by voltage v_1. Some of them pass through a narrow slit S_1 and are decelerated from P_1 to plate P_2 by voltage v_2. Some of these pass through slit S_2 into the space above P_2, with a kinetic energy determined by $v_1 - v_2$. A series of slits in the metal baffles b_1, b_2, etc., defines a circle of

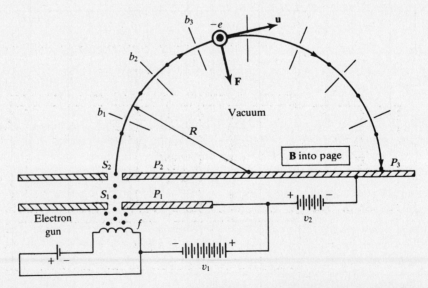

Fig. 28-5

radius R. Electrons emerging from S_2 enter a uniform magnetic field **B** normal to the paper, by which they are deflected in a circular path. For proper values of $v_1 - v_2$ and B they will pass through all slits and arrive at P_3. (*a*) For $v_1 = 600$ V, $v_2 = 200$ V, $R = 100$ mm, find the required value of B. (*b*) For $B = 0.4$ mT, what is R?

(*a*) Since the magnetic force is workless, an electron keeps the speed it had at S_2, which is given by energy conservation as

$$\frac{1}{2}mu^2 = e(v_1 - v_2) \qquad (28.12)$$

or

$$u = 1.186 \times 10^7 \text{ m/s}$$

The value of B must be such that the magnetic force constitutes the required centripetal force:

$$euB = \frac{mu^2}{R} \qquad \text{or} \qquad B = \frac{mu}{eR} \qquad (28.13)$$

Thus, $B = 0.6744$ mT. Note that B depends on e/m, not on e and m separately.

(*b*) From (*28.13*), $R = mu/eB$, and for $B = 4 \times 10^{-4}$ T (other quantities remaining the same), this gives $R = 168$ mm.

28.4. (*a*) A proton ($q = +e$, $m = 1.673 \times 10^{-27}$ kg) which has fallen through a difference in potential of 200 kV describes a circular path in a uniform magnetic field, $B = 0.6$ T. Compute R. (*b*) A doubly-ionized He atom (an α-particle: $q = 2e$, $m = 6.65 \times 10^{-27}$ kg), having fallen through a difference in potential of 100 kV in an apparatus similar to that of Fig. 28-5, describes a circular path of radius 250 mm. Compute B.

(*a*) From (*28.12*), the speed of the proton is 6.189×10^6 m/s. Then, from (*28.13*),

$$R = \frac{(1.673 \times 10^{-27})(6.189 \times 10^6)}{(1.602 \times 10^{-19})(0.6)} = 108 \text{ mm}$$

(*b*) The speed of the helium nucleus is found as $u = 3.104 \times 10^6$ m/s, and applying (*28.13*), $B = 0.258$ T.

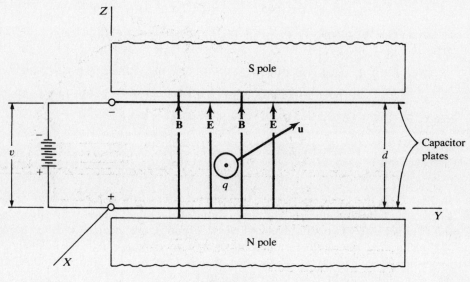

Fig. 28-6

28.5. A capacitor is placed between the poles of a large magnet, as shown in Fig. 28-6. In the space between the plates of the capacitor there is a uniform electric field **E** and also a uniform magnetic field **B**; the two fields are parallel and in the direction of Z. A small sphere, carrying charge $q = 3 \, \mu C$, has instantaneous velocity

$$\mathbf{u} = (4 \times 10^4)(0.766\mathbf{j} + 0.643\mathbf{k}) \text{ m/s}$$

If $d = 40$ mm, $v = 2000$ V, and $B = 1.5$ kT, what is the instantaneous force on the sphere?

Here,

$$\mathbf{E} = \frac{v}{d}\mathbf{k} = (5 \times 10^4)\mathbf{k} \text{ V/m}$$

and the Lorentz equation gives

$$\mathbf{F}_{tot} = (3 \times 10^{-6})[(5 \times 10^4)\mathbf{k} + (4 \times 10^4)(0.766\mathbf{j} + 0.643\mathbf{k}) \times (1.5\mathbf{k})]$$
$$= (3 \times 10^{-6})[(5 \times 10^4)\mathbf{k} + (4 \times 10^4)(0.766)(1.5)\mathbf{i}] = 0.1397\mathbf{i} + 0.15\mathbf{k} \quad \text{N}$$

which is a force in the XZ plane.

28.6. A charge $q = 40 \, \mu C$ moves with instantaneous velocity $\mathbf{u} = (5 \times 10^4)\mathbf{j}$ m/s through the uniform fields

$$\mathbf{E} = (6 \times 10^4)(0.52\mathbf{i} + 0.56\mathbf{j} + 0.645\mathbf{k}) \text{ V/m} \qquad \mathbf{B} = (1.7)(0.693\mathbf{i} + 0.6\mathbf{j} + 0.4\mathbf{k}) \text{ T}$$

Find the magnitude and direction of the instantaneous force on q.

$$\mathbf{F}_{tot} = (4 \times 10^{-5})[(6 \times 10^4)(0.52\mathbf{i} + 0.56\mathbf{j} + 0.645\mathbf{k}) + (5 \times 10^4)\mathbf{j} \times (0.693\mathbf{i} + 0.6\mathbf{j} + 0.4\mathbf{k})(1.7)]$$
$$= 2.61\mathbf{i} + 1.34\mathbf{j} - 0.81\mathbf{k} \quad \text{N}$$

From this, $F_{tot} = (2.61^2 + 1.34^2 + 0.81^2)^{1/2} = 3.04$ N. The direction cosines of $\mathbf{F}_{tot}$ are

$$\ell = \frac{2.61}{3.04} = 0.86 \qquad m = \frac{1.34}{3.04} = 0.44 \qquad n = \frac{-0.81}{3.04} = -0.27$$

28.7. As measured in an inertial coordinate system, a uniform magnetic field of 0.3 T exists parallel to the Z axis. A straight wire, of length 250 mm and having direction cosines

$$\ell = 0.45 \qquad m = 0.56 \qquad n = 0.6956$$

carries a constant current of 50 A. Find the magnitude and direction of the net force on the wire.

Write $\mathbf{B} = 0.3\mathbf{k}$ and $\mathbf{L} = (0.250)(0.45\mathbf{i} + 0.56\mathbf{j} + 0.6956\mathbf{k})$. Then, applying (28.5),

$$\mathbf{F} = 50[(0.250)(0.45\mathbf{i} + 0.56\mathbf{j} + 0.6956\mathbf{k}) \times (0.3\mathbf{k})] = 2.10\mathbf{i} - 1.688\mathbf{j} \quad \text{N}$$

Hence, $F = (2.1^2 + 1.688^2)^{1/2} = 2.694$ N; direction cosines of $\mathbf{F}$ are

$$\alpha_1 = \frac{2.1}{2.694} = 0.78 \qquad \alpha_2 = \frac{-1.688}{2.694} = -0.63 \qquad \alpha_3 = 0$$

28.8. In Fig. 28-7, the circle represents the face of the north pole of a large magnet. Near its surface the field **B** is uniform and normal out of the page; beyond the edge it is assumed that the field is zero. Two stiff insulated wires, a and b, are located close to the pole face in the XY plane; they are rigidly fastened together at an angle $\theta_b - \theta_a$. As indicated, a and b are in series and carry current I. For radius of pole face $R = 150$ mm, $I = 40$ A, $B = 0.4$ T, $\theta_a = 30°$, $\theta_b = 65°$, find the total force on the wires.

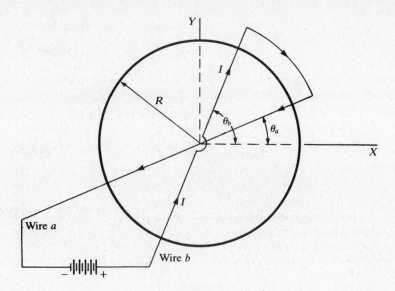

Fig. 28-7

The oriented length, in the direction of the current, of wire a is

$$\mathbf{L}_a = \mathbf{i}(-2R \cos \theta_a) + \mathbf{j}(-2R \sin \theta_a) = -0.26\mathbf{i} - 0.15\mathbf{j} \quad \text{m}$$

and so the force on wire a is

$$\mathbf{F}_a = I(\mathbf{L}_a \times \mathbf{B}) = 40 \begin{vmatrix} \mathbf{i} & \mathbf{j} & \mathbf{k} \\ -0.26 & -0.15 & 0 \\ 0 & 0 & 0.4 \end{vmatrix} = -2.4\mathbf{i} + 4.16\mathbf{j} \quad \text{N}$$

Similarly,

$$\mathbf{L}_b = \mathbf{i}(2R \cos \theta_b) + \mathbf{j}(2R \sin \theta_b) = 0.127\mathbf{i} + 0.272\mathbf{j} \quad \text{m}$$

$$\mathbf{F}_b = 40 \begin{vmatrix} \mathbf{i} & \mathbf{j} & \mathbf{k} \\ 0.127 & 0.272 & 0 \\ 0 & 0 & 0.4 \end{vmatrix} = 4.35\mathbf{i} - 2.03\mathbf{j} \quad \text{N}$$

Hence, the total force is $\mathbf{F} = \mathbf{F}_a + \mathbf{F}_b = 1.95\mathbf{i} + 2.13\mathbf{j}$ N, or $F = 28.9$ N.
    The problem might also have been solved by noting that

$$F_a = F_b = I(2R)B = 4.8 \text{ N}$$

and that the angle of $\mathbf{F}$ must be $\frac{1}{2}(\theta_a + \theta_b) = 47.5°$.

28.9. A parabolic section of wire, Fig. 28-8, is located in the XY plane and carries current $I = 12$ A. A uniform field, $B = 0.4$ T, making an angle of $60°$ with X, exists throughout the plane. Compute the total force on the wire between the origin and the point $x_1 = 0.25$ m, $y_1 = 1.00$ m.

    The equation of the wire is

$$\frac{y}{x^2} = \frac{y_1}{x_1^2} \qquad \text{or} \qquad y = 16x^2 \quad \text{(m)}$$

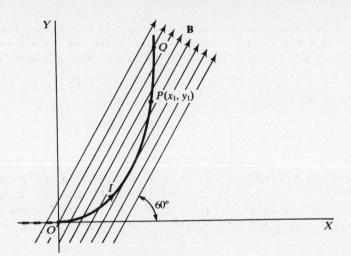

Fig. 28-8

Thus, along the wire, $dy = 32x\,dx$. By (28.4), the force on an element

$$d\mathbf{s} = \mathbf{i}\,dx + \mathbf{j}\,dy = \mathbf{i}\,dx + \mathbf{j}32x\,dx$$

of the wire is

$$d\mathbf{F} = 12 \begin{vmatrix} \mathbf{i} & \mathbf{j} & \mathbf{k} \\ dx & 32x\,dx & 0 \\ 0.4\cos 60° & 0.4\sin 60° & 0 \end{vmatrix} = \mathbf{k}(4.157 - 76.8x)\,dx \quad \text{(N)}$$

and

$$\mathbf{F} = \mathbf{k}\int_0^{0.25} (4.157 - 76.8x)\,dx = -1.36\mathbf{k}\ \text{N}$$

(For a method that avoids integration, see Problem 28.10.)

28.10. Referring to Fig. 28-9, the semicircular wire of radius R, supported at b_1 and b_2 and carrying current I, is in a uniform magnetic field parallel to Z. The angle between the YZ plane and the plane of the semicircle is arbitrary. Find the total force on the wire.

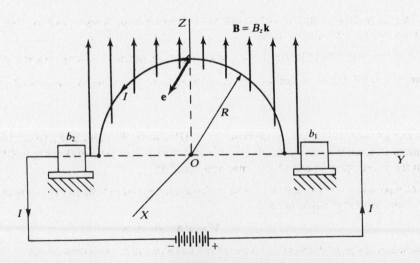

Fig. 28-9

Rather than integrate along the wire, as in Problem 28.9, we can exploit the fact (Example 28.1) that the net force on a current *loop* in a uniform magnetic field is zero. Hence we imagine current I also to flow along the Y axis from b_2 to b_1, completing the loop. The force on this straight segment would be

$$\mathbf{F}' = I(2R\mathbf{j} \times B_z\mathbf{k}) = 2IRB_z\mathbf{i}$$

The actual force on the semicircle must be such as to cancel $\mathbf{F}'$; i.e.,

$$\mathbf{F} = -\mathbf{F}' = -2iRB_z\mathbf{i}$$

28.11. A rectangular coil of n turns, Fig. 28-10, is located in the YZ plane; its area is ab and it carries current I in each turn. In the region there is a uniform magnetic field $\mathbf{B}$ having direction cosines β_1, β_2, β_3. Given that $a = 100$ mm, $b = 150$ mm, $n = 20$, $I = 12$ A, $\beta_1 = 0.49$, $\beta_2 = 0.56$, $\beta_3 = 0.668$, and $B = 0.4$ T, find the magnitude and direction of the torque on the coil.

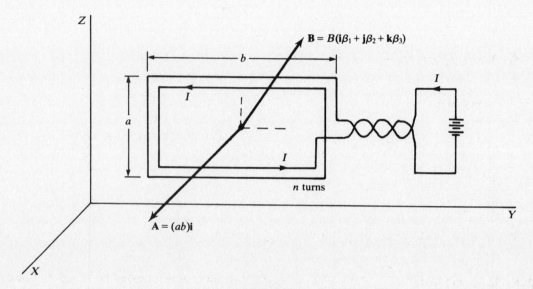

Fig. 28-10

We have $\mathbf{B} = (0.4)(0.49\mathbf{i} + 0.56\mathbf{j} + 0.668\mathbf{k})$ T, and in the given position, the coil has directed area $\mathbf{A} = (ab)\mathbf{i} = 0.015\mathbf{i}$ m^2. Then

$$\boldsymbol{\tau} = (20)(12)[0.015\mathbf{i} \times (0.4)(0.49\mathbf{i} + 0.56\mathbf{j} + 0.668\mathbf{k})] = -0.962\mathbf{j} + 0.806\mathbf{k} \quad \text{N} \cdot \text{m}$$

From this, $\tau = 1.255$ N $\cdot$ m, with direction cosines $\alpha_1 = 0$, $\alpha_2 = -0.766$, $\alpha_3 = 0.642$.

28.12. By analogy with the case of an electric dipole (Problems 23.7 and 23.8), define the *magnetic dipole moment* of a current-carrying planar coil and express the potential energy of the coil associated with its orientation in a uniform magnetic field $\mathbf{B}$.

Comparing $\boldsymbol{\tau} = \mathbf{P} \times \mathbf{E}$ for the electric dipole and $\boldsymbol{\tau} = nI(\mathbf{A} \times \mathbf{B})$ for the current-carrying coil, one defines the magnetic dipole moment as

$$\boldsymbol{\Pi} \equiv nI\mathbf{A} \quad (\text{A} \cdot \text{m}^2) \tag{28.14}$$

Then, reasoning as in Problem 23.8, one shows that the coil has potential energy

$$U_\theta = -\boldsymbol{\Pi} \cdot \mathbf{B} \quad (\text{J}) \tag{28.15}$$

28.13. A planar coil of 12 turns carries 15 A. The coil is oriented with respect to the uniform magnetic field

$$\mathbf{B} = 0.2\mathbf{i} + 0.3\mathbf{j} - 0.4\mathbf{k} \quad \text{T}$$

such that its directed area is $\mathbf{A} = 0.04\mathbf{i} = 0.05\mathbf{j} + 0.07\mathbf{k}$ m². Find (a) the dipole moment of the coil, (b) the potential energy of the coil in the given orientation, (c) the angle between the positive normal to the coil and the field, (d) the total magnetic flux "linking" (passing through) the coil.

(a) From (28.14),

$$\boldsymbol{\Pi} = (12)(15)(0.04\mathbf{i} - 0.05\mathbf{j} + 0.07\mathbf{k}) = 7.2\mathbf{i} - 9.0\mathbf{j} + 12.6\mathbf{k} \quad \text{A} \cdot \text{m}^2$$

(b) From (28.15),

$$U_\theta = -[(7.2)(0.2) + (-9.0)(0.3) + (12.6)(-0.4)] = +6.3 \text{ J}$$

(c) The desired angle θ is that between $\boldsymbol{\Pi}$ and $\mathbf{B}$. Since $U_\theta = -\Pi B \cos \theta$,

$$\cos \theta = -\frac{U_\theta}{\Pi B} = -\frac{6.3}{(7.2^2 + 9.0^2 + 12.6^2)^{1/2}(0.2^2 + 0.3^2 + 0.4^2)^{1/2}} = -0.6851$$

or $\theta = 133.24°$.

(d) The potential energy may be written as $U_\theta = -n I \mathbf{A} \cdot \mathbf{B} = -I(\mathbf{B} \cdot n\mathbf{A}) = -I\Phi$, where Φ is the total flux through the n turns, each of area A. Therefore,

$$\Phi = -\frac{U_\theta}{I} = -\frac{6.3}{15} = -0.42 \text{ Wb}$$

The negative value of Φ corresponds to the obtuse angle θ found in (c); that is, the flux is from the positive face of the coil to the negative face.

28.14. In Fig. 28-9, let β denote the angle between the YZ plane and the plane of the semicircle. What is the flux through the area bounded by the semicircular wire and the Y axis?

With the right-hand rule used to define the positive normal $\mathbf{e}$, the angle (in radians) between the uniform field and the given area is everywhere equal to $\beta + (\pi/2)$. Hence

$$\Phi = BA \cos\left(\beta + \frac{\pi}{2}\right) = B_z\left(\frac{\pi R^2}{2}\right)(-\sin \beta) = -\frac{B_z \pi R^2}{2}\sin \beta$$

Supplementary Problems

28.15. A charge $q = 300 \ \mu$C is moving with speed $u = 6$ km/s through a magnetic field $\mathbf{B}$; the angle between $\mathbf{u}$ and $\mathbf{B}$ is 32°. If the instantaneous force on the charge is 0.85 N, find the local value of B.
Ans. 0.89 T

28.16. In a long, air-core solenoid having a radius of 50 mm there is a uniform magnetic field, parallel to its axis, of 0.06 T. What is the flux through a cross-sectional area of the solenoid? *Ans.* 0.3927 mWb

28.17. Referring to Fig. 28-4, let $q = 500 \ \mu$C; $u_x = 5$ km/s, $u_y = -6$ km/s, $u_z = 8$ km/s; $B_x = -0.8$ T, $B_y = 0.4$ T, $B_z = -0.7$ T. Calculate the direction cosines of $\mathbf{F}$.
Ans. $\ell = 0.1340$, $m = -0.3887$, $n = 0.9115$

28.18. Refer to Problem 28.5 and Fig. 28-6. Suppose that the capacitor is rotated through an angle of 40° about a line parallel to Y, bringing the negative plate out of the page. All other quantities are unchanged. What is now the force on the sphere? *Ans.* $0.2343\mathbf{i} + 0.1149\mathbf{k}$ N

28.19. A long solenoid, Fig. 28-11, establishes a uniform field **B** parallel to its axis. An electron gun mounted inside fires a stream of electrons whose emergent velocity **u** has components $u_1 = u \cos \beta$ and $u_2 = u \sin \beta$, respectively normal and parallel to **B**. The electrons follow a helical path as indicated. Given $B = 1$ mT, $\beta = 30°$, $v_a = 600$ V, and $e/m = 1.76 \times 10^{11}$ C/kg for electrons, calculate the radius and the pitch of the helix.

Ans. $R = \dfrac{u \cos \beta}{(e/m)B} = \dfrac{\sqrt{2(m/e)v_a}}{B} \cos \beta = 71.5$ mm

$p = 2\pi R \tan \beta = 259.4$ mm

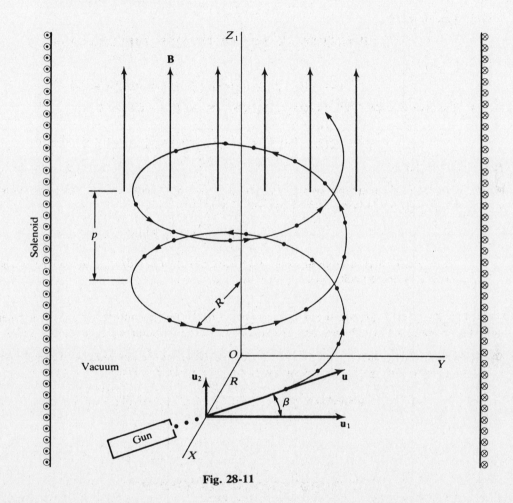

Fig. 28-11

28.20. A straight wire, 250 mm long and carrying a current $I = 50$ A, is in a uniform magnetic field **B** which makes an angle of 48° with the direction of I. The magnitude of the total force on the wire is 4.5 N. Find B.   *Ans.*   0.484 T

28.21. Refer to Problem 28.9 and Fig. 28-8. Without making any additional calculations, state the force on the section PQ of the wire.   *Ans.* $+1.36\mathbf{k}$ N

28.22. A triangular coil, Fig. 28-12, has 4 turns, carries a current of 6 A, and is in a uniform magnetic field

$$\mathbf{B} = 0.5\mathbf{i} + 0.45\mathbf{j} + 0.74\mathbf{k} \quad \text{T}$$

The vertices of the triangle lie on the axes at $x_1 = 0.20$ m, $y_1 = 0.15$ m, $z_1 = 0.25$ m. (a) Determine the vector area of the triangle. (*Hint*: The X component of **A** is the projected area of the triangle in the YZ plane; etc.) Find (b) the dipole moment of the coil, and (c) the potential energy of the coil.
Ans.   (a) $0.01875\mathbf{i} + 0.025\mathbf{j} + 0.015\mathbf{k}$ m²;   (b) $0.45\mathbf{i} + 0.6\mathbf{j} + 0.36\mathbf{k}$ A·m²;   (c) -0.7614 J

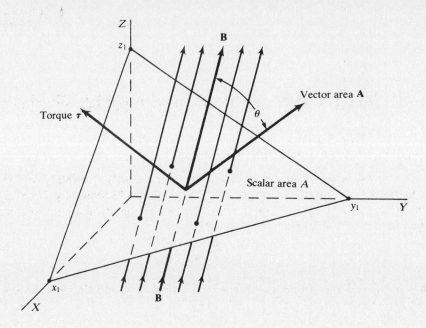

Fig. 28-12

28.23. (*a*) Verify the equivalences given in (*28.2*) and (*28.10*).   (*b*) Make a unit-dimensional check of (*28.13*).

Chapter 29

Sources of the Magnetic Field

29.1 MAGNETIC FIELD OF A MOVING CHARGE

Figure 29-1 shows a point charge q in motion with velocity $\mathbf{v}$ relative to inertial axes X, Y, Z. The instantaneous location of q with respect to the fixed observation point P is specified by the displacement vector $\mathbf{r}$ (or by $-\mathbf{r}$). Then at P there exists a magnetic field, whose instantaneous value as determined in the inertial frame is

$$\mathbf{B} = \frac{\mu_0 q}{4\pi r^3}(\mathbf{v} \times \mathbf{r}) \quad (\text{T}) \qquad (29.1)$$

The constant μ_0, called the *permeability of empty space*, is, in the SI, not an experimental, but a defined quantity (see Problem 29.6):

$$\mu_0 \equiv 4\pi \times 10^{-7} \text{ H/m}$$

Here, the *henry* (H) is the unit of inductance (see Chapter 31). It follows from (29.1) and (28.2) that

$$1 \text{ H} = 1 \frac{\text{N} \cdot \text{s}^2 \cdot \text{m}}{\text{C}^2} = 1 \frac{\text{Wb}}{\text{A}} \qquad (29.2)$$

In magnitude,

$$B = \frac{qv \sin \theta}{10r^2} \quad (\mu\text{T}) \qquad (29.3)$$

where θ is the angle between $\mathbf{v}$ and $\mathbf{r}$ (see Fig. 29-1). Observe that B is constant on circles centered on and perpendicular to the instantaneous line of motion of q, and that $B = 0$ along that line.

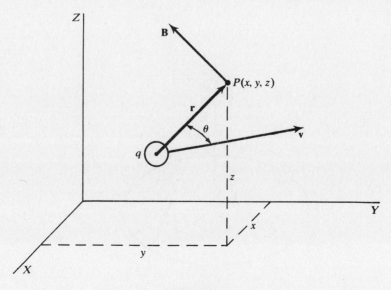

Fig. 29-1

255

29.2 MAGNETIC FIELD OF A CURRENT FILAMENT

When applied to an element $I\,d\mathbf{l}$ of a current filament, Fig. 29-2, (29.1) gives for the field produced at an external point P

$$d\mathbf{B} = \frac{\mu_0 I}{4\pi r^3}\,(d\mathbf{l} \times \mathbf{r}) \tag{29.4}$$

where $\mathbf{r}$ is the displacement from b, the site of the current element, to P. Relation (29.4) is the *Biot–Savart law*.

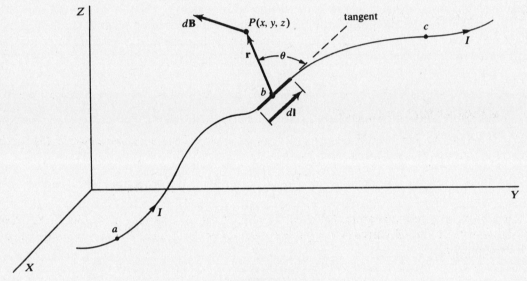

Fig. 29-2

The total field, $\mathbf{B}$, at P is obtained by integration of (29.4) over the entire current filament (e.g. over the entire length of a current-carrying wire). It is important to remember here that an actual current flows in a *closed* path, and a misleading result will be obtained if the integration is extended over only a part of the path. (See Problem 29.4.)

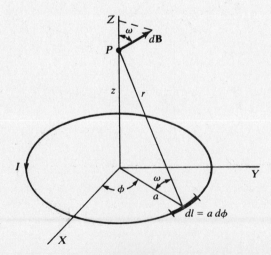

Fig. 29-3

EXAMPLE 29.1.   Let us calculate the field on the axis of a circular current loop of radius a.   Choosing axes as in Fig. 29-3, and noting that by symmetry the resultant field at P must be along Z, we have from (29.4):

$$dB_z = (dB)\cos\omega = \left(\frac{\mu_0 I}{4\pi r^3}\, dl\, r \sin 90°\right)\cos\omega$$

$$= \frac{\mu_0 I a^2}{4\pi r^3}\, d\phi = \frac{\mu_0 I a^2}{4\pi(a^2+z^2)^{3/2}}\, d\phi$$

Integration from $\phi = 0$ to $\phi = 2\pi$ then gives

$$B_z = B = \frac{\mu_0 I a^2}{2(a^2+z^2)^{3/2}} \tag{29.5}$$

In particular, at the center of the loop ($z = 0$),

$$B_{\text{center}} = \frac{\mu_0 I}{2a} \tag{29.6}$$

29.3 AMPERE'S CIRCUITAL LAW

Just as Coulomb's law, (23.3), can be expressed in integral form as Gauss's law, (24.3), so (29.4) can be rewritten as *Ampère's circuital law*:

$$\oint \mathbf{B} \cdot d\mathbf{s} = \mu_0 I \tag{29.7}$$

The line integral on the left is around an arbitrary closed path $\mathscr{C}$ (dashed line in Fig. 29-4).   On the right, I stands for the total current through any open surface S bounded by $\mathscr{C}$; i.e., in the case of a distributed current,

$$I = \int_S \mathbf{J} \cdot d\mathbf{S} \tag{29.8}$$

When $\mathscr{C}$ is planar, the following directional rule applies: If the fingers of the right hand are curled in the direction of integration along $\mathscr{C}$, the direction of positive I is indicated by the extended thumb.

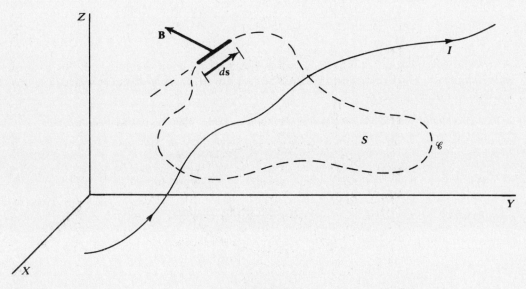

Fig. 29-4

Solved Problems

29.1. Referring to Fig. 29-1, let $q = 35\ \mu C$, $r = 50$ mm, $v = 2 \times 10^6$ m/s, and $\theta = 60°$. Find the magnitude of **B** at point P.

Applying (29.3),

$$B = \frac{(35 \times 10^{-6})(2 \times 10^6)(\sin 60°)}{10(50 \times 10^{-3})^2} = 2425\ \mu T$$

29.2. Again referring to Fig. 29-1, let the displacement from q to $P(x, y, z)$ be

$$\mathbf{r} = 50\mathbf{i} + 80\mathbf{j} + 70\mathbf{k} \quad mm$$

Let $q = 400\ \mu C$ and $\mathbf{v} = (3\mathbf{i} - 6\mathbf{j} + 9\mathbf{k}) \times 10^6$ m/s. Find B_x, B_y, B_z at P and also find angle θ.

The magnitude of **r** is $r = (50^2 + 80^2 + 70^2)^{1/2}(10^{-3}) = 0.1175$ m. Applying (29.1),

$$\mathbf{B} = \frac{10^{-7}(400 \times 10^{-6})}{(0.1175)^3}(10^6)(10^{-3})(3\mathbf{i} - 6\mathbf{j} + 9\mathbf{k}) \times (50\mathbf{i} + 80\mathbf{j} + 70\mathbf{k})$$

$$= (-2.813\mathbf{i} + 0.592\mathbf{j} + 1.332\mathbf{k})(10^{-2}) \quad T$$

whence

$$B_x = -28\,130\ \mu T \qquad B_y = 5920\ \mu T \qquad B_z = 13\,320\ \mu T$$

The magnitude of **B** is then

$$B = [(28\,130)^2 + (5920)^2 + (13\,320)^2]^{1/2} = 31\,682\ \mu T$$

and that of **v** is

$$v = (3^2 + 6^2 + 9^2)^{1/2}(10^6) = 11.225 \times 10^6 \text{ m/s}$$

Thus, from (29.3),

$$\sin \theta = \frac{10r^2B}{qv} = \frac{10(0.1175)^2(31\,682)}{(400 \times 10^{-6})(11.225 \times 10^6)} = 0.9737 \qquad \text{and} \qquad \theta = 76.8°$$

29.3. Make a unit-dimensional check of (29.6).

We have

$$B_{center} = \left|\frac{(H/m) \cdot A}{m}\right| = \left|H \cdot \frac{A}{m^2}\right| = \left|\frac{Wb}{A} \cdot \frac{A}{m^2}\right| = \left|\frac{Wb}{m^2}\right| = |T|$$

29.4. A straight wire, along Z in Fig. 29-5, carries a current I. Find the field at P due to section ab only (length, $\ell_1 + \ell_2$) (a) using the Biot–Savart law, (b) using Ampère's circuital law. (c) Explain the discrepancy.

(a) It is evident from (29.4) that $d\mathbf{B}$ at P due to any current element $I\,d\mathbf{l}$ has the direction $-\mathbf{i}$. Hence

$$d\mathbf{B} = \frac{\mu_0 I \sin \theta}{4\pi r^2} dz\ (-\mathbf{i})$$

But

$$r^2 = R^2 + z^2 \qquad \text{and} \qquad \sin \theta = \cos(\theta - 90°) = \frac{R}{r}$$

Thus,

$$d\mathbf{B} = \frac{\mu_0 IR}{4\pi(R^2 + z^2)^{3/2}} dz\ (-\mathbf{i})$$

and

$$\mathbf{B} = \frac{\mu_0 IR}{4\pi}(-\mathbf{i}) \int_{-\ell_2}^{+\ell_1} \frac{dz}{(R^2 + z^2)^{3/2}} = \frac{\mu_0 I}{4\pi R}\left[\frac{\ell_1}{(R^2 + \ell_1^2)^{1/2}} + \frac{\ell_2}{(R^2 + \ell_2^2)^{1/2}}\right](-\mathbf{i})$$

From the cylindrical symmetry of the problem it is evident that at all points of the circle $x^2 + y^2 = R^2$, **B** is tangential and of constant magnitude.

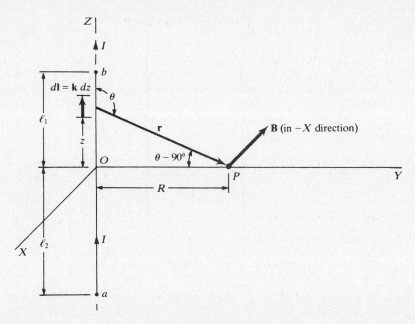

Fig. 29-5

(b) Application of (29.7) to the circle $x^2 + y^2 = R^2$, together with the symmetry considerations of (a), gives

$$B(2\pi R) = \mu_0 I \qquad \text{or} \qquad B = \frac{\mu_0 I}{2\pi R}$$

In particular, at point P,

$$\mathbf{B} = \frac{\mu_0 I}{2\pi R}(-\mathbf{i})$$

(c) Electric current cannot really appear abruptly at point a and disappear abruptly at point b; there must be a return path from b to a. If the integration in (a) were extended over the entire closed path, a different result for **B** would be obtained. Likewise, the problem no longer being cylindrically symmetric, a different result would also be found in (b); and these two new results would agree. For example, suppose that the path closes at infinity. Letting $\ell_1 \rightarrow \infty$ and $\ell_2 \rightarrow \infty$ in (a) and holding R fixed,

$$\frac{\ell_1}{(R^2 + \ell_1^2)^{1/2}} \rightarrow 1 \qquad \frac{\ell_2}{(R^2 + \ell_2^2)^{1/2}} \rightarrow 1$$

and so

$$\mathbf{B} \rightarrow \frac{\mu_0 I}{2\pi R}(-\mathbf{i})$$

This result agrees with that of (b), which, for an infinite straight wire, is correct as it stands.

29.5. Two very long, straight, parallel wires (Fig. 29-6) are separated by a distance d; they carry equal currents I in the same direction. Find the force per unit length between the wires. Is this force one of attraction or repulsion?

Consider the section $0 \leq y \leq 1$ m of wire 2. At each point of this section, wire 1 produces a magnetic field

$$\mathbf{B} = \frac{\mu_0 I}{2\pi d}(-\mathbf{k})$$

directed into the paper [see Problem 29.4(c)]. Then the force on the section is, by (28.5),

$$\mathbf{F} = I\left[1\mathbf{j} \times \frac{\mu_0 I}{2\pi d}(-\mathbf{k})\right] = \frac{\mu_0 I^2}{2\pi d}(-\mathbf{i}) \qquad (1)$$

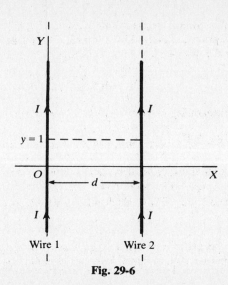

Fig. 29-6

By Newton's third law, the force per meter on wire 1 is the negative of the above. Thus the two wires attract each other with a force $\mu_0 I^2/2\pi d$ (N/m).

29.6. Explain why the permeability of empty space turns out to be exactly $4\pi \times 10^{-7}$ (in SI units).

Equation (*1*) of Problem 29.5 is used to define the ampere, as follows: When d is set at 1 m and F is measured to be 2×10^{-7} N/m (a convenient value), then $I = 1$ A, exactly. This forces the value $\mu_0 = 4\pi \times 10^{-7}$.

29.7. An air-core solenoid of length L and radius r, Fig. 29-7, is wound with $n' = n/L$ turns per unit length and carries current I. (*a*) Determine the magnetic field at the arbitrary axial point P. (*b*) What can be said about the field on axis when L becomes very large as compared to r?

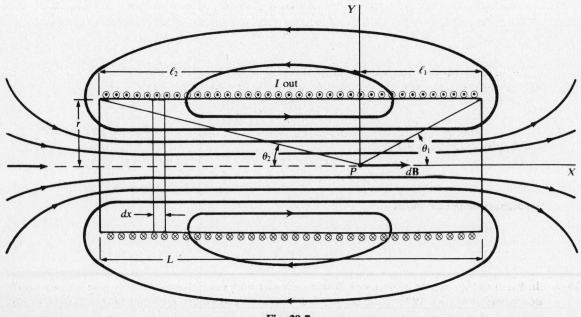

Fig. 29-7

(a) Take a longitudinal section of the solenoid as the XY plane, with P as the origin. Divide the solenoid into a succession of infinitesimal rings of width dx, each carrying current $n'I\,dx$. Then, by (29.5), a ring located a distance x from P contributes

$$d\mathbf{B} = \frac{\mu_0(n'I\,dx)r^2}{2(r^2 + x^2)^{3/2}}\,\mathbf{i}$$

to the field at P. Integrating from $x = -\ell_2$ to $x = \ell_1$, where $\ell_1 + \ell_2 = L$, we find

$$\mathbf{B} = \frac{\mu_0 n'I}{2}\left[\frac{\ell_1}{(r^2 + \ell_1^2)^{1/2}} + \frac{\ell_2}{(r^2 + \ell_2^2)^{1/2}}\right]\mathbf{i} = \frac{\mu_0 n'I}{2}(\cos\theta_1 + \cos\theta_2)\mathbf{i}$$

where θ_1 and θ_2 are the half-angles subtended at P by the two ends of the solenoid.

(b) Let $L \to \infty$ in such a way that both $\ell_1 \to \infty$ and $\ell_2 \to \infty$. Moreover, let $n \to \infty$ in such a way that $n/L = n'$ remains fixed. Then $\cos\theta_1 \to 1$, $\cos\theta_2 \to 1$, and

$$\mathbf{B} \to \mu_0 n'I\mathbf{i}$$

i.e. the field along the axis of the solenoid becomes uniform. As a matter of fact (see Problem 29.15), the field *everywhere* within the infinite solenoid has this same constant value.

29.8. Figure 29-8 shows the cross section of a long metal rod of radius $R = 30$ mm. The rod carries a current $I = 5$ kA directed out of the paper. Find B at point P ($r = 20$ mm) and at the surface of the rod, assuming that the permeability of the metal may be approximated as μ_0.

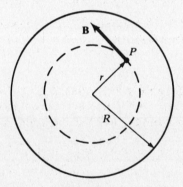

Fig. 29-8

Assuming uniform current density over the cross section, the lines of $\mathbf{B}$ will be circles concentric with the surface of the rod. The current through the dashed circle, of area πr^2, is

$$\frac{\pi r^2}{\pi R^2}I = \frac{r^2}{R^2}I$$

and Ampère's circuital law gives

$$B(2\pi r) = \mu_0 \frac{r^2}{R^2}I \qquad \text{or} \qquad B = \left(\frac{\mu_0 I}{2\pi R^2}\right)r$$

Substituting numerical values,

$$B_P = 0.0222 \text{ T} \qquad B_{\text{surface}} = \frac{3}{2}B_P = 0.0333 \text{ T}$$

29.9. In the Bohr hydrogen atom (see Chapter 40), the single electron orbits the nucleus in a circle of radius $a \approx 5.3 \times 10^{-11}$ m, making $f \approx 6.6 \times 10^{15}$ revolutions each second. Estimate the magnetic field at the nucleus.

The revolving electron is equivalent to a circular loop of current I, where

$$I = ef = (1.6 \times 10^{-19})(6.6 \times 10^{15}) = 1.06 \times 10^{-3} \text{ A}$$

Then, from (29.6),

$$B_{\text{nucleus}} = \frac{(4\pi \times 10^{-7})(1.06 \times 10^{-3})}{2(5.3 \times 10^{-11})} \approx 13 \text{ T}$$

Supplementary Problems

29.10. Point charges $q_1 = +800 \ \mu\text{C}$ and $q_2 = -500 \ \mu\text{C}$ have respective velocities $v_1 = 7 \times 10^6$ m/s and $v_2 = 4 \times 10^6$ m/s along the Y axis, as shown in Fig. 29-9. At the instant q_1 and q_2 are at points $y_1 = 0.30$ m, $y_2 = 0.45$ m, respectively, find the magnetic field established by the charges (*a*) at the origin O, (*b*) at the point $P(0, 0.15, 0.35)$.   *Ans.*   (*a*) $0 + 0 = 0$;   (*b*) $(3.55 - 0.7146)\mathbf{i}$ mT

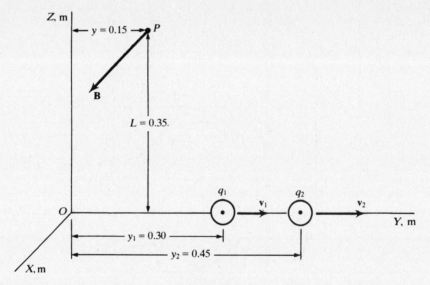

Fig. 29-9

29.11. The solenoid of Fig. 29-7 is wound with $n = 300$ turns; $L = 300$ mm; $r = 50$ mm; $I = 12$ A. Find B at the center of the solenoid.   *Ans.*   14.31 mT

29.12. Five very long, straight, insulated wires are closely bound together to form a small cable. Currents carried by the wires are $I_1 = 20$ A, $I_2 = -6$ A, $I_3 = 12$ A, $I_4 = -7$ A, $I_5 = 18$ A (negative currents are opposite in direction to the positive). Find B at a distance of 10 cm from the cable.   *Ans.*   74 μT

29.13. A long, straight, nonconducting string, painted with charge at a density of 40 μC/m, is pulled along its length at a speed of 300 m/s. What is the magnetic field at a normal distance of 5 mm from the moving string?   *Ans.* $B = 0.48 \ \mu$T

29.14. A nonconducting circular disk rotates about its axis, with rim speed v. If the disk carries uniform surface charge σ (C/m^2), find the magnetic field at the center. [*Hint:* Use (29.6) and integrate.]
Ans. $B_{\text{center}} = \mu_0 \sigma v / 2$

29.15. By applying Ampère's circuital law to an appropriate contour, show that **B** is uniform within a very long solenoid *of arbitrary cross section*.   *Ans.* $\mathbf{B} = \mu_0 n' I \mathbf{e}$, where **e** is the unit vector along the axis.

Chapter 30

Faraday's Law of Induced Electromotive Force

30.1 INDUCED EMF

Figure 30-1 shows the instantaneous configuration of a flexible conducting loop, referred to inertial axes X, Y, Z; the loop is translating, rotating, and perhaps changing shape, through a magnetic field $\mathbf{B}$. This field may vary in magnitude and direction from point to point and may also be changing with time.

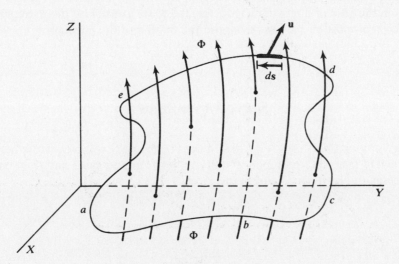

Fig. 30-1

As a joint result of the motion and the magnetic field every charge in the wire experiences a force. Specifically, a charge q in an element of length ds, which element, at the instant considered, has velocity $\mathbf{u}$, experiences a force

$$\mathbf{F} = q(\mathbf{E}_i + \mathbf{u} \times \mathbf{B}) \qquad (30.1)$$

In the *Lorentz equation*, (30.1), $\mathbf{E}_i$ represents an *induced electric field* which is present when, for example, $\mathbf{B}$ is changing with time. The instantaneous *induced electromotive force* (*emf*) in the loop is defined as the work which the (nonconservative) Lorentz force would do in taking a positive unit charge one time around the closed path. That is,

$$v_i = \frac{1}{q} \oint \mathbf{F} \cdot d\mathbf{s} = \oint (\mathbf{E}_i + \mathbf{u} \times \mathbf{B}) \cdot d\mathbf{s} \quad (\text{V}) \qquad (30.2)$$

Observe that any conservative force that might be included in $\mathbf{F}$ would integrate to zero, making no contribution to v_i. Hence we omitted from (30.1) any electrostatic field that might be present.

It was shown by the experiments of Faraday (and it follows from Maxwell's equations) that the right-hand side of (30.2) is equal to $-d\Phi/dt$, the time rate of change of the magnetic flux Φ threading the loop. Hence we write $v_i = -d\Phi/dt$, or for a loop consisting of N turns,

$$v_i = -N \frac{d\Phi}{dt} \quad \text{(V)} \tag{30.3}$$

Equation (30.3) is *Faraday's law*; it is valid regardless of the nature of the factor or factors responsible for the change in magnetic flux.

30.2 LENZ'S LAW

In the case of a planar loop, let the direction of positive flux and the positive direction around the loop be related by the usual right-hand rule. Then the minus sign in Faraday's law means that an *increasing* flux through the loop induces an emf in the *negative* direction around the loop. Frequently, and particularly for more complicated geometries, we use (30.3) only to find the magnitude of v_i, and separately infer the direction of v_i from *Lenz's law*, a consequence of the conservation of energy. Lenz's law states: (1) the current that would flow in response to the induced emf must produce a magnetic field the flux of which opposes the flux *change* that is inducing the emf; or (2) in the case of a moving conductor, the force exerted by the external magnetic field on the induced current must be such as to oppose the motion of the conductor.

Solved Problems

30.1. A rectangular loop of N turns, Fig. 30-2, moves to the right with velocity **u** between the poles of a permanent magnet. Assuming **B** uniform between pole faces and (for convenience) zero beyond, compute the induced emf in the loop.

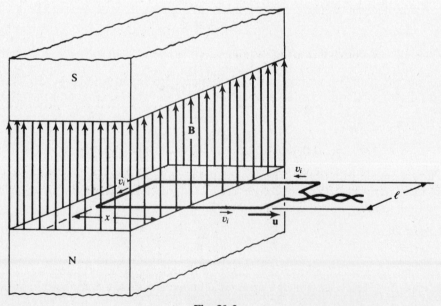

Fig. 30-2

The total flux linking the loop is $N\Phi = NB\ell x$. Noting that x is decreasing, we have for the instantaneous emf

$$v_i = -\frac{d}{dt}(N\Phi) = -NB\ell\dot{x} = +NB\ell u$$

as long as $x > 0$. The plus sign on v_i indicates that the emf is in the positive (counterclockwise) direction around the loop. This is confirmed by Lenz's law: a counterclockwise current gives rise to an upward-directed magnetic field within the loop, and hence to an increase in flux through the loop. This increase counters the decreasing flux from the magnet.

30.2. Suppose that the permanent magnet in Problem 30.1 is replaced by an electromagnet carrying an alternating current, which establishes a magnetic field $B = B_0 \sin \omega t$. Everything else remaining the same, derive an expression for the induced voltage.

Now the total flux is $N\Phi = N(B_0 \sin \omega t)\ell x$, so that the instantaneous emf is

$$v_i = -\frac{d}{dt}(N\Phi) = -NB_0\ell(\omega x \cos \omega t + \dot{x} \sin \omega t) = NB_0\ell(u \sin \omega t - \omega x \cos \omega t)$$

If the loop moves with *constant* velocity, then, in the above expression, $x = x_0 + ut$.

30.3. In Problem 30.2, let $B_0 = 0.1$ T, $\omega = 400$ rad/s, $\ell = 150$ mm, $N = 10$, $u = 5$ m/s = constant, $x_0 = 300$ mm. Evaluate v_i at times corresponding to (*a*) $\omega t = 0$, (*b*) $\omega t = \pi/4$, (*c*) $\omega t = \pi/2$, (*d*) $\omega t = 3\pi/4$.

(*a*) -18 V (clockwise); (*b*) -11.77 V; (*c*) $+0.75$ V; (*d*) $+12.01$ V.

30.4. A planar loop consisting of $N = 10$ turns of flexible wire is placed inside a long solenoid having n' turns per meter and carrying current $I = I_0 \sin \omega t$. The area A enclosed by the loop can be changed by, say, pulling the sides apart; however, the plane of the loop always remains normal to the axis of the solenoid. For $n' = 3000$ m^{-1}, $I = 12 \sin 150t$ (A), and assuming that

$$A = 0.15 \text{ m}^2 \qquad \frac{dA}{dt} = 2 \text{ m}^2/\text{s}$$

at $t = 0.005$ s, compute the emf induced in the loop at this instant.

From Problem 29.7(*b*),

$$B = \mu_0 n' I = \mu_0 n' I_0 \sin \omega t$$

within the solenoid, so that the flux linkage with the loop is

$$N\Phi = NBA = NA\mu_0 n' I_0 \sin \omega t$$

and

$$v_i = -\frac{d}{dt}(N\Phi) = -\mu_0 I_0 n' N\left(A\omega \cos \omega t + \frac{dA}{dt}\sin \omega t\right)$$

Substituting the data,

$$v_i = -(4\pi \times 10^{-7})(12)(3000)(10)[(0.15)(150)\cos(150)(0.005) + 2\sin(150)(0.005)] = -8.064 \text{ V}$$

Thus the induced emf is 8.064 V, directed as indicated in Fig. 30-3. This direction is in accordance with Lenz's law, since at the given instant both B and A are increasing, and so Φ is increasing.

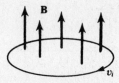

Fig. 30-3

30.5. Suppose that in Problem 30.4 the deformable loop is replaced by a rigid circular loop, of radius r, with its center on the axis of the solenoid. All other conditions remain the same. Find (a) the induced emf v_i in the loop, (b) the induced electric field E_i at any point on the loop.

(a) Setting

$$A = \pi r^2 \qquad \frac{dA}{dt} = 0$$

in (1) of Problem 30.4, we obtain

$$v_i = -\mu_0 I_0 n' N\pi r^2 \omega \cos \omega t$$

(b) Since the loop is stationary, $\mathbf{u} = 0$ at each point of it, and (30.2) becomes

$$v_i = \oint \mathbf{E}_i \cdot d\mathbf{s} = E_i(2\pi r)$$

since, by symmetry, $\mathbf{E}_i$ is constant in magnitude and tangential around the circle. Thus

$$E_i = \frac{v_i}{2\pi r} = \frac{-\mu_0 I_0 n' N r\omega}{2} \cos \omega t \qquad (1)$$

For certain values of t, E_i as given by (1) is negative. This means simply that, at those times, $\mathbf{E}_i$ (like v_i) is directed clockwise around the loop (see Fig. 30-3).

30.6. Make a unit-dimensional check of (1) of Problem 30.5.

Recall that $\omega = |s^{-1}|$, the radian being a dimensionless unit. Thus, $\cos \omega t$, along with N, is a pure number. Furthermore (Section 29.1),

$$\mu_0 = \left| \frac{N \cdot s^2}{C^2} \right|$$

Hence
$$E_i = \left| \frac{N \cdot s^2}{C^2} \right| \left| \frac{C}{s} \right| |m^{-1}| |m| |s^{-1}| = \left| \frac{N}{C} \right|$$

or $|V/m|$.

30.7. The shaded square in Fig. 30-4 represents one face of a large electromagnet. The magnet establishes a magnetic field, $B = B_0 \sin \omega t$, uniform over and normal to the pole faces; we assume that $B = 0$ beyond the edges. Wires ab, bc, and cd represent portions of a rectangular loop. They are attached to a rigid nonconducting framework, not shown. Metal rod ad can slide (parallel to bc) in contact with wires ab and dc. It is made to move such that $\ell_2 = \ell_0 \sin \alpha t$, where $0 < \ell_0 < \ell_1$. At the same time, the framework moves with constant

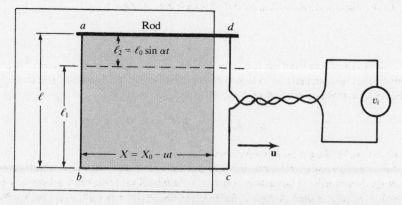

Fig. 30-4

velocity **u** as indicated.   (*a*) Derive an expression for the instantaneous emf, v_i, induced in the loop *abcda*.   (*b*) In which direction does current flow in the rod at the moment when *ab* passes the edge of the pole piece?

(*a*)   The instantaneous flux threading the loop is

$$\Phi = B\ell x = (B_0 \sin \omega t)(\ell_1 + \ell_0 \sin \alpha t)(x_0 - ut)$$

wherein B_0, ω, ℓ_1, ℓ_0, α, x_0, u are constants.   Hence

$$v_i = -\frac{d\Phi}{dt} = -B_0[(\omega \cos \omega t)(\ell_1 + \ell_0 \sin \alpha t)(x_0 - ut) + (\sin \omega t)(\ell_0 \alpha \cos \alpha t)(x_0 - ut)$$
$$- u(\sin \omega t)(\ell_1 + \ell_0 \sin \alpha t)]$$

(*b*)   When $x = 0$, i.e. when $t = \tau \equiv x_0/u$, we have from (*a*):

$$v_i(\tau) = u(B_0 \sin \omega \tau)(\ell_1 + \ell_0 \sin \alpha \tau) \equiv uB(\tau)\,\ell(\tau)$$

Since u and ℓ are positive, the sign of $v_i(\tau)$ will be that of $B(\tau)$, which is determined by its phase, $\omega\tau$.   Thus, if $B(\tau) > 0$ (i.e. if **B** is out of the page in Fig. 30-4 as the framework leaves the field), then the emf is counterclockwise around the loop, driving current from d to a in the rod.   If $B(\tau) < 0$, current flows from a to d.

These same conclusions also follow from Lenz's law, when the force on wire *ab* is computed.

30.8.   A straight wire of length L moves with constant velocity **U** (no rotation) through a uniform magnetic field **B**, Fig. 30-5.   Find the voltage difference between the ends of the wire.

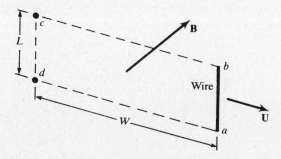

Fig. 30-5

Method 1.

In a certain time interval, the wire sweeps out the interior of a parallelogram *abcda*, where the fixed points c and d are the initial positions of the ends b and a.   Applying (*30.2*) counterclockwise around the contour, under the assumption that $E_i = 0$ everywhere, we have:

$$v_i = \oint (\mathbf{u} \times \mathbf{B}) \cdot d\mathbf{s} = \int_a^b (\mathbf{U} \times \mathbf{B}) \cdot d\mathbf{s}$$

where the second equality follows from the fact that $\mathbf{u} = 0$ along sides *bc*, *cd*, and *da*, while $\mathbf{u} = \mathbf{U}$ along *ab*.   Thus the entire induced emf appears as a voltage difference between a and b.   Since **U** and **B** are constants, we can evaluate the integral as follows:

$$v_i = v_{ba} = (\mathbf{U} \times \mathbf{B}) \int_a^b d\mathbf{s} = (\mathbf{U} \times \mathbf{B}) \cdot \mathbf{L} \tag{1}$$

Here, **L** is the vector length of the wire, from a to b.

For the directions of **U**, **B**, and **L** indicated in Fig. 30-5, v_{ba} is negative, which means that end a is at higher voltage than end b.   The induced current in the wire flows from b to a (check by Lenz's law); that is, it flows from low voltage to high voltage.   In the external circuit, it would (as usual) flow from high voltage to low voltage.   Thus the moving wire is equivalent to a battery with emf v_{ba}.

Method 2.

In Fig. 30-5, let $\mathbf{W}$ be the vector from d to a, so that

$$\frac{d\mathbf{W}}{dt} = \mathbf{U}$$

Then the directed area of the parallelogram may be written as $\mathbf{A} = \mathbf{W} \times \mathbf{L}$, and the instantaneous flux through the parallelogram is $\Phi = \mathbf{B} \cdot \mathbf{A} = \mathbf{B} \cdot (\mathbf{W} \times \mathbf{L})$.

Faraday's law now gives (remember that $\mathbf{B}$ and $\mathbf{L}$ are constants)

$$v_i = -\frac{d\Phi}{dt} = -\mathbf{B} \cdot \left(\frac{d\mathbf{W}}{dt} \times \mathbf{L}\right) = -\mathbf{B} \cdot (\mathbf{U} \times \mathbf{L}) = \mathbf{B} \cdot (\mathbf{L} \times \mathbf{U}) \qquad (2)$$

Formula (2) is equivalent to formula (1) above (see Problem 30.16). Observe that this method required no assumptions about $\mathbf{E}_i$.

30.9. In Fig. 30-6, the circular ring indicates a cross section of a long air-core solenoid, of radius $R = 0.10$ m. It has $n' = 2000$ turns per meter and carries an alternating current

$$I = I_0 \sin \omega t = 25 \sin 300t \quad (A)$$

The dashed line indicates a concentric circular wire of radius $r = 0.12$ m. Compute (a) the induced emf v_i in the wire and (b) the induced electric field $\mathbf{E}_i$ at any point in the wire. (c) If the circular wire were removed, would $\mathbf{E}_i$ remain unchanged at all points on the dashed circle?

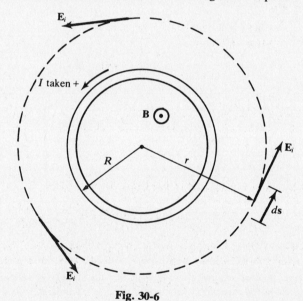

Fig. 30-6

(a) Outside the solenoid $\mathbf{B} = 0$, so that the flux linking the wire is the flux through the cross section of the solenoid: $\Phi_s = BA = (\mu_0 n' I)(\pi R^2) = \pi R^2 \mu_0 n' I_0 \sin \omega t$. Then

$$v_i = -\frac{d\Phi_s}{dt} = -\pi R^2 \mu_0 n' I_0 \omega \cos \omega t = -0.5922 \cos 300t \quad (V)$$

This result might have been obtained at once by setting $N = 1$ and $r = R$ in Problem 30.5(a).

(b) As in Problem 30.5(b),

$$E_i = \frac{v_i}{2\pi r} = -0.7854 \cos 300t \quad (V/m)$$

Here the minus sign implies that when I is positive and increasing, $\mathbf{E}_i$ is in the opposite direction to that indicated in Fig. 30-6.

(c) Yes.

30.10. Outside the solenoid of Problem 30.9 there is a wire PQ fixed in the XY plane, as shown in Fig. 30-7.   Throughout the entire region there exists a uniform magnetic field **B**, parallel to X.   The external source establishing this field (a large electromagnet, for example) has a uniform velocity **u** in the negative direction of Z.   That is, $\mathbf{B} = B_x\mathbf{i}$ and $\mathbf{u} = -u_z\mathbf{k}$.   Derive an expression for the emf in the wire.

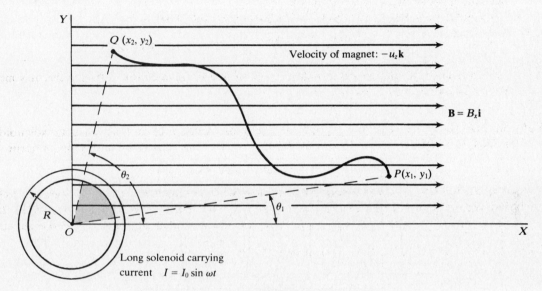

Fig. 30-7

Let us apply Faraday's law to the loop $OPQO$.   The only flux through the loop is that through the shaded subregion, due to the solenoid.   (The moving field has no Z component.)   Therefore

$$\Phi = \frac{\theta_2 - \theta_1}{2\pi}\,\Phi_s$$

where Φ_s is the solenoid flux calculated in Problem 30.9(a), and

$$v_i = v_{PO} + v_{QP} + v_{OQ} = -\frac{d\Phi}{dt} = -\frac{\theta_2 - \theta_1}{2\pi}\frac{d\Phi_s}{dt}. \qquad (1)$$

The straight segment OP is, in effect, moving through the uniform field **B** with velocity $+u_z\mathbf{k}$.   Thus we can calculate v_{PO} from (1) of Problem 30.8, with $\mathbf{U} = +u_z\mathbf{k}$, $\mathbf{B} = B_x\mathbf{i}$, and $\mathbf{L} = x_1\mathbf{i} + y_1\mathbf{j}$:

$$v_{PO} = (u_z\mathbf{k} \times B_x\mathbf{i}) \cdot (x_1\mathbf{i} + y_1\mathbf{j}) = u_zB_xy_1 \qquad (2)$$

Similarly,

$$v_{OQ} = (u_z\mathbf{k} \times B_x\mathbf{i}) \cdot (-x_2\mathbf{i} - y_2\mathbf{j}) = -u_zB_xy_2 \qquad (3)$$

Then, from (1), (2), and (3),

$$v_{QP} = -\frac{\theta_2 - \theta_1}{2\pi}\frac{d\Phi_s}{dt} + u_zB_x(y_2 - y_1)$$

$$= -\frac{1}{2}R^2\mu_0 n'I_0\omega(\theta_2 - \theta_1)\cos\omega t + u_zB_x(y_2 - y_1)$$

where $d\Phi_s/dt$ has been evaluated from Problem 30.9(a).

Supplementary Problems

30.11. A long air-core solenoid, C_1 in Fig. 30-8, has n' turns per meter, total resistance R_1, and radius r. Another coil, C_2, consisting of N turns of insulated wire wound around the solenoid, is connected to a resistor; let the total resistance of the C_2-circuit be R_2. A short time after switch S is closed, the current in C_1 has reached the steady value $I_1 = v/R_1$. However, while current is building up in C_1, an emf is induced in C_2 and charge flows through that circuit. Find the total charge Q which flows through C_2, and make a unit-dimensional check of your result. (*Hint*: Relate the magnitude of Q to the final value of Φ.)

Ans. $\quad Q = \dfrac{\mu_0 \pi r^2 N n' v}{R_1 R_2}$ (C)

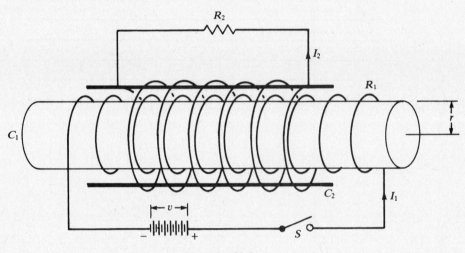

Fig. 30-8

30.12. Refer to Problem 30.9 and Fig. 30-6. (*a*) Find the magnitude of the induced electric field at any point on a circle of radius r ($r < R$) concentric with a cross section. (*b*) For the data of Problem 30.9, along with $r = 0.08$ m and $300t = 0.8\pi$, evaluate the result of (*a*).

Ans. (*a*) $E_i = -\dfrac{1}{2} r \mu_0 n' I_0 \omega \cos \omega t$; (*b*) 0.61 V/m

30.13. A rectangular loop, Fig. 30-9, has N turns, area $\ell_1 \ell_2$, and it rotates with constant angular velocity $\dot{\theta}$ in a magnetic field as shown. This field (established by an electromagnet, not shown) is uniform in space but varies with time according to $B_z = B_0 \sin \omega t$. (*a*) Write the induced voltage as a function of time, given that $\theta = 0$ at $t = 0$. (*b*) For $N = 100$, $B_0 = 0.2$ T, $\ell_1 \ell_2 = 0.04$ m^2, $\theta = 1800$ rev/min, and $\omega = 300$ rad/s, evaluate v_i at $t = (\pi/600)$ s.
Ans. (*a*) $v_i = N B_0 \ell_1 \ell_2 (\dot{\theta} \sin \omega t \sin \dot{\theta} t - \omega \cos \omega t \cos \dot{\theta} t)$; (*b*) 125.82 V

30.14. Problem 30.13 is changed as follows: B_z is constant and the rectangle undergoes angular SHM, $\theta = \theta_0 \sin \beta t$. (*a*) Determine v_i and (*b*) evaluate it at $t = (\pi/600)$ s, given $B_z = 0.2$ T, $\ell_1 \ell_2 = 0.04$ m^2, $N = 100$, $\theta_0 = (\pi/4)$ rad, and $\beta = 200$ rad/s.
Ans. (*a*) $v_i = +N B_z \ell_1 \ell_2 \theta_0 \beta [\sin(\theta_0 \sin \beta t)] \cos \beta t$; (*b*) 39.52 V

30.15. In Fig. 30-10, the large ring indicates a cross section of a long air-core solenoid having n' turns per unit length and carrying an alternating current $I = I_0 \sin \omega t$. (*a*) Find $\mathbf{E}_i$, the induced electric field, at a distance $r = (x^2 + y^2)^{1/2} < R$ from the center. (*b*) A point charge q moves in the XY plane with velocity $\mathbf{u} = u_x \mathbf{i} + u_y \mathbf{j}$, as indicated in Fig. 30-10. Find the force on the charge.
Ans. (*a*) $\mathbf{E}_i = -(\frac{1}{2}\mu_0 n' I_0 \omega \cos \omega t)(-y\mathbf{i} + x\mathbf{j})$
(*b*) $\mathbf{F} = \mu_0 n' I_0 q [(u_y \sin \omega t + \frac{1}{2} y \omega \cos \omega t)\mathbf{i} - (u_x \sin \omega t + \frac{1}{2} x \omega \cos \omega t)\mathbf{j}]$

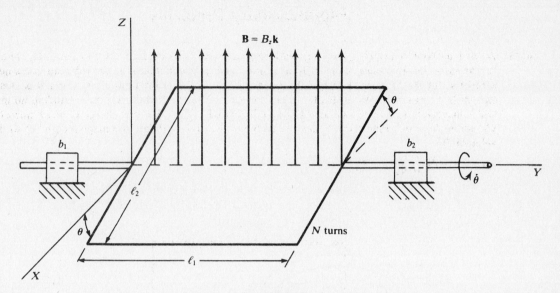

Fig. 30-9

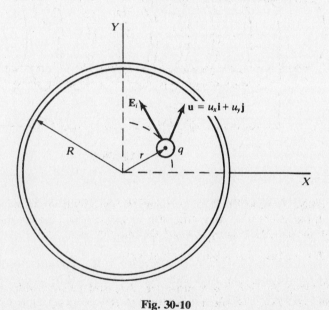

Fig. 30-10

30.16. Performing the calculations in terms of rectangular components, show that

$$(\mathbf{U} \times \mathbf{B}) \cdot \mathbf{L} = \mathbf{B} \cdot (\mathbf{L} \times \mathbf{U})$$

for arbitrary vectors **B**, **L**, **U**.

Chapter 31

Inductance

31.1 SELF-INDUCTANCE OF A COIL

In Fig. 31-1, current I establishes a magnetic field in and around coil C as indicated. Each line of flux threads (or "links") some or all of the turns. The total linkage is $N\Phi$, where N is the total number of turns and Φ is the *average* flux linking a turn.

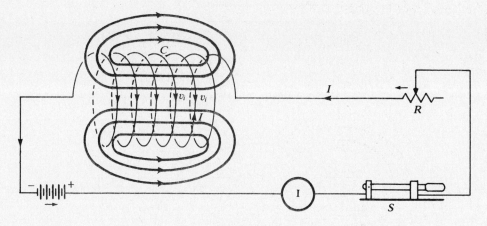

Fig. 31-1

If Φ changes (as by sliding the contact on R), an electromotive force

$$v_i = -N \frac{d\Phi}{dt}$$

is induced in the coil. For an increase in Φ (decrease in R), $d\Phi/dt$ is positive and v_i is negative; that is, in the direction opposite to that of I (see Section 30.2). For Φ decreasing (increasing R), v_i is positive; that is, in the direction of I.

Assuming C in empty space and no magnetic material nearby, the flux established, and thus the flux linkage, is directly proportional to I:

$$N\Phi = LI \qquad \text{or} \qquad L = \frac{N\Phi}{I} \qquad (31.1)$$

The proportionality factor L is referred to as the *self-inductance* (or just *inductance*). The SI unit of inductance, the henry, was defined in (29.2).

Differentiating (31.1) with respect to time gives for the (back-) emf in the inductor

$$v_i = -N \frac{d\Phi}{dt} = -L\frac{dI}{dt} \qquad (31.2)$$

Suppose that upon closing switch S in Fig. 31-1 the current in the circuit is built up from an initial value zero to a final value I. Then the work done (by the battery) in driving charge through C, against the induced emf, is

$$\mathscr{E} = \int (-v_i)\, dq = \int \left(L\frac{dI}{dt}\right)(I\, dt) = L \int_0^I I\, dI = \tfrac{1}{2} LI^2 \quad (\text{J}) \qquad (31.3)$$

This amount of energy is actually stored in the final magnetic field around the inductor; i.e., principally in the space enclosed by the turns of C.   (See Problem 31.3.)

31.2 MUTUAL INDUCTANCE OF TWO COILS

In Fig. 31-2, *part* of the flux established by I_1 in coil C_1 threads *some* of the turns of coil C_2.   The total linkage of flux from C_1 with C_2 is $N_2\Phi_{12}$, where Φ_{12} is the *average number* of lines from C_1 linking each of the N_2 turns.   Since the flux established in C_1 is directly proportional to I_1,

$$N_2\Phi_{12} = M_{12}I_1 \qquad \text{or} \qquad M_{12} = \frac{N_2\Phi_{12}}{I_1} \qquad (31.4)$$

where the proportionality factor, M_{12}, is constant, assuming that the coils are fixed relative to each other.   Similarly, interchanging the roles of C_1 and C_2,

$$N_1\Phi_{21} = M_{21}I_2 \qquad \text{or} \qquad M_{21} = \frac{N_1\Phi_{21}}{I_2} \qquad (31.5)$$

It can be shown that

$$M_{12} = M_{21} = M \qquad (31.6)$$

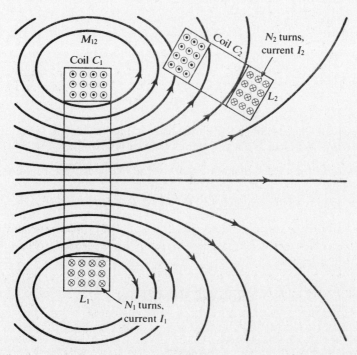

Fig. 31-2

Thus, the two coils have a single *mutual inductance*, M; like self-inductance, mutual inductance is measured in henries.

If the two currents I_1 and I_2 are changing with time, the induced emf in either coil will be made up of a part due to the change in its own current—as given by (31.2)—and a part due to the current change in the other coil—as given by Faraday's law and either (31.4) or (31.5).   Explicitly,

$$(v_i)_1 = -L_1\frac{dI_1}{dt} - M\frac{dI_2}{dt} \qquad (v_i)_2 = -L_2\frac{dI_2}{dt} - M\frac{dI_1}{dt} \qquad (31.7)$$

In the first equation (31.7), M has replaced M_{21}; in the second, M_{12}. Of course, if only one of the currents varies, there is only one term on the right in either equation (31.7).

EXAMPLE 31.1. In applying (31.7), attention must be given to the signs of the inductances. Under our convention that the direction of positive I is the direction of positive self-induced emf, self-inductance L is always positive. However, depending on the geometry, mutual inductance M may be positive or negative. Consider, for instance, the setup of Fig. 31-3, in which the positive current directions are as indicated. An *increase* in I_1 would, by the right-hand rule, produce an increase in flux (from left to right) through C_2, thereby inducing in C_2 a negative emf v_{12} (that is, v_{12} is opposite in direction to I_2). We are therefore compelled to take $M_{12} = M$ as *positive* in the relation

$$v_{12} = -M_{12}\frac{dI_1}{dt}$$
$$(-)\ \ (-)\ (+)\ (+)$$

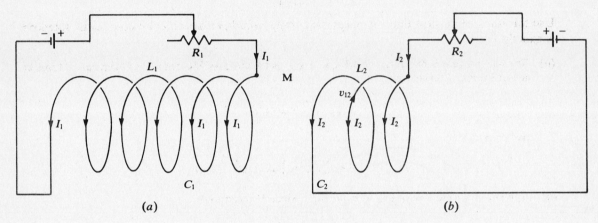

Fig. 31-3

But now suppose the C_2-circuit rotated through 180° about an axis normal to the plane of Fig. 31-3. Voltage v_{12} will now be in the *same* direction as I_2, and M must therefore be regarded as *negative*.

For the energy stored in the joint magnetic field of the two coils, we obtain from (31.7):

$$\mathscr{E}_{tot} = \int \left[-(v_i)_1 I_1 - (v_i)_2 I_2\right] dt = \frac{1}{2}\left(L_1 I_1^2 + L_2 I_2^2 + 2MI_1 I_2\right) \qquad (31.8)$$

Solved Problems

31.1. (a) In a certain coil a flux linkage of 3.5 "weber-turns" is established by a current of 10 A. Determine the inductance of the coil. (b) In (a), I is made to decrease at the rate of 100 A/s. What is the voltage of self-induction and what is its direction? (c) For a current of 15 A, how much energy is stored in the coil?

(a) It is given that $N\Phi = 3.5$ Wb in (31.1). Hence,

$$L = \frac{3.5\ \text{Wb}}{10\ \text{A}} = 0.35\ \text{H}$$

(b)
$$v_i = -(0.35)(-100) = +35\ \text{V}, \quad \text{in the direction of the current}$$

(c)
$$\mathscr{E} = \frac{1}{2}(0.35)(15)^2 = 39.375\ \text{J}$$

31.2. (a) In a certain coil the flux is changing at the rate $N\dot{\Phi} = 30$ Wb/s, while I increases at the rate of 60 A/s; compute L and v_i. (b) If instead $\dot{I} = -100$ A/s, what is v_i?

(a) By (31.2), $30 = L\,(60)$, or $L = 0.5$ H, and $v_i = -N\dot{\Phi} = -30$ V, opposite in direction to I.

(b) $$v_i = L\dot{I} = -(0.5)(-100) = 50 \text{ V}, \text{ in the direction of } I$$

31.3. A very long air-core solenoid is wound with n' turns per meter of wire carrying current I. (a) Calculate the self-inductance per meter, L', of the solenoid, and from it determine $\mathcal{E}'$, the work per meter required to establish current I. (b) In Maxwell's theory, the energy stored in a volume V in which a magnetic field exists is given by

$$\mathcal{E}_V = \frac{1}{2\mu_0} \int_V B^2 \, dv \qquad (31.9)$$

Use this integral to show that the magnetic energy stored per meter in the core of the solenoid is exactly $\mathcal{E}'$.

(a) The flux per turn is $\Phi = BA = \mu_0 n' IA$, where A is the cross-sectional area, and there are n' turns in a 1-m length. Hence, by (31.1),

$$L' = \frac{n'\Phi}{I} = \mu_0 n'^2 A \quad \text{(H/m)}$$

and, from (31.3),

$$\mathcal{E}' = \frac{1}{2} L' I^2 = \frac{1}{2}\mu_0 n'^2 A I^2 \quad \text{(J/m)}$$

(b) Applying (31.9) to 1 m of the core, we have, since B is constant,

$$\mathcal{E}_V = \frac{1}{2\mu_0} B^2 V = \frac{1}{2\mu_0}(\mu_0^2 n'^2 I^2)(A \times 1) = \frac{1}{2}\mu_0 n'^2 A I^2 = \mathcal{E}'$$

31.4. The air-core toroidal coil shown in Fig. 31-4 has a circular cross section and is uniformly wound with $N = 500$ turns of insulated wire. Its average radius is $R = 100$ mm, and the cross-sectional radius is $\rho = 20$ mm. Compute approximately the self-inductance of the coil.

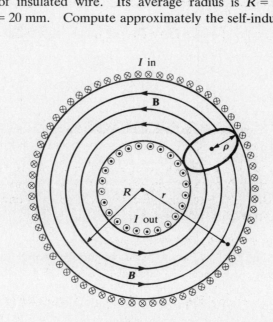

Fig. 31-4

The flux lines are concentric circles.   Applying Ampère's circuital law around a circle of radius r, we obtain

$$B(2\pi r) = \mu_0 NI \qquad \text{or} \qquad B = \frac{\mu_0 NI}{2\pi r}$$

Thus, B varies as $1/r$ over the cross section of the coil.   However, if R is large with respect to ρ (as it is for the values given), we may write

$$\frac{1}{r} \approx \frac{1}{R} = \text{constant}$$

over the cross section, so that

$$B \approx \mu_0 n'I = \text{constant} \qquad (n' \equiv N/2\pi R)$$

To this approximation, the total flux linkage is

$$N\Phi = NBA = (2\pi Rn')(\mu_0 n'I)(\pi \rho^2) = 2\pi^2 \mu_0 n'^2 R\rho^2 I$$

For the numerical data, $L = 0.6283$ mH.

31.5. Suppose that the toroid in Problem 31.4 has an iron core and that the iron has a permeability $\mu = 100\mu_0$ (assumed constant, which is never exactly the case).   Compute L, and the energy stored for $I = 15$ A.

Increasing the permeability by a factor of 100 increases the field, and consequently the inductance, by that same factor.   Thus, $L = 62.83$ mH and

$$\mathscr{E} = \frac{1}{2}LI^2 = \frac{1}{2}(62.83 \times 10^{-3})(15)^2 = 7.07 \text{ J}$$

31.6. (a) With a current of 10 A in C_1, Fig. 31-2, there is a flux linkage of 0.5 "weber-turns" (see Problem 31.1) with C_2.   What is the mutual inductance?   (b) With a current of 10 A in C_2, what is the linkage with C_1?   (c) If I_1 is increasing at the rate of 200 A/s, what electromotive force v_2 is induced in C_2?

(a)   From (31.4),

$$M = M_{12} = \frac{0.5}{10} = 0.05 \text{ H} = 50 \text{ mH}$$

(b)   From (31.5), $N_1\Phi_{21} = (0.05)(10) = 0.5$ Wb.   That is, for $I_1 = I_2$, the linkages are equal.

(c)

$$v_2 = -M\frac{dI_1}{dt} = -(0.05)(200) = -10 \text{ V}$$

31.7. Suppose that coils C_1 and C_2, Fig. 31-2, have self-inductances $L_1 = 200$ mH and $L_2 = 120$ mH, and mutual inductance $M = 50$ mH.   For $I_1 = 20$ A and $I_2 = 15$ A, compute the total energy of inductance.

By (31.8),

$$\mathscr{E}_{\text{tot}} = \frac{1}{2}[(0.200)(20)^2 + (0.120)(15)^2 + 2(0.050)(20)(15)] = 68.5 \text{ J}$$

31.8. As is seen from Fig. 31-3, I_1 and I_2 can be independently increased, decreased, or held constant by varying resistances R_1 and R_2.   Let $L_1 = 50$ mH, $L_2 = 40$ mH, $M = +15$ mH. (a) I_1 is made to increase at the rate of 120 A/s; I_2 is held constant.   Compute the voltage induced in each coil.   (b) I_1 is made to decrease at the rate of 120 A/s; I_2 is held constant. Compute v_1 and v_2.   (c) I_1 is increased at the rate of 120 A/s and I_2 is decreased at the rate of 200 A/s.   Compute v_1 and v_2.

Use (31.7).

(a)
$$v_1 = -L_1 \frac{dI_1}{dt} + 0 = -(0.050)(120) = -6 \text{ V}$$

$$v_2 = 0 - M \frac{dI_1}{dt} = -(0.015)(120) = -1.8 \text{ V}$$

(b)
$$v_1 = -L_1 \frac{dI_1}{dt} + 0 = -(0.050)(-120) = +6 \text{ V}$$

$$v_2 = 0 - M \frac{dI_1}{dt} = -(0.015)(-120) = +1.8 \text{ V}$$

(c)
$$v_1 = -L_1 \frac{dI_1}{dt} - M \frac{dI_2}{dt} = -(0.050)(120) - (0.015)(-200) = -3 \text{ V}$$

$$v_2 = -L_2 \frac{dI_2}{dt} - M \frac{dI_1}{dt} = -(0.040)(-200) - (0.015)(120) = +6.2 \text{ V}$$

Supplementary Problems

31.9. In the circuit of Fig. 31-1, I decreases at the rate of 64 A/s, producing an induced emf of +8 V.   (a) Find L.   (b) If $I = 12$ A, what is the energy of inductance?   *Ans.*   (a) 125 mH;   (b) 9 J

31.10. The coil of Fig. 31-1 has 100 turns and $L = 125$ mH.   (a) For $I = 10$ A, find the average flux linking a turn.   (b) For $I = 15$ A, find the energy stored.   *Ans.*   (a) 12.5 mWb;   (b) 14.06 J

31.11. In Fig. 31-1, the battery is replaced by an alternator, providing current

$$I = I_0 \sin \omega t = 25 \sin 400t \quad \text{(A)}$$

For $L = 60$ mH, determine the induced voltage.   *Ans.* $v_i = -600 \cos 400t$ (V)

31.12. In Problem 31.11, find an expression for the terminal voltage of the alternator.
Ans. $v_a = I_0(R^2 + L^2\omega^2)^{1/2} \sin(\omega t + \delta)$, where $\tan \delta = \omega L/R$ and where R includes the resistance of the coil.   (Check that R and ωL are dimensionally the same.)

31.13. Refer to Problem 31.8 and Fig. 31-3.   (a) For $I_1 = 25$ A and $I_2 = 15$ A, find the total magnetic energy stored.   (b) If C_2 is reversed in direction but located so that the magnitude of M remains the same, what is the new value of the stored energy?   *Ans.*   (a) 25.75 J;   (b) 14.5 J

31.14. Two coils, such as those in Fig. 31-2, have self-inductances L_1 and L_2 and mutual inductance M.   If connected in series, they are equivalent to a single coil.   What would be the self-inductance of that coil? *Ans.* $L_1 + 2M + L_2$

31.15. In Fig. 31-5, coil C_1 is a long solenoid, of radius r and having n'_1 turns per unit length.   Coil C_2 consists of N_2 turns near the mid-portion of C_1.   (a) Express the mutual inductance of the coils.   (b) Evaluate M, given $n'_1 = 3$ mm^{-1}, $N_2 = 500$, $r = 20$ mm.   *Ans.*   (a) $M = \mu_0 n'_1 N_2 \pi r^2$;   (b) 2.369 mH

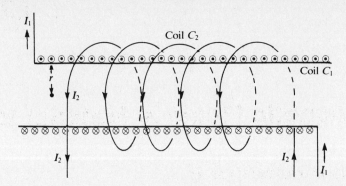

Fig. 31-5

31.16. Two coils are stated to have self-inductances of 5 mH and 18 mH, and a mutual inductance of 10 mH. Are these values consistent with one another? [*Hint*: (*31.9*) implies that total energy of induction is always positive.]   *Ans.*   Necessarily, $L_1 L_2 > M^2$; the stated values are inconsistent.

Chapter 32

Magnetic Fields in Material Media

32.1 THE THREE MAGNETIC VECTORS

Magnetic phenomena in empty space can be described in terms of a single vector quantity, the magnetic induction **B** (Chapters 28 through 31).   In a material medium, **B** is still the vector field that accounts for the force on a moving charge, via the definition (*28.1*).   Now, however, we introduce two new vectors, **H** and **M**, such that

$$\mathbf{B} = \mu_0\mathbf{H} + \mu_0\mathbf{M} \qquad\qquad (32.1)$$

H, called the *magnetic field strength* or the *magnetic intensity*, is defined by the Biot–Savart law or Ampère's law in the form

$$d\mathbf{H} = \frac{I}{4\pi r^3}(d\mathbf{l} \times \mathbf{r}) \qquad \text{or} \qquad \oint \mathbf{H} \cdot d\mathbf{s} = I$$

where I denotes the true current in the medium, i.e. the macroscopic flow of charge through the medium.   In vacuum, **H** is simply $\mathbf{B}/\mu_0$.   Thus **H** (or $\mu_0\mathbf{H}$) is the part of **B** that is independent of the microscopic properties of the medium.   The SI unit of **H** is the A/m.

M, called the *magnetization* or the *magnetic polarization* or the *magnetic dipole moment per unit volume*, is just what the third name implies: the sum of the magnetic dipole moments (Problem 28.12)—due to electronic spin, nuclear spin, and orbital motions of electrons—per unit volume of the material.   In empty space, $\mathbf{M} = 0$.   Thus **M** (or $\mu_0\mathbf{M}$) is the part of **B** that depends on the microscopic properties of the medium.   Values of **M** can be given in A/m or (see Problem 32.5) in $J/T \cdot m^3$.   We note the following equivalences:

$$1\,\frac{A}{m} = 1\,\frac{N}{Wb} = 1\,\frac{Wb}{H \cdot m} = 1\,\frac{J}{T \cdot m^3} \qquad\qquad (32.2)$$

32.2 MAGNETIC SUSCEPTIBILITY; PERMEABILITY

For an isotropic medium in which **B**, **H**, and **M** are all in the same direction, (*32.1*) may be written

$$B = \mu_0(H + M) = \mu_0(H + \chi_m H) = \mu H \qquad \text{where} \qquad \mu = \mu_0(1 + \chi_m) \qquad (32.3)$$

The dimensionless quantity χ_m is called the *magnetic susceptibility* of the medium; and μ, which like μ_0 carries the units H/m, is the *permeability* of the material.   The ratio $\mu_{\text{rel}} \equiv \mu/\mu_0$, a pure number, is called the *relative permeability*.

Materials are classified according to their susceptibilities, as follows:

Diamagnetic materials, for which χ_m is a very small *negative* constant.

Paramagnetic materials, for which χ_m is small and positive, and is inversely proportional to the absolute temperature.

Ferromagnetic materials, for which χ_m is positive and may be much greater than 1.   Moreover, χ_m depends in a complicated manner on H, so that M is not proportional to H in these materials.

32.3 MAGNETIC CIRCUITS

Consider a toroidal iron-core solenoid of such dimensions that **B** and **H** may be taken as uniform over a cross section. Ampère's law for **H** gives $H\ell = NI$, where ℓ is the mean circumferential length of the core, and N is the total number of turns in the winding. The flux in the core is given by

$$\Phi = BA = \mu HA$$

where A is the cross-sectional area. Thus the relation between winding current and core flux is

$$NI = \Phi \frac{\ell}{\mu A} \qquad \text{or} \qquad \text{mmf} = \Phi \mathcal{R} \qquad\qquad (32.4)$$

where we have titled NI the *magnetomotive force* (mmf) and defined $\ell/\mu A$ to be the *reluctance* $\mathcal{R}$ of the core. It is customary to specify mmf in "ampere-turns," which are, of course, the same as A; reluctance is measured in reciprocal henries (H^{-1}).

As (32.4) has the form of Ohm's law, with

$$\text{mmf} \leftrightarrow \text{emf} \qquad \text{flux} \leftrightarrow \text{current} \qquad \text{reluctance} \leftrightarrow \text{resistance}$$

we can view the core as the analog of a resistive electric circuit. However, for ferromagnetic substances, the analogy breaks down in one important respect: whereas $R = \ell/\sigma A$ (Section 26.2) is independent of emf and is known beforehand, $\mathcal{R} = \ell/\mu A$ depends, through μ, on the mmf. Consequently, (32.4) becomes a nonlinear equation in H, which must be solved by iteration or graphically.

32.4 ENERGY DENSITY

The general expression for the energy per unit volume in a magnetic field is

$$\varepsilon = \frac{1}{2} \mathbf{B} \cdot \mathbf{H} \quad (\text{J/m}^3) \qquad\qquad (32.5)$$

For a medium in which **B** and **H** are parallel, this becomes

$$\varepsilon = \frac{1}{2} BH = \frac{1}{2\mu} B^2 = \frac{\mu}{2} H^2 \quad (\text{J/m}^3) \qquad\qquad (32.6)$$

[compare (31.9)].

Solved Problems

32.1. A toroidal coil has $N = 1200$ turns; average length of core, $\ell = 80$ cm; cross-sectional area, $A = 60$ cm^2; current, $I = 1.5$ A. Compute B, H, total flux Φ, and ε. Assume an empty core.

It has already been seen (Problem 31.4) that **B** is approximately uniform over any cross section, of magnitude

$$B = \mu_0 n' I = (4\pi \times 10^{-7})\left(\frac{1200}{0.80}\right)(1.5) = 2.8274334 \text{ mT}$$

Then,

$$H = \frac{B}{\mu_0} = n'I = 2250 \text{ A/m} \qquad \Phi = BA = 16.964598 \ \mu\text{Wb} \qquad \varepsilon = \frac{1}{2} BH = 3.1808626 \text{ J/m}^3$$

32.2. Repeat Problem 32.1 for a bismuth core ($\chi_m = -2 \times 10^{-6}$).

H, which depends only on I, is the same as in the empty core; thus, $H = 2250$ A/m. The permeability of Bi is

$$\mu = \mu_0(1 + \chi_m) = (4\pi \times 10^{-7})(1 - 2 \times 10^{-6}) = 1.2566345 \times 10^{-6} \text{ H/m}$$

as compared to $\mu_0 = 1.256637 \times 10^{-6}$ H/m for empty space. Therefore,

$$B = \mu H = 2.8274276 \text{ mT} \qquad \Phi = BA = 16.964566 \ \mu\text{Wb} \qquad \varepsilon = \frac{1}{2} BH = 3.180856 \text{ J/m}^3$$

It is seen that the presence of the diamagnetic material brings about a very slight reduction in the values of B, Φ, and ε, as compared to the free-space values. For a typical paramagnetic material, e.g., iron ammonium alum ($\chi_m = 7 \times 10^{-4}$), the values would slightly exceed the free-space values.

32.3. The flux density in an air-core toroidal coil is B_1 for an mmf N_1I_1. In a coil of the same length wound on a toroidal ring of constant susceptibility χ_m, the flux density is B_2 for an mmf N_2I_2. Show that (a)

$$\frac{B_2}{B_1} = (1 + \chi_m) \frac{N_2I_2}{N_1I_1}$$

(b) the presence of the magnetic material in effect increases the coil mmf by $\chi_m N_1 I_1$.

(a)

$$\frac{B_2}{B_1} = \frac{\mu}{\mu_0} \frac{H_2}{H_1} = (1 + \chi_m) \frac{N_2I_2}{N_1I_1}$$

(b) For $N_2I_2 = N_1I_1$, part (a) gives $B_2 = (1 + \chi_m)B_1 = B_1 + \chi_m B_1$. The extra field $\chi_m B_1$ in the air core corresponds to an extra mmf of

$$\chi_m \times (\text{mmf to establish } B_1) = \chi_m N_1 I_1$$

32.4. The ring in Problem 32.3 is made of a good grade of iron for which, for $NI = 12\,000$ ampere-turns and a mean radius $R = 150$ mm, $B = 1.8$ T. Compute H, M (assumed constant), χ_m, and the equivalent ampere-turns contributed by the magnetic dipoles of the iron.

$$H = \frac{NI}{2\pi R} = \frac{12\,000}{2\pi(0.150)} = 12.732 \text{ kA/m}$$

From $B = \mu H$,

$$\mu = \frac{1.8}{12.732 \times 10^3} = 1.4137 \times 10^{-4} \text{ H/m}$$

Then, from $\mu = \mu_0(1 + \chi_m)$, $\chi_m = 111.5$ and

$$M = \chi_m H = (111.5)(12.732 \times 10^3) = 1420 \text{ kA/m}$$

By Problem 32.3(b), the number of ampere-turns contributed by the magnetic dipoles of the iron is

$$\chi_m(NI) = (111.5 \times 12\,000) = 1.338 \times 10^6$$

which is more than 100 times the mmf of the coil current. That is, to establish $B = 1.8$ T in an air core would require

$$12\,000 + 1.338 \times 10^6 = 1.35 \times 10^6$$

ampere-turns.

32.5. (a) Show that magnetic dipole moments can be given in J/T. (b) In the Bohr hydrogen atom (Chapter 40), the orbital angular momentum of the electron is quantized in units of $h/2\pi$, where $h = 6.626 \times 10^{-34}$ J · s is *Planck's constant*. Calculate, in J/T, the smallest allowed magnitude of the atomic dipole moment. (This quantity is known as the *Bohr magneton*.)

(a) Equation (28.15), $U_\theta = -\mathbf{\Pi} \cdot \mathbf{B}$, gives the unit J/T for $\mathbf{\Pi}$, since U_θ is in J and $\mathbf{B}$ is in T.   On the atomic or subatomic scale, dipole moments show up through their contributions to the atomic energy.

(b) The dipole moment, $\mathbf{\Pi}$, is directly proportional to the angular momentum, mvr, of the electron in its circular orbit; in fact,

$$\Pi = IA = \left(\frac{e}{2\pi r/v}\right)(\pi r^2) = \frac{e}{2m}(mvr)$$

Thus, the Bohr magneton is given by

$$\Pi_{min} = \frac{e}{2m}\left(\frac{h}{2\pi}\right) = \frac{1.602 \times 10^{-19}\ C}{2(9.109 \times 10^{-31}\ kg)}\left(\frac{6.626 \times 10^{-34}\ J \cdot s}{2\pi}\right)$$

$$= 9.27 \times 10^{-24}\ \frac{C \cdot J \cdot s}{kg} = 9.27 \times 10^{-24}\ J/T$$

32.6. The number of atoms per cubic meter of iron is about 8.5×10^{28}.   Assuming that each has a magnetic dipole moment of one Bohr magneton, and that for $NI = 3000$ ampere-turns on the ring of Problem 32.4 one-quarter of the dipoles are aligned with $\mathbf{H}$, compute M, χ_m, μ, B.

Taking the value of the Bohr magneton from Problem 32.5, we have:

$$M = \frac{1}{4}(8.5 \times 10^{28})(9.27 \times 10^{-24}) = 1.97 \times 10^5\ J/T \cdot m^3 = 197\ kA/m$$

$$H = \frac{3000}{2\pi(0.150)} = 3.1831\ kA/m$$

$$\chi_m = \frac{M}{H} = 61.9$$

$$\mu = \mu_0(1 + \chi_m) = 7.9 \times 10^{-5}\ H/m$$

$$B = \mu H = 0.25\ T$$

32.7. Show that the energy stored in a linear magnetic circuit may be written as

$$\mathscr{E} = \frac{1}{2}\Phi^2 \mathscr{R} \tag{32.7}$$

From (31.1) and (32.4),

$$\mathscr{E} = \frac{1}{2}LI^2 = \frac{1}{2}N\Phi I = \frac{1}{2}\Phi(\text{mmf}) = \frac{1}{2}\Phi^2 \mathscr{R}$$

Alternatively, assuming a constant energy density as given by (32.6),

$$\mathscr{E} = \varepsilon V = \left(\frac{1}{2\mu}B^2\right)(\ell A) = \frac{1}{2}(BA)^2\left(\frac{\ell}{\mu A}\right) = \frac{1}{2}\Phi^2 \mathscr{R}$$

Note that (32.7) is *not* the analog of $P = I^2R$ for a resistive electric circuit.   There is no energy loss associated with reluctance in a linear magnetic circuit.

32.8. Data regarding the magnetic circuit shown in Fig. 32-1 are: cross-sectional areas $A_1 = 1200\ mm^2$, $A_2 = 800\ mm^2$; lengths $\ell_1 = 210\ mm$, $\ell_2 = 430\ mm$, and length of air gap $\ell_a = 2\ mm$; relative permeability of lower leg of iron $(\mu_{rel})_1 = 200$, for remaining iron $(\mu_{rel})_2 = 300$; total turns $N = 1000$; $I = 2.5\ A$.   Compute (a) the mmf, (b) total reluctance of the circuit, (c) total flux threading the circuit, (d) flux density in each part.

(a) $\text{mmf} = NI = 2500$ ampere-turns

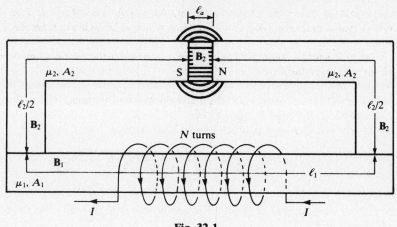

Fig. 32-1

(b) Reluctances in series or parallel are combined like electrical resistances in series or parallel. Thus,

$$\mathscr{R}_1 = \frac{\ell_1}{\mu_1 A_1} = \frac{210 \times 10^{-3}}{(200 \times 4\pi \times 10^{-7})(1200 \times 10^{-6})} = 0.6963 \ \mu\text{H}^{-1}$$

$$\mathscr{R}_2 = \frac{\ell_2}{\mu_2 A_2} = 1.426 \ \mu\text{H}^{-1}$$

$$\mathscr{R}_a = \frac{\ell_a}{\mu_0 A_2} = 1.989 \ \mu\text{H}^{-1}$$

$$\mathscr{R}_{\text{tot}} = \mathscr{R}_1 + \mathscr{R}_2 + \mathscr{R}_a = 4.11 \ \mu\text{H}^{-1}$$

where we have assumed the air gap to have effective area A_2 (no fringing). Observe that the reluctance of the air gap is almost equal to that of the entire iron path.

(c)
$$\Phi = \frac{\text{mmf}}{\mathscr{R}_{\text{tot}}} = \frac{2500}{4.11} = 608 \ \mu\text{Wb}$$

(d)
$$B_1 = \frac{\Phi}{A_1} = \frac{608 \times 10^{-6}}{1200 \times 10^{-6}} = 0.507 \ \text{T}$$

$$B_2 = B_a = \frac{\Phi}{A_2} = \frac{608 \times 10^{-6}}{800 \times 10^{-6}} = 0.76 \ \text{T}$$

Supplementary Problems

32.9. Verify the conversion of units

$$1 \ \frac{\text{C} \cdot \text{J} \cdot \text{s}}{\text{kg}} = 1 \ \frac{\text{J}}{\text{T}}$$

made in Problem 32.5(b).

32.10. Verify the following equivalences for the units of reluctance:

$$\mathscr{R} = \left| \frac{\text{A}^2}{\text{N} \cdot \text{m}} \right| = \left| \frac{\text{A}^2}{\text{J}} \right| = \left| \frac{\text{C}^2}{\text{kg} \cdot \text{m}^2} \right| = \left| \frac{\text{A}}{\text{Wb}} \right| = \left| \frac{1}{\text{H}} \right|$$

32.11. A wound iron ring has 1200 turns, a cross-sectional area of 24 cm^2 and an average radius $R = 16$ cm.   It is found by experiment that, for $I = 4$ A, the flux is $\Phi = 1.4$ mWb.   Assuming M directly proportional to H (that is, assuming μ constant, which it rarely is), determine (a) the magnetic flux density in the iron, (b) the magnetic field strength in the iron, (c) the permeability of the iron (assumed constant), (d) the magnetic susceptibility of the iron, (e) the magnetic dipole moment per unit volume, (f) the relative permeability of the iron, (g) the additional current that would be required to produce the same flux density in an air core, (h) the self-inductance of the ring, (i) the energy per unit volume in the magnetic field, (j) the total energy stored.

Ans.   (a) $B = 0.5833$ T   (e) $M = 459\,500$ A/m   (h) $L = 0.42$ H
　　　　(b) $H = 4774.65$ A/m   (f) $\mu_{rel} = 97.244$ (i) $\varepsilon = 1392.5$ J/m^3
　　　　(c) $\mu = 1.222 \times 10^{-4}$ H/m   (g) $\Delta I = 384.89$ A   (j) $\mathscr{E} = 3.36$ J
　　　　(d) $\chi_m = 96.244$

32.12. In Problem 32.8, what percent of the total magnetic energy is stored in the air gap?

Ans. $\mathscr{R}_a/\mathscr{R}_{tot} = 48.4\%$

Chapter 33

Time Response of Simple Electric Circuits

33.1 THE SERIES R-L-C CIRCUIT

The type of circuit to be treated here is illustrated in Fig. 33-1.

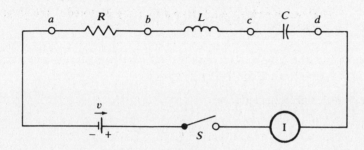

Fig. 33-1

When switch S is closed, the source (a battery) begins to deliver charge q, thus establishing a current I in the circuit. Given the values of R, L, C and emf v (assumed constant), we have to find expressions for q and I as functions of time. A more complete analysis also includes a determination of voltages v_{ab}, v_{bc}, v_{cd}; energies stored in L and C; and the power delivered by the source—all as functions of time.

Applying Kirchhoff's voltage relation to the circuit, we have

$$v = v_{ab} + v_{bc} + v_{cd} = RI + L\frac{dI}{dt} + \frac{q}{C}$$

where the three terms on the right have been respectively obtained from (26.5), (31.2), and (25.9). Writing $\dot{q}$ for I, we put this equation in the form

$$L\ddot{q} + R\dot{q} + \frac{1}{C}q = v \tag{33.1}$$

A solution of the linear, second-order differential equation (33.1), subject to appropriate initial conditions, gives q as a function of t, and from this all other results can be obtained.

The solution of (33.1) will have the form

$$q(t) = c_1 q_1(t) + c_2 q_2(t) + Q(t) \tag{33.2}$$

in which $q_1(t)$ and $q_2(t)$ are independent solutions of (33.1) with v replaced by zero. The combination $c_1 q_1(t) + c_2 q_2(t)$ is called the *transient solution*; the values of the constants c_1 and c_2 are determined from the initial conditions. As time becomes large, the transient solution goes to zero (except in the case $R = 0$), leaving the *steady-state solution*, $Q(t) = Cv$, which represents the ultimate charge on the capacitor.

The governing differential equation for a two- or one-element circuit is obtained from (33.1) by simply omitting the term(s) corresponding to the absent element(s). In particular, for an R-L circuit (no capacitance), $Q(t) = (v/R)t$ in (33.2). This corresponds to a steady-state current $\dot{Q} = v/R$, i.e. Ohm's law.

33.2 ELECTROMECHANICAL ANALOGIES

An electric circuit and a mechanical system may be identical in the sense that the governing differential equations for the two have exactly the same mathematical form. This allows the time behavior of the one to be inferred from that of the other.

Several mechanical systems analogous to simple R-L-C circuits are considered in the Solved Problems.

Solved Problems

33.1. Give a full analysis of the simple R-C circuit (Fig. 33-1, with L absent). Assume v constant, and $q = 0$ at $t = 0$ (the instant the switch is closed).

For the first-order differential equation

$$R\dot{q} + \frac{1}{C}q = v \tag{1}$$

the transient function is $q_1(t) = e^{-t/RC}$, as may be verified by substitution in

$$R\dot{q}_1 + \frac{1}{C}q_1 = 0$$

Hence, $q(t) = c_1 e^{-t/RC} + Cv$. The initial condition gives

$$0 = c_1 + Cv \qquad \text{or} \qquad c_1 = -Cv$$

Thus, finally,

$$q(t) = Cv(1 - e^{-t/RC}) \tag{2}$$

The remaining results follow from (2):

$$\text{current:} \quad I = \dot{q} = (v/R)e^{-t/RC}$$

$$\text{voltage across resistor:} \quad v_{ab} = RI = ve^{-t/RC}$$

$$\text{voltage across capacitor:} \quad v_{ab} = q/C = v(1 - e^{-t/RC})$$

$$\text{power from battery:} \quad P = Iv = (v^2/R)e^{-t/RC}$$

$$\text{energy in capacitor:} \quad \mathscr{E}_C = \int v_{cd}\,dq = \frac{1}{C}\int q\,dq = \frac{q^2}{2C} = \frac{1}{2}Cv^2(1 - e^{-t/RC})^2$$

At $t \to \infty$, the various quantities approach their steady-state values at a rate which depends on the size of RC; the smaller RC, the faster the approach. We have:

$$q \to Cv \qquad I \to 0 \qquad v_{ab} \to 0 \qquad v_{cd} \to v \qquad P \to 0 \qquad \mathscr{E}_C \to \frac{1}{2}Cv^2$$

33.2. The parameter RC in (2) of Problem 33.1 is called the *time constant* of the R-C circuit. (*a*) The time constant must have the dimensions of time, to make the exponent in (2) a pure number. Verify that this is so by means of a unit-dimensional analysis. (*b*) What should be the size of the time constant if the current in the circuit is to decrease by 50% in the first millisecond?

(*a*)

$$RC = |\Omega|\,|\text{F}| = \left|\frac{\text{V}}{\text{A}}\right|\left|\frac{\text{C}}{\text{V}}\right| = \left|\frac{\text{C}}{\text{A}}\right| = \left|\frac{\text{A}\cdot\text{s}}{\text{A}}\right| = |\text{s}|$$

(b) From $I = I_0 e^{-t/RC}$, we obtain, with time measured in ms,

$$\frac{1}{2} = e^{-1/RC} \qquad \text{or} \qquad RC = \frac{1}{\ln 2} = 1.44 \text{ ms}$$

This could be attained with, say, $R = 144 \ \Omega$, $C = 10 \ \mu\text{F}$.

33.3. Give a full analysis of the simple R-L circuit (Fig. 33-1, with C absent). Assume v constant, and $I = 0$ at $t = 0$ (the instant the switch is closed).

From (33.1),

$$L\ddot{q} + R\dot{q} = v \qquad \text{or} \qquad L\dot{I} + RI = v$$

As this equation is formally identical to (1) of Problem 33.1, and as the initial conditions correspond, the solution is yielded by (2) of Problem 33.1 as

$$I(t) = \frac{v}{R}(1 - e^{-Rt/L})$$

From this, and the further condition that $q = 0$ at $t = 0$,

$$\text{charge:} \qquad q = \int_0^t I \, dt = \frac{v}{R} t - \frac{vL}{R^2}(1 - e^{-Rt/L})$$

$$\text{voltage across resistor:} \qquad v_{ab} = RI = v(1 - e^{-Rt/L})$$

$$\text{voltage across inductor:} \qquad v_{bc} = v - v_{ab} = v e^{-Rt/L}$$

$$\text{power from battery:} \qquad P = Iv = \frac{v^2}{R}(1 - e^{-Rt/L})$$

$$\text{energy in inductor:} \qquad \mathcal{E}_L = \frac{1}{2} L I^2 = \frac{Lv^2}{2R^2}(1 - e^{-Rt/L})^2$$

The time constant of the R-L circuit is L/R [see Problem 33.8(a)].

33.4. Give a full analysis of the simple L-C circuit (Fig. 33-1, with R absent). Assume v constant, $q = q_0$ at $t = 0$ (C is charged before being connected in the circuit), and $I = 0$ at $t = 0$.

When the circuit has zero resistance (an idealized case), its behavior is essentially different from the exponential decay found in Problems 33.1 and 33.3. The governing differential equation, (33.1), is now

$$L\ddot{q} + \frac{1}{C} q = v \tag{1}$$

Setting $v = 0$ in (1), we find that the "transient" functions are

$$q_1(t) = \sin \omega t \qquad q_2(t) = \cos \omega t$$

where $\omega \equiv (LC)^{-1/2}$. These functions, and their derivatives, do not die out as time increases; rather they represent undamped oscillations (SHM). The solution of (1) is then of the form

$$q(t) = c_1 \sin \omega t + c_2 \cos \omega t + Cv$$

The initial conditions require

$$q_0 = c_2 + Cv \qquad \text{and} \qquad 0 = \omega c_1$$

whence $c_1 = 0$, $c_2 = q_0 - Cv$, and

$$q(t) = (q_0 - Cv) \cos \omega t + Cv \tag{2}$$

From (2),

$$\text{current:} \qquad I = \dot{q} = -\omega(q_0 - Cv) \sin \omega t$$

voltage across capacitor: $v_{cd} = \dfrac{q}{C} = \left(\dfrac{q_0}{C} - v\right)\cos \omega t + v$

voltage across inductor: $v_{bc} = v - v_{cd} = -\left(\dfrac{q_0}{C} - v\right)\cos \omega t$

power from battery: $P = Iv = -\omega v(q_0 - Cv)\sin \omega t$

energy in capacitor: $\mathscr{E}_C = \dfrac{q^2}{2C} = \dfrac{1}{2C}[(q_0 - Cv)\cos \omega t + Cv]^2$

Note that I is an alternating current, just as if it were the (steady-state) current produced by an ac generator running at angular velocity ω.

33.5. Briefly analyze the R-L-C circuit of Fig. 33-1 (v constant). Assume the same initial conditions as in Problem 33.4, and suppose the circuit parameters to be such that $R^2 < 4L/C$.

Depending on the relative values of R, L, and C, the transient response of the R-L-C circuit can range between the extremes of exponential decay and undamped oscillation. The relationship assumed above will be seen to lead to damped oscillations (the so-called *underdamped case*).

The circuit equation

$$L\ddot{q} + R\dot{q} + \frac{1}{C}q = v \tag{1}$$

will have transient solutions of the form

$$q_1(t) = e^{-\alpha t}\sin \omega t$$

$$q_2(t) = e^{-\alpha t}\cos \omega t$$

Substituting in (1) (with $v = 0$), we evaluate the positive real constants α and ω as

$$\alpha = \frac{R}{2L} \qquad \omega = \left(\frac{1}{LC} - \frac{R^2}{4L^2}\right)^{1/2}$$

Then $q(t) = c_1 q_1(t) + c_2 q_2(t) + Cv$, which, after c_1 and c_2 are evaluated from the initial conditions, becomes

$$q = \frac{q_0 - Cv}{\omega}e^{-\alpha t}(\alpha \sin \omega t + \omega \cos \omega t) + Cv \tag{2}$$

From (2), the current is found as

$$I = \dot{q} = \frac{q_0 - Cv}{\omega}(\alpha^2 + \omega^2)e^{-\alpha t}\sin \omega t = \frac{v_L(0)}{\omega L}e^{-Rt/2L}\sin \omega t \tag{3}$$

where $v_L(0)$ is the voltage across the inductor the instant after the switch is closed. The current is an exponentially-damped sinusoid; see Fig. 33-2.

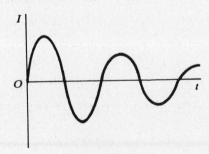

Fig. 33-2

33.6. Show that systems (a), (b). (c) of Fig. 33-3 are equivalent.

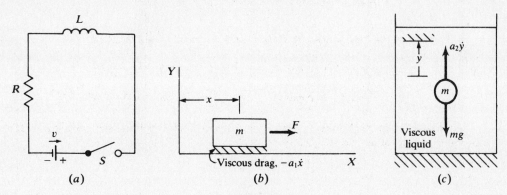

Fig. 33-3

System (a) is just the $R\text{-}L$ circuit treated in Problem 33.3. In (b) the block, of mass m, moves along a horizontal line under a constant applied force, F, and a viscous drag, $-a_1\dot{x}$. In (c) a sphere falls through a viscous fluid under the force of gravity, mg, and the viscous drag, $-a_2\dot{y}$. (The coefficients a_1 and a_2 are positive, and buoyant force on the sphere has been neglected.)

The differential equations for the three systems are

$$(a) \quad L\ddot{q} + R\dot{q} = v \qquad (b) \quad m\ddot{x} + a_1\dot{x} = F \qquad (c) \quad m\ddot{y} + a_2\dot{y} = mg$$

all of which have the same mathematical form. Hence, the solution to (b) that satisfies the initial conditions $x = \dot{x} = 0$ at $t = 0$ may be found by substitution of x, F, a_1, m for q, v, R, L, respectively, in the expression for the charge obtained in Problem 33.3:

$$x = \frac{F}{a_1} t - \frac{Fm}{a_1^2} (1 - e^{-a_1 t/m})$$

Likewise, the kinetic energy of the block at time t is

$$\mathscr{E}_m = \frac{1}{2} m\dot{x}^2 = \frac{mF^2}{2a_1^2} (1 - e^{-a_1 t/m})^2$$

In the same way—substituting y, mg, a_2, m for q, v, R, L—we solve (b) under the initial conditions $y = \dot{y} = 0$ at $t = 0$:

$$y = \frac{mg}{a_2} t - \frac{m^2 g}{a_2^2} (1 - e^{-a_2 t/m})$$

33.7. Find criteria for the underdamped motion of the mechanical systems shown in Fig. 33-4.

In system (a), the sphere, of mass m, experiences the downward force of gravity, mg; the upward spring force, $-k(x - \ell_0)$, where ℓ_0 is the natural length of the spring; and a viscous drag, $-a_1\dot{x}$ $(a_1 > 0)$. Its equation of motion is therefore

$$m\ddot{x} = mg - k(x - \ell_0) - a_1\dot{x}$$

or

$$(a) \quad m\ddot{x} + a_1\dot{x} + kx = mg + k\ell_0$$

In system (b), the disk is acted on by three torques: $\tau_1 = fr$, where force f is applied to the crank handle as shown;

$$\tau_2 = -r_1(a\dot{\theta}) \equiv -b\dot{\theta}$$

due to the viscous brake force; and $\tau_3 = -\kappa\theta$, the restoring torque in the spring. Thus its equation of motion is

$$I\ddot{\theta} = fr - b\dot{\theta} - \kappa\theta$$

or

$$(b) \quad I\ddot{\theta} + b\dot{\theta} + \kappa\theta = fr$$

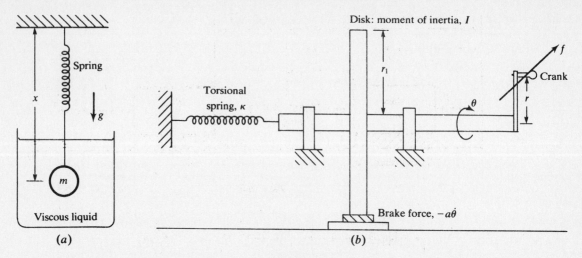

Fig. 33-4

It is seen that both (a) and (b) are analogs of the R-L-C circuit, whose differential equation is

$$L\ddot{q} + R\dot{q} + \frac{1}{C}q = v$$

Hence, by comparison with Problem 33.5, the criteria for underdamped motion are

$$(a) \quad a_1^2 < 4mk \qquad (b) \quad b^2 < 4I\kappa$$

Supplementary Problems

33.8. (a) In Problem 33.3, verify that $L/R = |s|$. (b) In Problem 33.5, verify that $\omega = |s^{-1}|$.

33.9. Consider an R-C circuit with no applied emf. Find the charge on the condensor at any time, if the initial charge was q_0.   *Ans.* $q = q_0 e^{-t/RC}$

33.10. An undriven L-C circuit has parameters $L = 1.5$ H, $C = 4$ μF. The initial charge on the condensor is 4 mC. Find (a) the frequency of oscillation, (b) the current as a function of time.
Ans. (a) $f = 65$ Hz; (b) $I = -1.633 \sin 408.25t$ (A)

33.11. Show that energy is conserved in the circuit of Problem 33.4.
Ans. $\mathscr{E}_L + \mathscr{E}_C = \dfrac{q_0^2}{2C} + \displaystyle\int_0^t 4P\,dt$

33.12. Figure 33-5 shows a mechanical system consisting of two spring-coupled masses; the spring has stiffness k and natural length ℓ_0. Devise an electrical analog of this system.
Ans.   See Fig. 33-6, in which it is assumed that the mutual inductance is zero.

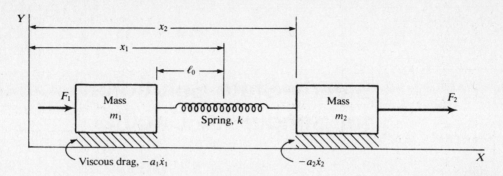

Fig. 33-5

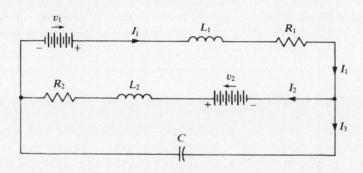

Fig. 33-6

33.13. (a) Applying Kirchhoff's laws, show that differential equations for the electrical system in Fig. 33-6 (assume no mutual inductance) are

$$v_1 = L_1\ddot{q}_1 + R_1\dot{q}_1 + \frac{1}{C}(q_1 - q_2) \qquad v_2 = L_2\ddot{q}_2 + R_2\dot{q}_2 + \frac{1}{C}(q_2 - q_1) \qquad (1)$$

(b) Applying the simple laws of mechanics, show that the equations of motion for the mechanical system in Fig. 33-5 are

$$F_1 = m_1\ddot{x}_1 + a_1\dot{x}_1 + k(x_1 - x_2) \qquad F_2 = m_2\ddot{x}_2 + a_2\dot{x}_2 + k(x_2 - x_1) \qquad (2)$$

where F_1 and F_2 are applied forces, k = constant of coil spring, ℓ_0 = unstretched length of spring, a_1 and a_2 are coefficients of viscous drag on the blocks.

Comparing (1) and (2), we see that, mathematically, the two systems are identical, with coordinates x_1 and x_2 corresponding to charges q_1 and q_2, respectively; $\dot{x}_1$ to $\dot{q}_1$, $\ddot{x}_1$ to $\ddot{q}_1$, etc.; forces F_1 and F_2 to voltages v_1 and v_2; masses m_1 and m_2 to inductances L_1 and L_2; spring constant k to $1/C$; coefficients of viscous drag a_1 and a_2 to resistances R_1 and R_2.   (c) Show that kinetic energy $\mathscr{E}_1$ of masses m_1 and m_2, energy $\mathscr{E}_2$ stored in L_1 and L_2, energy $\mathscr{E}_3$ of the spring, and energy $\mathscr{E}_4$ stored in capacitor C (Fig. 33-6) are given by

$$\mathscr{E}_1 = \frac{1}{2}m_1\dot{x}_1^2 + \frac{1}{2}m_2\dot{x}_2^2 \qquad \mathscr{E}_3 = \frac{k}{2}(x_2 - x_1)^2$$

$$\mathscr{E}_2 = \frac{1}{2}L_1\dot{q}_1^2 + \frac{1}{2}L_2\dot{q}_2^2 \qquad \mathscr{E}_4 = \frac{1}{2C}(q_2 - q_1)^2$$

Note the correspondence between $\mathscr{E}_1$ and $\mathscr{E}_2$; likewise, between $\mathscr{E}_3$ and $\mathscr{E}_4$.

Chapter 34

Steady-State Solutions
for Simple AC Circuits

34.1 SERIES CIRCUIT

Referring to the circuit in Fig. 34-1, assume that a sinusoidal voltage, $u_a = v_a \sin \omega t$, is maintained between points P_1 and P_2 by an ac generator ("alternator"), where u_a = instantaneous value of the voltage, v_a = maximum value or amplitude of the voltage wave, and $\omega = 2\pi f$ with f = frequency in Hz.

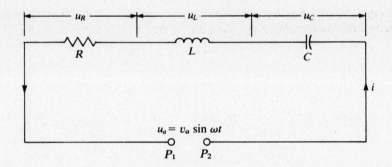

Fig. 34-1

The instantaneous values of voltage across resistor R, inductor L, and capacitor C are expressed in terms of instantaneous current i and instantaneous charge q by

$$u_R = Ri = R\dot{q} \qquad u_L = L\frac{di}{dt} = L\ddot{q} \qquad u_C = \frac{q}{C}$$

These must sum to u_a, which gives as the differential equation of the circuit

$$L\ddot{q} + R\dot{q} + \frac{1}{C}q = v_a \sin \omega t \qquad (34.1)$$

Like *(33.1)*, *(34.1)* has a transient and a steady-state solution. We are concerned here only with the latter, which we assume to be of the form

$$q = -\frac{v_a}{\omega Z}\cos(\omega t - \phi) \qquad \text{or} \qquad i = \frac{v_a}{Z}\sin(\omega t - \phi) \qquad (34.2)$$

Substitution of *(34.2)* into *(34.1)* determines the unknown constants Z and ϕ as

$$Z = \left[R^2 + \left(\omega L - \frac{1}{\omega C}\right)^2\right]^{1/2} \equiv [R^2 + (X_L - X_C)^2]^{1/2}$$

$$\tan \phi = \left(\omega L - \frac{1}{\omega C}\right)\bigg/ R \equiv (X_L - X_C)/R \qquad (34.3)$$

where we have defined the *impedance* Z (Ω) and the *phase angle* ϕ (rad) in terms of the resistance R (Ω), the *inductive reactance* $X_L \equiv \omega L$ (Ω), and the *capacitative reactance* $X_C \equiv (\omega C)^{-1}$ (Ω). The steady-state values of the current and the three voltage drops may now be written as

292

$$i = \frac{v_a}{Z} \sin(\omega t - \phi) \equiv I \sin(\omega t - \phi)$$

$$u_R = IR \sin(\omega t - \phi) \equiv v_R \sin(\omega t - \phi)$$

$$u_L = IX_L \cos(\omega t - \phi) \equiv v_L \cos(\omega t - \phi)$$

$$u_C = -IX_C \cos(\omega t - \phi) \equiv -v_C \cos(\omega t - \phi)$$

(34.4)

where the amplitudes of the four waves have been indicated as $I \equiv v_a/Z$ (A), $v_R \equiv IR$ (V), $v_L \equiv IX_L$ (V), and $v_C \equiv IX_C$ (V)

Relations (34.4) are conveniently pictured in *the rotating vector diagram*, Fig. 34-2. The various waves are represented by vectors from the origin, $\mathbf{I}$, $\mathbf{v}_R$, $\mathbf{v}_L$, $\mathbf{v}_C$, $\mathbf{v}_a$, whose lengths are respectively I, v_R, v_L, v_C, v_a. Vectors $\mathbf{v}_R$, $\mathbf{v}_L$, $\mathbf{v}_C$, $\mathbf{v}_a$ respectively make the fixed angles 0°, 90°, 270°, ϕ vector $\mathbf{I}$. If the five vectors are regarded as a rigid unit that rotates counterclockwise about O with constant angular velocity ω, it is seen that the Y component of each vector gives the instantaneous value of the corresponding wave. Note that $\mathbf{I}$ and $\mathbf{v}_R$ are "in phase"; $\mathbf{v}_L$ "leads" $\mathbf{I}$ by 90°; $\mathbf{v}_C$ "lags" $\mathbf{I}$ by 90°; and $\mathbf{v}_a$ "leads" or "lags" $\mathbf{I}$ according as ϕ is positive or negative. Note also that the circuit equation (34.1) is reflected in the diagram through the vector equality

$$\mathbf{v}_a = \mathbf{v}_R + \mathbf{v}_L + \mathbf{v}_C$$

(34.5)

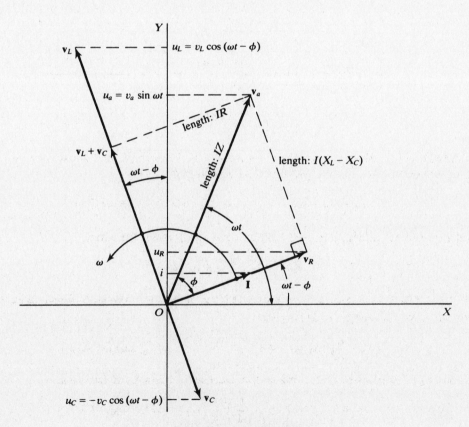

Fig. 34-2. Series Circuit

The power delivered by the alternator has instantaneous value iu. Of more interest, under steady-state conditions, is the average power delivered, P_{avg}, where the average is taken over a long period of time, or, equivalently, over one cycle of the applied voltage. It is shown in Problem 34.1 that

$$P_{\text{avg}} = I_{\text{rms}} v_{a,\text{rms}} \cos \phi = (I_{\text{rms}})^2 R = \frac{1}{2} IR^2 \qquad (34.6)$$

where

$(I_{\text{rms}})^2 \equiv$ average value of i^2 over one cycle $= I^2/2$

$(v_{a,\text{rms}})^2 \equiv$ average value of u_a^2 over one cycle $= v_a^2/2$

[compare (17.2)]. Because of the way it enters (34.6), $\cos \phi = R/Z$ is known as the *power factor*. In terms of the vector diagram, an alternative expression is

$$P_{\text{avg}} = \frac{1}{2} \mathbf{I} \cdot \mathbf{v}_a \qquad (34.7)$$

34.2 PARALLEL CIRCUIT

In the circuit of Fig. 34-3, an alternating voltage $u_a = v_a \sin \omega t$ is applied between terminals P_1 and P_2. As this is the voltage across each circuit element, the three current equations are

$$R i_R = u_a \qquad L \frac{di_L}{dt} = u_a \qquad \frac{1}{C} i_C = \frac{du_a}{dt} \qquad (34.8)$$

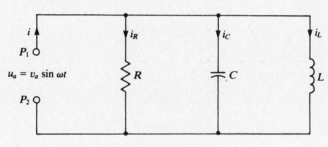

Fig. 34-3

The steady-state solutions of (34.8) are

$$i_R = \frac{v_a}{R} \sin \omega t \equiv I_R \sin \omega t$$

$$i_L = -\frac{v_a}{\omega L} \cos \omega t \equiv -\frac{v_a}{X_L} \cos \omega t \equiv -I_L \cos \omega t \qquad (34.9)$$

$$i_C = v_a \omega C \cos \omega t \equiv \frac{v_a}{X_C} \cos \omega t \equiv I_C \cos \omega t$$

where the reactances X_L and X_C are as defined in Section 34.1. For the total current, we have

$$i = i_R + i_L + i_C$$

$$= v_a \left[\frac{1}{R} \sin \omega t + \left(\frac{1}{X_C} - \frac{1}{X_L} \right) \cos \omega t \right] \equiv I \sin (\omega t + \phi) \qquad (34.10)$$

where $I \equiv v_a/Z$, and the impedance and phase angle are defined by

$$\frac{1}{Z} \equiv \left[\left(\frac{1}{R} \right)^2 + \left(\frac{1}{X_C} - \frac{1}{X_L} \right)^2 \right]^{1/2}$$

$$\tan \phi \equiv \left(\frac{1}{X_C} - \frac{1}{X_L} \right) \Big/ \frac{1}{R} \qquad (34.11)$$

In the vector diagram for this parallel circuit, Fig. 34-4, the five rotating vectors are $\mathbf{v}_a$, $\mathbf{I}_R$, $\mathbf{I}_L$, $\mathbf{I}_C$, $\mathbf{I}$, where $\mathbf{I}_R$ is in phase with $\mathbf{v}_a$, $\mathbf{I}_C$ leads $\mathbf{v}_a$ by 90°, $\mathbf{I}_L$ lags $\mathbf{v}_a$ by 90°, and $\mathbf{I}$ leads or lags $\mathbf{v}_a$ by $|\phi|$. As before, Y components give instantaneous values. The vector equality

$$\mathbf{I} = \mathbf{I}_R + \mathbf{I}_L + \mathbf{I}_C \tag{34.12}$$

reflects conservation of charge in the circuit.

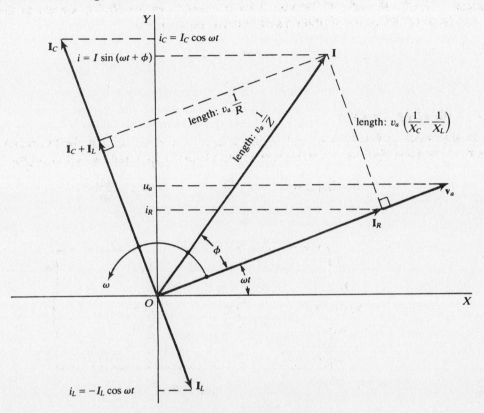

Fig. 34-4.   Parallel Circuit

The formulas for average power from the alternator, analogous to (34.6) and (34.7), are

$$P_{\text{avg}} = I_{\text{rms}} v_{a,\text{rms}} \cos\phi = (I_{R,\text{rms}})^2 R \tag{34.13}$$

where now $\cos\phi = Z/R$. As before,

$$P_{\text{avg}} = \frac{1}{2}\mathbf{I}\cdot\mathbf{v}_a \tag{34.14}$$

Solved Problems

34.1. Express the steady-state, average power input to the circuit of Fig. 34-1.

During one period of the applied voltage—say, from $t = 0$ to $t = 2\pi/\omega \equiv T$—conservation of energy requires that the energy into the circuit, $P_{\text{avg}}T$, be stored in the inductor and capacitor or dissipated in the resistor. Thus

$$P_{\text{avg}}T = \frac{1}{2}Li^2\bigg|_0^T + \frac{q^2}{2C}\bigg|_0^T + \int_0^T i^2 R\, dt$$

But, in the steady state, i and q are periodic of period T, and so the first two terms on the right-hand side vanish, giving

$$P_{avg} = R\left(\frac{1}{T}\int_0^T i^2\,dt\right) \equiv R(I_{rms})^2$$

Since i is a sinusoid, we can easily show that

$$I_{rms} = \frac{1}{\sqrt{2}}I = \frac{1}{\sqrt{2}}\frac{v_a}{Z}$$

from which the rest of (34.6) follows at once.

34.2. A sinusoidal voltage of frequency $f = 60$ Hz and peak value 150 V is applied to a series R-L circuit, where $R = 20\ \Omega$ and $L = 40$ mH. (a) Compute the period T, ω, X_L, Z, ϕ. (b) Compute the amplitudes I, v_R, v_L and find the instantaneous values i, u_R, u_L at $t = T/6$. (c) Check that

$$u_a = u_R + u_L \qquad v_a = (v_R^2 + v_L^2)^{1/2}$$

at $t = T/6$. (d) Compute I_{rms}, $v_{a,rms}$, and the average power into the circuit.

(a)
$$T = \frac{1}{f} = \frac{1}{60}\ \text{s}$$

$$\omega = 2\pi f = 377\ \text{rad/s}$$

$$X_L = \omega L = (377)(0.040) = 15.08\ \Omega$$

$$Z = (R^2 + X_L^2)^{1/2} = 25.05\ \Omega$$

$$\phi = \arctan\frac{X_L}{R} = \arctan 0.754 = 37°$$

(b)
$$I = \frac{v_a}{Z} = \frac{150}{25.05} = 6\ \text{A} \qquad v_R = IR = 120\ \text{V} \qquad v_L = IX_L = 90.5\ \text{V}$$

At $t = T/6$ or $\omega t = \pi/3 = 60°$,

$$i = I \sin(\omega t - \phi) = 6 \sin 23° = 2.344\ \text{A}$$

$$u_R = iR = (2.344)(20) = 46.88\ \text{V}$$

$$u_L = v_L \cos(\omega t - \phi) = 90.5 \cos 23° = 83.29\ \text{V}$$

(c) At $\omega t = 60°$, $u_a = 150 \sin 60° = 130\ \text{V} = (46.88\ \text{V}) + (83.29\ \text{V})$. Moreover (at all times),

$$[(120)^2 + (90.5)^2]^{1/2} = 150.3\ \text{V} \approx 150\ \text{V}$$

(d)
$$I_{rms} = \frac{6}{\sqrt{2}} = 4.24\ \text{A} \qquad v_{a,rms} = \frac{150}{\sqrt{2}} = 106.07\ \text{V}$$

$$P_{avg} = (I_{rms})^2 R = (18)(20) = 360\ \text{W}$$

34.3. In Problem 34.2, the frequency is changed to 1200 Hz; what is now the average power into the circuit?

With the frequency increased by a factor of 20,

$$X_L = 20 \times 15.08 = 301.6\ \Omega$$

$$Z = [(20)^2 + (301.6)^2]^{1/2}\ \Omega$$

Then, since P_{avg} varies inversely as Z^2,

$$P_{avg} = \frac{(20)^2 + (15.08)^2}{(20)^2 + (301.6)^2}(360) = 2.46\ \text{W}$$

34.4. In the circuit of Fig. 34-1, let $R = 20\ \Omega$, $L = 0.16$ H, $C = 30\ \mu$F, and let the applied voltage be $u = 250 \sin 400t$ (V).   (a) Compute X_L, X_C, Z, ϕ, I, v_R, v_L, v_C.   (b) Verify that

$$v_a = [v_R^2 + (v_L - v_C)^2]^{1/2}$$

(c) Write expressions for i, u_R, u_L, u_C and show that $u_a = u_R + u_L + u_C$.   (d) Compute P_{avg}.

(a) $X_L = (400)(0.16) = 64\ \Omega$ $I = \dfrac{250}{27.817} = 8.987$ A

$X_C = \dfrac{1}{(400)(30 \times 10^{-6})} = 83.333\ \Omega$ $v_R = (8.987)(20) = 179.75$ V

$Z = [(20)^2 + (64 - 83.333)^2]^{1/2} = 27.817\ \Omega$ $v_L = (8.987)(64) = 575.19$ V

$\phi = \arctan \dfrac{64 - 83.333}{20} = -0.7685$ rad $v_C = (8.987)(83.333) = 748.94$ V

and $[(179.75)^2 + (575.19 - 748.94)^2]^{1/2} = 250.0$

(b) $i = 8.987 \sin(400t + 0.7685)$ (A) $u_L = 575.19 \cos(400t + 0.7685)$ (V)

$u_R = 179.75 \sin(400t + 0.7685)$ (V) $u_C = -748.94 \cos(400t + 0.7685)$ (V)

(c)   Using the trigonometric identity

$$A \sin \theta + B \cos \theta = (A^2 + B^2)^{1/2} \sin(\theta + \delta)\qquad \text{where}\qquad \tan \delta = \frac{B}{A}$$

and the last result of part (a),

$$u_R + u_L + u_C = 179.75 \sin(400t + 0.7685) + (575.19 - 748.94)\cos(400t + 0.7685)$$
$$= 250.0 \sin(400t + 0.7685 + \delta) = u_a$$

where the last equality follows from

$$\delta = \arctan \frac{575.19 - 748.94}{179.75} = \arctan \frac{64 - 83.333}{20} = -0.7685$$

(d) $P_{avg} = (I_{rms})^2 R = \left(\dfrac{8.987}{\sqrt{2}}\right)^2 (20) = 807.66$ W

34.5. For a certain angular frequency, ω_0, of the applied voltage, the R-L-C circuit will have unity power factor (a condition known as *voltage resonance*).   Because

$$P_{avg} = I_{rms} v_{a,rms} \cos \phi = \left(\frac{v_{a,rms}}{R}\right)^2 \cos^2 \phi$$

the circuit consumes maximum power, for a given v_a, at voltage resonance.   Rework Problem 34.4(a) at voltage resonance.

We have $\cos \phi = 1$ only if

$$X_L = X_C \qquad \text{or}\qquad \omega L = \frac{1}{\omega C}\qquad \text{or}\qquad \omega = \omega_0 \equiv \sqrt{\frac{1}{LC}}$$

Then $\omega_0 = \sqrt{\dfrac{1}{(0.16)(30 \times 10^{-6})}} = 456.435$ rad/s

and $X_L = (456.435)(0.16) = 73.03\ \Omega = X_C$ $I = \dfrac{250}{20} = 12.5$ A

$Z = R = 20\ \Omega$ $v_R = v_a = 250$ V

$\phi = 0$ $v_L = (12.5)(73.03) = 912.87$ V $= v_C$

Note that the maximum voltage between terminals of the inductor (or the capacitor) is $912.87/250 = 3.65$ times the maximum applied voltage.

34.6. Suppose that in Fig. 34-3 the applied voltage is $v_a = 150 \sin 400t$ (V), $R = 20\ \Omega$, $L = 0.06$ H, $C = 18\ \mu$F. Give a complete analysis of the circuit.

$$X_L = \omega L = (400)(0.060) = 24\ \Omega$$

$$X_C = \frac{1}{\omega C} = \frac{1}{(400)(18 \times 10^{-6})} = 138.889\ \Omega$$

$$\frac{1}{Z} = \left[\frac{1}{20^2} + \left(\frac{1}{138.889} - \frac{1}{24}\right)^2\right]^{1/2} \qquad \text{or} \qquad Z = 16.467\ \Omega$$

$$I_R = \frac{150}{20} = 7.5\ \text{A} \qquad I_L = \frac{150}{24} = 6.25\ \text{A} \qquad I_C = \frac{150}{138.889} = 1.08\ \text{A}$$

$$I = \frac{150}{16.467} = 9.109\ \text{A}$$

Alternatively, $I = [I_R^2 + (I_C - I_L)^2]^{1/2}$.

Instantaneous values of current are:

$$i_R = 7.5 \sin 400t \quad \text{(A)} \qquad i_L = -6.25 \cos 400t \quad \text{(A)} \qquad i_C = 1.08 \cos 400t \quad \text{(A)}$$

The phase angle is given by

$$\tan \phi = R\left(\frac{1}{X_C} - \frac{1}{X_L}\right) = -0.6893 \qquad \text{or} \qquad \phi = -0.6035\ \text{rad}$$

From (34.10), $i = 9.109 \sin (400t - 0.6035)$ (A). The average power taken by the circuit is

$$P_{\text{avg}} = (I_{R,\text{rms}})^2 R = \left(\frac{7.5}{\sqrt{2}}\right)^2 (20) = 562.5\ \text{W}$$

Supplementary Problems

34.7. An alternating voltage $u_a = 150 \sin 6283.2t$ (V) is applied to a series R-C circuit, with $R = 20\ \Omega$ and $C = 8\ \mu$F. Find (a) f, (b) X, (c) Z, (d) power factor, (e) I, (f) v_R, (g) v_C.
Ans. (a) 1000 Hz; (b) 19.894 Ω; (c) 28.21 Ω; (d) 0.709; (e) 5.317 A; (f) 106.34 V; (g) 105.78 V

34.8. Show that (a) X_L and X_C have the dimensions of resistance, (b) $(1/LC)^{1/2} = |s^{-1}|$.

34.9. A simple series circuit contains a resistance $R = 20\ \Omega$ and inductance $L = 0.06$ H. The applied voltage is

$$u_a = 260 \sin 400t \quad \text{(V)}$$

Find (a) Z, (b) I, (c) ϕ, (d) i, (e) P_{avg}.
Ans. (a) 31.24 Ω; (b) 8.32 A; (c) 50.194°; (d) 8.32 $\sin (400t - 0.8760)$ (A); (e) 692.2 W

34.10. Data on the series circuit shown in Fig. 34-5 are: applied voltage, $u_a = 300 \sin 400t$ (V); $R_1 = 15\ \Omega$; $R_2 = 10\ \Omega$; $L = 0.16$ H; $C = 60\ \mu$F. Compute (a) X_L, (b) X_C, (c) Z, (d) I, (e) $\cos \phi$, (f) v_{R_1}, (g) v_{R_2}, (h) v_L (i) v_C, (j) v_1, (k) v_2, (l) P_{avg}.
Ans. (a) 64 Ω; (b) 41.667 Ω; (c) 33.523 Ω; (d) 8.949 A; (e) 0.74577; (f) 134.24 V; (g) 89.49 V; (h) 572.74 V; (i) 372.88 V; (j) $[v_{R_1}^2 + (v_L - v_C)^2]^{1/2} = 240.76$ V; (k) 218.98 V; (l) 1.0011 kW

34.11. For Problem 34.10, find the resonant angular frequency and the corresponding values of Z, I, v_L, and v_C.
Ans. $\omega = 322.75$ rad/s; $Z = 25\ \Omega$; $I = 12$ A; $v_L = v_C = 619.68$ V

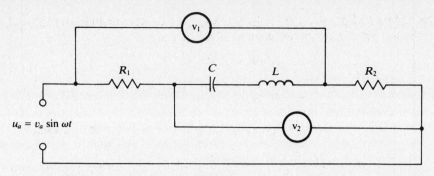

$u_a = v_a \sin \omega t$

Fig. 34-5

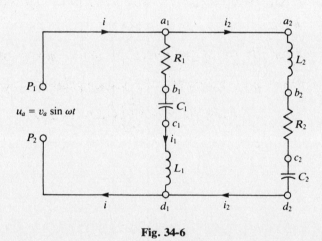

Fig. 34-6

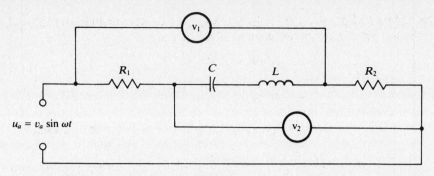

Fig. 34-7

34.12. In the circuit of Fig. 34-3, let $u_a = 320 \sin 400t$ (V), $C = 45$ μF, $L = 0.12$ H, $R = 40$ Ω. Determine (a) X_L, (b) X_C, (c) Z, (d) I_R, (e) I_L, (f) I_C, (g) I, (h) ϕ, (i) i_R, (j) i_L, (k) i_C, (l) i, (m) P_{avg}.

    Ans.   (a) 48 Ω;   (b) 55.555 Ω;   (c) 39.7456 Ω;   (d) 8 A;   (e) 6.666 A;   (f) 5.76 A;   (g) 8.051 A;
             (h) $-6.466°$;   (i) $8 \sin 400t$ (A);   (j) $-6.666 \cos 400t$ (A);   (k) $5.76 \cos 400t$ (A);
             (l) $8.051 \sin (400t + 0.113)$ (A); (m) 1.280 kW

34.13. Consider the series-parallel network of Fig. 34-6. A voltage $u_a = v_a \sin \omega t$ is applied between points P_1 and P_2. (a) Show that branch $a_1 d_1$ and branch $a_2 d_2$ may be treated as separate R-L-C circuits. (b) Draw the rotating vector diagram for the entire network, and from it obtain expressions for the maximum total current I and the total impedance Z.

    Ans.   (b) See Fig. 34-7. From $\mathbf{I} = \mathbf{I}_1 + \mathbf{I}_2$,

$$\mathbf{I} \cdot \mathbf{I} = \mathbf{I}_1 \cdot \mathbf{I}_1 + \mathbf{I}_2 \cdot \mathbf{I}_2 + 2\mathbf{I}_1 \cdot \mathbf{I}_2 \qquad \text{or} \qquad I^2 = I_1^2 + I_2^2 + 2I_1 I_2 \cos (\phi_1 - \phi_2)$$

Now substitute $I = v/Z$, $I_1 = v_a/Z_1$, $I_2 = v_a/Z_2$ to obtain

$$\frac{1}{Z^2} = \frac{1}{Z_1^2} + \frac{1}{Z_2^2} + \frac{2}{Z_1 Z_2} \cos (\phi_1 - \phi_2)$$

Chapter 35

Reflection, Refraction, and Polarization of Light

35.1 LAWS OF REFLECTION AND REFRACTION

A *ray* of light is a line whose direction gives the direction of flow of radiant energy. In an isotropic medium, rays are straight lines, along which energy travels at speed $v = c/n$, where n is the *refractive index* of the medium. Because $n > 1$, the speed is less than the speed in vacuum, $c \approx 3 \times 10^8$ m/s. The refractive index is a function of the wavelength (in vacuum) of the light.

Let a ray in medium 1 be incident on the interface with a medium 2, making angle θ_1 with the normal to the interface (Fig. 35-1). Then, in general, there will be a reflected ray in medium 1 and a refracted ray in medium 2 such that

   (i)   The three rays and the normal all lie in a common plane, the *plane of incidence*.
  (ii)   The angle of incidence equals the angle of reflection: $\theta_1 = \theta_r$.
 (iii)   The directions of the incident and refracted rays are related by *Snell's law*:

$$n_1 \sin \theta_1 = n_2 \sin \theta_2$$

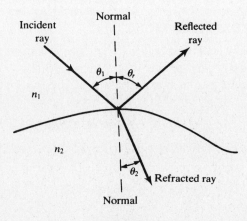

Fig. 35-1

If $n_1 > n_2$ and θ_1 exceeds the *critical angle* θ_c, where

$$\sin \theta_c = \frac{n_2}{n_1}$$

then there will be no refracted ray—a phenomenon called *total reflection* or, from the point of view of medium 1, *total internal reflection*.

35.2 POLARIZATION

An electromagnetic or other transverse wave is *polarized* whenever the disturbance lacks cylindrical symmetry about the direction of propagation (i.e. about the ray direction).

If all along a ray of an electromagnetic wave the electric field lies in a fixed plane, the wave is said to be *plane* or *linearly polarized*. Equivalently, at a given point on the ray, **E** oscillates along a fixed straight line, and the lines corresponding to all the points are parallel, sweeping out a plane.

In the case of *circularly polarized* light, the electric field has perpendicular components of equal amplitudes with a phase difference of 90°. However, if the electric field has perpendicular components of unequal amplitudes and some fixed phase difference, the light is *elliptically polarized*.

In *unpolarized* or *natural* light, the electric field has perpendicular components with equal amplitudes but with a phase difference that undergoes many random fluctuations during the time it is observed. In other words, an unpolarized wave can be considered as the superposition of two, perpendicular, *incoherent*, plane polarized waves of the same amplitude.

Polarization by birefringence. Certain anisotropic materials (e.g. calcite) display *two* indices of refraction, one for field oscillations parallel to a certain direction (the *optic axis*) and the other for oscillations perpendicular to that direction. A ray of unpolarized light incident on the surface of such a birefringent medium is usually split into two refracted rays, linearly polarized at right angles to each other. Snell's law holds for one of the refracted beams, the *ordinary ray*, but not for the other, the *extraordinary ray*. A birefringent crystal of such thickness that the two emerging waves differ in phase by 180° is called a *half-wave plate*; or, if they differ by 90°, a *quarter-wave plate*.

Polarization by reflection. Light reflected from a smooth dielectric surface (glass) is completely linearly polarized normal to the plane of incidence if the angle of incidence is given by *Brewster's law*:

$$\tan \theta_{1p} = \frac{n_2}{n_1}$$

35.3 INTENSITY OF POLARIZED LIGHT

A *perfect polarizer* is a plate that transmits all light polarized along a given transverse axis and absorbs all light polarized normal to this axis. If the incident wave has intensity I_0 (W/m^2) and is linearly polarized at angle θ to the transmission axis, the transmitted intensity is given by *Malus's law* as

$$I = I_0 \cos^2 \theta \quad (\text{W/m}^2)$$

When unpolarized light of intensity I_0 is incident on a perfect polarizer, the transmitted intensity is $\frac{1}{2}I_0$.

Solved Problems

35.1. In what direction does the fish in Fig. 35-2 see the setting sun? The index of refraction of water is $n_2 = 4/3$ and of air is $n_1 \approx 1$.

The rays of the setting sun come in nearly tangent to the surface of the water. By Snell's law, with $\theta_1 = 90°$,

$$1 = \frac{4}{3} \sin \theta_2 \qquad \text{or} \qquad \theta_2 = \arcsin \frac{3}{4} = 48.6°$$

Note that θ_2 is the critical angle for the reversed ray (from the fish to the sun). The fish perceives the sun at $90° - \theta_2 = 41.4°$ above the horizontal.

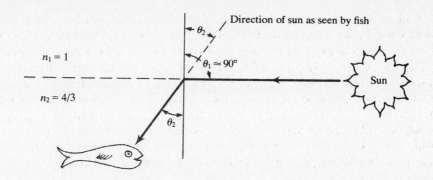

Fig. 35-2

35.2. Assume that the index of refraction of a glass sphere is $n_1 = 1.76$. For rays originating within the sphere, find the critical angle if the sphere is immersed in (*a*) air ($n_2 = 1$), (*b*) water ($n_2 = 1.33$).

$$\sin \theta_c = \frac{n_2}{n_1}$$

and so

(*a*)
$$\theta_c = \arcsin \frac{1}{1.76} = 34.6°$$

(*b*)
$$\theta_c = \arcsin \frac{1.33}{1.76} = 49.1°$$

35.3. Two identical beakers, one filled with water ($n = 1.361$) and the other filled with mineral oil ($n = 1.47$), are viewed from above. Which beaker appears to contain the greater depth of liquid, and what is the ratio of the apparent depths?

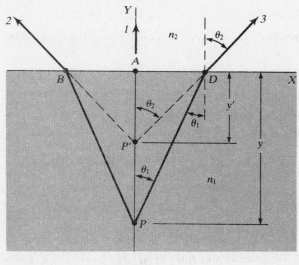

Fig. 35-3

Figure 35-3 indicates the bottom of a beaker by point P. For an observer looking in the $-Y$ direction, the image of P, labeled P', will be determined by a vertical ray *1* and the nearly-vertical ray *3* (or *2*). Since θ_1 and θ_2 are small, Snell's law gives

$$n_1\theta_1 \approx n_2\theta_2 \qquad \text{or} \qquad \frac{\theta_1}{\theta_2} \approx \frac{n_2}{n_1}$$

But

$$\overline{AD} = y\tan\theta_1 = y'\tan\theta_2 \qquad \text{or} \qquad \frac{\tan\theta_1}{\tan\theta_2} \approx \frac{\theta_1}{\theta_2} = \frac{y'}{y}$$

Consequently, for given y and n_2, y' is inversely proportional to n_1. The water-filled beaker appears deeper, by a factor of

$$\frac{1.47}{1.361} = 1.08$$

35.4. A ray of light is incident on the left vertical face of a glass cube of refractive index $n_2 = 1.6$, as shown in Fig. 35-4. The plane of incidence is the plane of the page, and the cube is surrounded by water ($n_1 = 1.33$). What is the largest angle of incidence θ_1 for which total internal reflection occurs at the top surface?

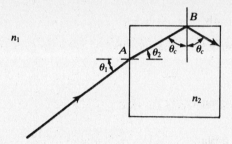

Fig. 35-4

Applying Snell's law at A,

$$n_1\sin\theta_1 = n_2\sin\theta_2 \tag{1}$$

But $\theta_2 = 90° - \theta_c$, and so

$$\cos\theta_2 = \sin\theta_c = \frac{n_1}{n_2} \tag{2}$$

Elimination of θ_2 between (1) and (2) gives

$$\sin\theta_1 = \sqrt{\left(\frac{n_2}{n_1}\right)^2 - 1} = 0.6684$$

whence $\theta_1 = 42.0°$.

35.5. Light is incident at angle θ_1 on an isosceles triangular prism of apex angle α and refractive index $n > 1$ (Fig. 35-5). (a) Obtain an expression for the minimum angular deviation, δ_{min}, produced by the prism. (b) Show that if α is sufficiently small, $\delta \approx (n-1)\alpha$, independent of θ_1.

(a) Because the prism is symmetrical and because light paths are reversible, it is easy to see that the deviation δ will be a minimum when $\theta_1 = \theta_3 = \kappa$. Under this condition, Snell's law gives at the two interfaces:

$$\sin\kappa = n\sin\theta_2 \qquad \text{and} \qquad n\sin(\alpha - \theta_2) = \sin\kappa$$

which imply $\theta_2 = \alpha/2$. Then,

$$\sin\kappa = n\sin\frac{\alpha}{2}$$

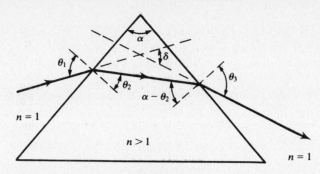

Fig. 35-5

But $\delta_{min} = \theta_1 + \theta_3 - \alpha = 2\kappa - \alpha$, or

$$\kappa = \frac{\alpha + \delta_{min}}{2}$$

Hence

$$\sin\frac{\alpha + \delta_{min}}{2} = n\sin\frac{\alpha}{2}$$

This equation implicitly determines δ_{min}.

(b)   When α is very small, so is δ_{min}; and the result of (a) becomes

$$\frac{\alpha + \delta_{min}}{2} \approx n\frac{\alpha}{2} \qquad \text{or} \qquad \delta_{min} \approx (n-1)\alpha$$

Furthermore, for α small enough, all deviations δ are close to δ_{min}, so that $\delta \approx (n-1)\alpha$.

35.6.   A ray of light strikes a plane mirror at an angle of incidence 45°, as shown in Fig. 35-6.   After reflection, the ray passes through a prism of refractive index 1.50, whose apex angle is 4°.   Through what angle must the mirror be rotated if the total deviation of the ray is to be 90°?

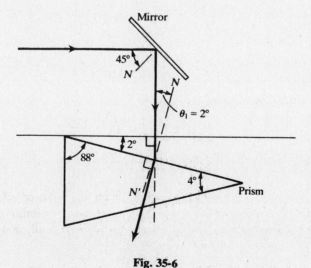

Fig. 35-6

By Problem 35.5(b), the prism will, to the first order, cause a fixed deviation,

$$\delta = (1.50 - 1)(4°) = 2°$$

whatever the orientation of the mirror.   In the given orientation, the mirror by itself produces the desired 90° deviation; hence it must be rotated so as to cancel the additional 2°.   Thus, it must be rotated through angle of $\frac{1}{2}(2°) = 1°$, counterclockwise in Fig. 35-6.

35.7. Four perfect polarizing plates are stacked so that the axis of each is turned 30° clockwise with respect to the preceding plate; the last plate is crossed with the first. How much of the intensity of an incident unpolarized beam of light is transmitted by the stack?

The first plate transmits 1/2 of the incident intensity. Each succeeding plate makes a vector resolution at angle 30°, transmitting the fraction $\cos 30°$ of the amplitude, or $\cos^2 30° = 3/4$ of the intensity, leaving the preceding plate. The fraction of the initial intensity transmitted by the stack is then

$$\left(\frac{1}{2}\right)\left(\frac{3}{4}\right)^3 = 0.211$$

35.8. At what angle β above the horizon is the sun when a person observing its rays reflected in water ($n_2 = 1.33$) finds them linearly polarized along the horizontal? See Fig. 35-7.

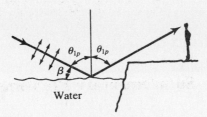

Water

Fig. 35-7

In order for the reflected rays to be linearly polarized, the angle of incidence must be Brewster's angle.

$$\tan \theta_{1p} = \frac{n_2}{n_1} = \frac{1.33}{1.00} \qquad \text{or} \qquad \theta_{1p} \approx 53°$$

and $\beta = 90° - \theta_{1p} \approx 37°$.

35.9. Prove that a particle undergoing two simple harmonic vibrations of the same frequency, at right angles and out of phase, traces an elliptical path.

Suppose the vibrations take place along the X and Y axes, with a phase difference α; i.e.

$$x = A \sin \omega t \qquad y = B \sin (\omega t - \alpha)$$

Then

$$\frac{y}{B} = \sin \omega t \cos \alpha + \cos \omega t \sin \alpha = \frac{x}{A} \cos \alpha + \sqrt{1 - \frac{x^2}{A^2}} \sin \alpha$$

and so

$$\left(\frac{y}{B} - \frac{x}{A} \cos \alpha\right)^2 = \left(1 - \frac{x^2}{A^2}\right) \sin^2 \alpha$$

which gives upon expansion

$$\frac{x^2}{A^2} + \frac{y^2}{B^2} - \frac{2xy}{AB} \cos \alpha = \sin^2 \alpha \qquad (1)$$

This is the equation of an ellipse whose major and minor axes are inclined to the X and Y axes. See Problem 35.16.

35.10. A quarter-wave plate is made from a material whose indices of refraction for light of free-space wavelength $\lambda_0 = 589$ nm are $n_\perp = 1.732$ and $n_\parallel = 1.456$. What is the minimum necessary thickness of the plate for this wavelength?

The *optical path length* (c times the travel time) of the ordinary wave in a plate of thickness ℓ is $n_\perp \ell$ and the optical path length of the extraordinary ray is $n_\parallel \ell$. Since the two rays must emerge from the plate with a 90° phase difference, the optical paths must differ by $(k + \frac{1}{4})\lambda_0$, $k = 0, 1, 2, \ldots$. The minimum thickness thus satisfies

$$\frac{\lambda_0}{4} = \ell(n_\perp - n_\parallel) \qquad \text{or} \qquad \ell = \frac{\lambda_0}{n_\perp - n_\parallel} = \frac{589 \text{ nm}}{1.732 - 1.546} = 3170 \text{ nm} = 0.00317 \text{ mm}$$

35.11. Derive Malus's law, $I = I_0 \cos^2 \theta$.

The incident plane polarized wave, $\mathbf{E} = \mathbf{E}_0 \sin \omega t$, is equivalent to two plane polarized waves,

$$\mathbf{E}_\parallel = (\mathbf{E}_0 \cos \theta) \sin \omega t \qquad \mathbf{E}_\perp = (\mathbf{E}_0 \sin \theta) \sin \omega t$$

respectively along and perpendicular to the transmission axis. Since intensity is proportional to the square of the amplitude, and since only $\mathbf{E}_\parallel$ is transmitted,

$$\frac{I}{I_0} = \frac{(\mathbf{E}_0 \cos \theta) \cdot (\mathbf{E}_0 \cos \theta)}{\mathbf{E}_0 \cdot \mathbf{E}_0} = \cos^2 \theta$$

Supplementary Problems

35.12. A ray of light is incident at an angle of 60° on one surface of a glass plate 40 mm thick and of refractive index 1.50. The medium on either side of the plate is air. Find the transverse displacement between the incident and emergent rays.   *Ans.* $20(\sqrt{3} - 1/\sqrt{2}) = 20.5$ mm

35.13. A cylindrical tin can open at the top has a height of 0.4 m. In the center of its bottom surface is a tiny black dot and the can is completely filled with water ($n = 1.33$). Calculate the radius of the smallest opaque circular disk that would prevent the dot from being seen, if the disk were floated centrally on the surface of the water.   *Ans.*   0.46 m

35.14. A prism having an apex angle 4° and refractive index 1.50 is located in front of a vertical plane mirror as shown in Fig. 35-8. A horizontal ray of light is incident on the prism. (*a*) What is the angle of incidence at the mirror? (*b*) Through what total angle is the ray deviated?
Ans.   (*a*) 2°;   (*b*) −178°

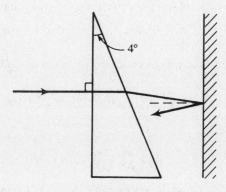

Fig. 35-8

35.15. Find the direction of polarization when at some point the electric field is given by

$$E_x = 20 \cos \omega t \quad \text{(V/m)} \qquad E_y = 40 \cos (\omega t + \pi) \quad \text{(V/m)}$$

Ans.   Linearly polarized at −63.4° with +X.

35.16. In (*1*) of Problem 35.9, show that an appropriate change of variables

$$x = x' \cos \psi - y' \sin \psi$$

$$y = x' \sin \psi + y' \cos \psi$$

yields the ellipse graphed in Fig. 35-9.

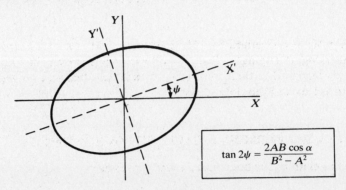

$$\tan 2\psi = \frac{2AB \cos \alpha}{B^2 - A^2}$$

Fig. 35-9

35.17. Unpolarized light of intensity I' is incident upon a stack of two filters whose transmission axes make an angle θ. Express the intensity I of the emerging beam.   *Ans.* $I = \frac{1}{2}I' \cos^2 \theta$

35.18. Find the transmitted intensity when circularly polarized light of intensity I' is incident on a perfect polarizer.   *Ans.* $I'/2$

35.19. Linearly polarized light is incident at Brewster's angle on the surface of a medium.   What can be said about the refracted and reflected beams if the incident beam is polarized (*a*) parallel to the plane of incidence?   (*b*) perpendicular to the plane of incidence?
Ans.   (*a*) At Brewster's angle the parallel component is completely refracted; thus no light is reflected at all.   (*b*) Some of the incident light is reflected and some is refracted.   Both the reflected and the refracted beams will be polarized perpendicular to the plane of incidence.

Chapter 36

Geometrical Optics

Geometrical optics is concerned with the effects on light rays of mirrors (plane, concave, convex) and lenses (converging, diverging). In what follows, it is assumed that all lenses are *thin*, i.e. their thickness is small compared to their radii of curvature, and that all rays are *paraxial*, i.e. they make small angles with the axis of the optical system.

36.1 GAUSSIAN LENS FORMULA; MAGNIFICATION FORMULA

The two basic relations for both mirrors and lenses are

$$\frac{1}{d_o} + \frac{1}{d_i} = \frac{1}{f} \qquad (36.1)$$

$$\frac{d_o}{d_i} = -\frac{\ell_o}{\ell_i} \qquad (36.2)$$

where

$f \equiv$ focal length, cm
$d_o \equiv$ distance of object, cm
$d_i \equiv$ distance of image, cm
$\ell_o \equiv$ linear size of object, cm
$\ell_i \equiv$ linear size of image, cm

The five quantities are conventionally given in centimeters. Each quantity carries an algebraic sign, as specified in Table 36-1. Recall that light rays diverge from a *real object* and converge towards a *virtual object*. Contrarily, light rays converge towards a *real image* (which could be projected directly on a screen), but they diverge from a *virtual image* (which could not be projected directly on a screen).

Table 36-1

Quantity	Sign	
	+	−
f	converging lens, concave mirror	diverging lens, convex mirror
d_o	real object	virtual object
d_i	real image	virtual image
ℓ_o	erect object	inverted object
ℓ_i	erect image	inverted image

36.2 RAY TRACING

The position, size, and kind of image formed can always be determined graphically, by drawing the paths of certain rays. This method will be employed side by side with (36.1) and (36.2) in the solved problems below.

Solved Problems

36.1. Describe the image that a plane mirror forms of a real object (an illuminated material body).

In Fig. 36-1, $P_1 P_2$ represents a real object at distance d_o from the reflecting surface of mirror M.

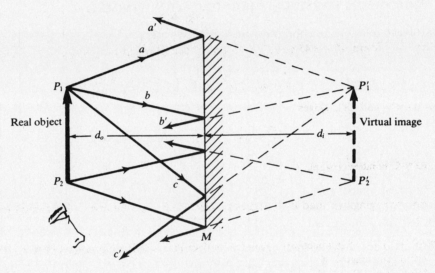

Fig. 36-1

Draw any number of "real rays," as a, b, c, from P_1 to the mirror. Corresponding reflected rays (constructed by applying the law of reflection) are a', b', c'. Projected backward as dashed lines, the *virtual rays* intersect at P_1', which locates one point on the virtual image $P_1' P_2'$. When rays from any point on the real object are treated in the same way, a corresponding point is located on $P_1' P_2'$. Hence, the entire image can be located.

The image is erect. As can be shown from simple geometry, $\overline{P_1 P_2}$ (length of object, ℓ_o) is equal to $\overline{P_1' P_2'}$ (length of image, ℓ_i). Likewise d_o and d_i are equal in magnitude. An eye placed as shown would see an arrow $P_1' P_2'$ *behind* the mirror.

For a plane mirror, $f = \infty$. Thus, (36.1) gives $d_i = -d_o < 0$ and (36.2) gives $\ell_i = \ell_o > 0$ (d_o and ℓ_o are positive, by Table 36-1).

36.2. Describe the image formed by a concave mirror when a real object is situated outside the center of curvature, C.

In Fig. 36-2, real rays a, b, c are drawn from a point P_1 on the real object $P_1 P_2$. For convenience a is drawn parallel to the optic axis, b through F, and c through C. From the geometry of the mirror and the law of reflection, reflected ray a' passes through F, b' is parallel to the optic axis, and c' returns through C along the path of c. The intersection of the real rays a', b', c' at P_1' locates one point on the real image $P_1' P_2'$. Rays a, b, c may seem very special. However, any (paraxial) ray from P_1, after reflection, passes through P_1', as may be shown by an application of the law of reflection. Note that the image is inverted.

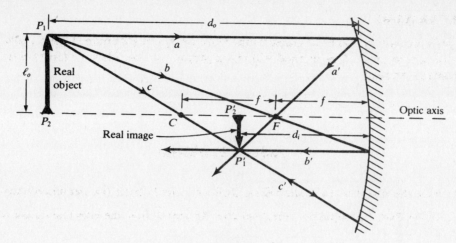

Fig. 36-2

Let $f = +20$ cm, $d_o = +45$ cm, $\ell_o = +5$ cm. Then (*36.1*) gives

$$\frac{1}{45} + \frac{1}{d_i} = \frac{1}{20} \qquad \text{or} \qquad d_i = +36 \text{ cm}$$

(a real image), and (*36.2*) gives

$$\frac{45}{36} = -\frac{5}{\ell_i} \qquad \text{or} \qquad \ell_i = -4 \text{ cm}$$

(an inverted, minified image).

36.3. Describe the image formed by a concave mirror when a real object is situated inside the focal point, *F*.

See Fig. 36-3. Constructing rays as in Problem 36.2, we find the image to be erect, virtual (behind the mirror), and magnified.

Letting $f = +20$ cm, $d_o = +15$ cm, $\ell_o = +5$ cm, we have from (*36.1*) and (*36.2*):

$$d_i = -60 \text{ cm} \quad \text{(virtual)} \qquad \ell_i = +20 \text{ cm} \quad \text{(erect, magnified)}$$

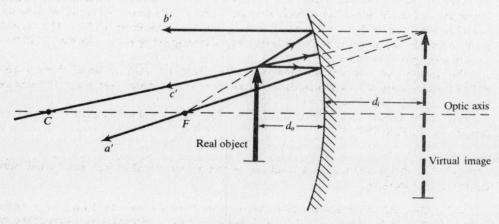

Fig. 36-3

36.4. Describe the image formed by a thin converging (positive) lens when a real object is situated outside the front focal point, *F*.

In Fig. 36-4, real rays a, b, c from P_1 on real object P_1P_2 are, for convenience, drawn parallel to the optic axis, through F, and through the center of the lens. Refracted ray a' passes through the back focal point F', b' is parallel to the optic axis, and c' is the continuation of c. These rays, of course, follow the law of refraction. The intersection of real rays a', b', c' locates point P_1' on the real image $P_1'P_2'$. Other points on $P_1'P_2'$ can be located in the same way. The image is inverted.

Letting $f = +20$ cm, $d_o = +30$ cm, $\ell_o = +6$ cm, we find from (36.1) and (36.2):

$$d_i = +60 \text{ cm} \quad (\text{real}) \qquad \ell_i = -12 \text{ cm} \quad (\text{inverted})$$

For the assumed values, the image is magnified.

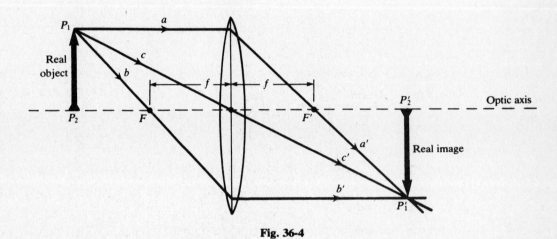

Fig. 36-4

36.5. Repeat Problem 36.4 for a real object inside F.

Figure 36-5 gives the ray construction; the image is virtual, erect, and magnified. Letting $f = +15$ cm, $d_o = +10$ cm, $\ell_o = 2$ cm in (36.1) and (36.2), we find:

$$d_i = -30 \text{ cm} \quad (\text{virtual}) \qquad \ell_i = +6 \text{ cm} \quad (\text{erect, magnified})$$

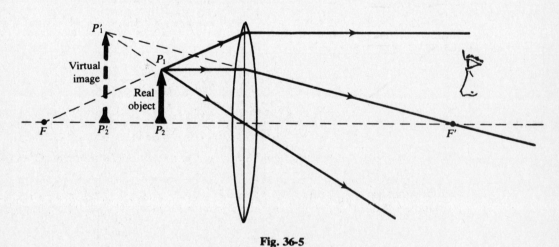

Fig. 36-5

36.6. Describe the image of a real object formed by a thin diverging (negative) lens.

Figure 36-6 gives the ray construction, which is independent of the relationship between d_o and f. The image is virtual, erect, and minified.

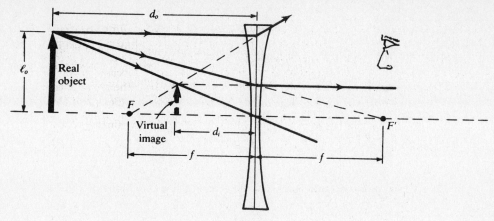

Fig. 36-6

Checking with $f = -30$ cm, $d_o > 0$, $\ell_o = 10$ cm,

$$\frac{1}{d_o} + \frac{1}{d_i} = -\frac{1}{30} \qquad \text{or} \qquad d_i = -\frac{30 d_o}{30 + d_o} \quad \text{(virtual)}$$

and
$$\frac{30 + d_o}{30} = -\frac{10}{\ell_i} \qquad \text{or} \qquad \ell_i = \frac{10}{1 + (d_o/30)} \quad \text{(erect, minified)}$$

36.7. Describe the image formed by a plane mirror of a virtual object.

A virtual object is an intercepted real image (produced by some auxiliary optical system). Thus, in Fig. 36-7, the auxiliary converging lens L forms the real image I_1 which serves as a virtual object for plane mirror M. The ray construction shows the image to be real (the reflected rays a', b', c' converge towards it), erect (same orientation as I_1), and of the same size and distance from the mirror as I_1.

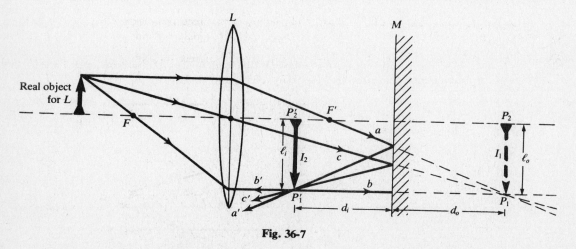

Fig. 36-7

We can check the result using (*36.1*) and (*36.2*). Assuming a focal length of +15 cm for the lens, a real object at a distance of 45 cm from the lens, and a separation of 18 cm between lens and mirror, we find from (*36.1*):

$$\frac{1}{45} + \frac{1}{x} = \frac{1}{15} \qquad \text{or} \qquad x = 22.5 \text{ cm}$$

i.e. I_1 is $22.5 - 18 = 4.5$ cm behind M. Then, using $d_o = -4.5$ cm (virtual object), $\ell_o > 0$, and $f = \infty$ in (*36.1*) and (*36.2*), we obtain

$$d_i = +4.5 \text{ cm} \quad \text{(real, same distance)} \qquad \ell_i = \ell_o \quad \text{(erect, same size)}$$

We could say that the net effect of M has been to move I_1 9.0 cm closer to the lens.

36.8. Describe the image formed by a positive lens of a virtual object.

In Fig. 36-8, the positive lens, L_2, intercepts the real image I_1 produced by a lens L_1 (not shown); I_1 is the virtual object for L_2. Applying (*36.1*), with $d_o < 0$ (virtual object) and $f > 0$, we obtain

$$-\frac{1}{|d_o|} + \frac{1}{d_i} = \frac{1}{f} \qquad \text{or} \qquad d_i = \left(\frac{|d_o|}{|d_o| + f}\right)f$$

which shows that the image is real and situated inside the back focal point of the lens. Moreover, (*36.2*) gives, for $\ell_o > 0$,

$$\ell_i = \left(\frac{f}{|d_o| + f}\right)\ell_o$$

and so the image is erect and minified. These conclusions are borne out by Fig. 36-8.

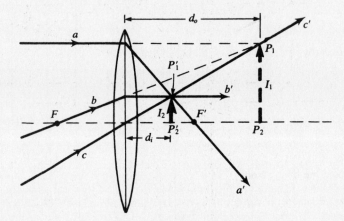

Fig. 36-8

36.9. Describe the image formed by a negative lens when a virtual object is situated inside the back focal point.

In Fig. 36-9, the negative lens, L_2, intercepts the real image I_1 produced by a lens L_1 (not shown). I_1 is the virtual object for L_2. Applying (*36.1*), with $f < d_o < 0$, we obtain

$$d_i = \frac{d_o f}{d_o - f} > 0 \quad \text{(real)}$$

and (*36.2*), with $\ell_o > 0$, gives

$$\ell_i = -\frac{\ell_o f}{d_o - f} > 0 \quad \text{(erect)}$$

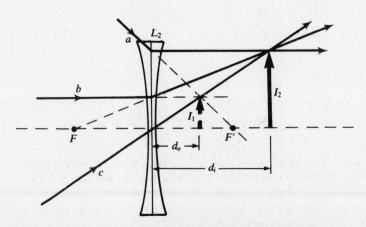

Fig. 36-9

36.10. Figure 36-10 shows a combination of two thin lenses (a *compound lens*). A real object is presented to L_1. (*a*) Find the image distance d_i in terms of the object distance d_o and the parameters of the system. (*b*) Show that as $s \to 0$ the system becomes equivalent to a single thin lens of focal length f, where

$$\frac{1}{f} = \frac{1}{f_1} + \frac{1}{f_2}$$

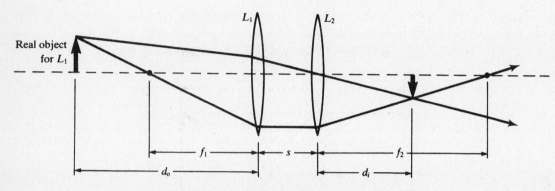

Fig. 36-10

(*a*) First locate the image from L_1 alone:

$$\frac{1}{d_o} + \frac{1}{x} = \frac{1}{f_1} \qquad \text{or} \qquad x = \frac{f_1 d_o}{d_o - f_1}$$

This image serves as the (real or virtual) object for L_2 alone:

$$\frac{1}{s - x} + \frac{1}{d_i} = \frac{1}{f_2}$$

whence

$$d_i = \frac{f_2(s - x)}{s - x - f_2} = f_2 \frac{s - [f_1 d_o/(d_o - f_1)]}{s - [f_1 d_o/(d_o - f_1)] - f_2} \qquad (1)$$

Depending on the relative sizes of d_o and the parameters, d_i may be positive (real image) or negative (virtual image).

Observe that no assumptions have been made about the signs of f_1 and f_2; thus (*1*) holds for any pair of thin lenses.

(*b*) As $s \to 0$, (*1*) becomes

$$d_i = \frac{f_2 f_1 d_o/(d_o - f_1)}{[f_1 d_o/(d_o - f_1)] + f_2} \qquad \text{or} \qquad \frac{1}{d_i} = \frac{1}{f_2} + \frac{1}{f_1} - \frac{1}{d_o} \qquad \text{or} \qquad \frac{1}{d_o} + \frac{1}{d_i} = \frac{1}{f}$$

which last is the Gaussian formula for a thin lens of focal length $f = (f_1^{-1} + f_2^{-1})^{-1}$.

Supplementary Problems

36.11. Referring to Fig. 36-2, show by ray construction and also by relation (*36.1*) that if a real object is placed in the position of $P_1' P_2'$, a real image will be formed in the position of $P_1 P_2$.

36.12. In Fig. 36-3, let $f = 35$ cm. A virtual object, of size 6 cm, is located behind the mirror at a distance of 70 cm. Describe the image formed.
   Ans. A real erect image, of size 2 cm, is formed 23.33 cm in front of the mirror.

36.13. A convex (negative) mirror has a focal length of -25 cm. A real object is placed in front at $d_o = 30$ cm. Locate the image. Is it erect?   *Ans.*   An erect virtual image 13.64 cm behind mirror.

36.14. A virtual object is located behind the mirror of Problem 36.13, at a distance of 35 cm. Locate the image.   *Ans.*   A virtual image 87.5 cm behind the mirror.

36.15. The radius of curvature of a positive mirror is 40 cm, $d_i = 50$ cm, $\ell_i = 10$ cm. Find f, d_o, ℓ_o.
      Ans. $f = +20$ cm; $d_o = 33.333$ cm; $\ell_o = 6.67$ cm

36.16. In Fig. 36-6, a virtual object is located inside F' on the emergent side of the negative lens. Describe the image.   *Ans.*   The image is real, erect, and on the emergent side.

Chapter 37

Interference and Diffraction of Light

37.1 INTERFERENCE

Unless amplitudes are very large, sinusoidal light waves obey the principle of superposition (Section 20.4). However, the *intensity*—the time-average flux of power in the direction of propagation—of the sum of two waves may not be equal to the sum of the two intensities, a phenomenon known as *interference*. Interference occurs only between waves from *coherent* sources, sources that have a fixed phase difference between them. If the sources are *incoherent*, the two intensities simply add.

Let observation point P be located at distances r_1 and r_2, respectively, from two coherent sources of light of wavelength λ. For simplicity, assume that the waves from the two sources are plane polarized in the same plane. Then, by a calculation similar to that in Problem 20.9, it can be shown that the resultant intensity, I, at P is given in terms of the intensities, I_1 and I_2, at P from the two sources, by

$$I = I_1 + I_2 + I_{12} \qquad \text{where} \qquad I_{12} \equiv 2\sqrt{I_1 I_2}\cos\phi \qquad (37.1)$$

The term I_{12}, which measures the interference, depends on the difference in phase, ϕ, between the two waves at P, which in turn is determined by the *path difference* $\Delta r \equiv r_1 - r_2$ and the *phase difference* $\Delta\phi' \equiv \phi_1' - \phi_2'$ of the two sources:

$$\phi = \frac{2\pi}{\lambda}\Delta r + \Delta\phi' \qquad (37.2)$$

In the special case $I_1 = I_2 = I_0$ (37.1) becomes

$$I = 4I_0 \cos^2\frac{\phi}{2} \qquad (37.3)$$

Interference is *constructive* at points P where $I_{12} > 0$. In particular, there will be an *intensity maximum* if

$$\cos\phi = +1 \qquad \text{or} \qquad \phi = 2\pi m \quad (m = 0, \pm 1, \pm 2, \ldots)$$

Thus, in the Young's double-slit arrangement, Fig. 37-1(a), where the two slits act as in-phase sources ($\Delta\phi' = 0$) of equal intensities, intensity maxima are located by

$$\Delta r = m\lambda \qquad \text{or} \qquad d\sin\theta = m\lambda \qquad \text{or} \qquad y = \frac{mD\lambda}{d}$$

for θ small; (37.3) gives the value of the maximum intensity as $4I_0$.

Interference is *destructive* at points P where $I_{12} < 0$. In particular, *intensity minima* are found where

$$\cos\phi = -1 \qquad \text{or} \qquad \phi = 2\pi(m + \tfrac{1}{2}) \quad (m = 0, \pm 1, \pm 2, \ldots)$$

In Fig. 37-1(a), these minima are located by

$$\Delta r = (m + \tfrac{1}{2})\lambda \qquad \text{or} \qquad d\sin\theta = (m + \tfrac{1}{2})\lambda \qquad \text{or} \qquad y = \frac{(m + \tfrac{1}{2})D\lambda}{d}$$

that is, they are midway between the maxima; the value of the minimum intensity is 0. Both the

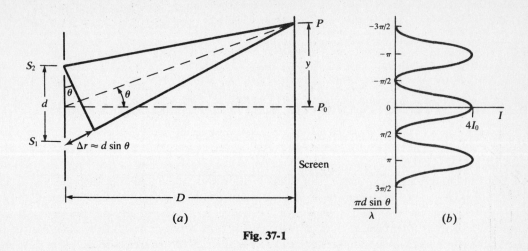

Fig. 37-1

maxima and minima are shown in Fig. 37-1(b), which is the graph of (37.3) for the double-slit arrangement.

The generalization of (37.3) to an N-slit arrangement is

$$I = I_0 \frac{\sin^2 (N\phi/2)}{\sin^2 (\phi/2)} \qquad (37.4)$$

where I_0 is the intensity at P from one slit, and ϕ is the phase difference between waves arriving at P from adjacent slits. As in the case $N = 2$, there are equal intensity maxima, of magnitude $N^2 I_0$, whose angular positions satisfy

$$d \sin \theta = m\lambda \quad (m = 0, \pm 1, \pm 2, \ldots)$$

As was shown in Problem 20.12, electromagnetic waves will undergo a 180° phase shift when reflected by an optically denser medium. Thus, interference effects are possible when light is reflected from (or transmitted through) two closely spaced interfaces.

37.2 DIFFRACTION

For plane light waves (i.e. the wavefronts are planes) incident normally on an opaque surface with an aperture, *Huygens' principle* states that the radiation field on the far side of the aperture is the same as would be produced by identical sources distributed uniformly over the area of the aperture and oscillating coherently in phase. These secondary sources would illuminate regions within the geometrical shadow of the barrier; hence the name *diffraction* ("breaking" or "bending"). The phenomenon is simplest at large distances behind the barrier, where it is called *Fraunhofer diffraction*. If the aperture has the form of a long narrow slit, of width w [Fig. 37-2(a)], the intensity distribution in Fraunhofer diffraction is given by

$$I = I_0 \frac{\sin^2 (\phi/2)}{(\phi/2)^2} \qquad (37.5)$$

where $\phi \equiv (2\pi/\lambda)w \sin \theta$ is the phase difference, at observation angle θ, between the two edges of the slit. Figure 37-2(b) shows the graph of (37.5).

A *diffraction grating* consists of a large number, N, of parallel slits, ruled lines, or grooves. For normal incidence, its Fraunhofer pattern can be shown to be the N-slit interference pattern, (37.4), modulated by the single-slit diffraction pattern, (37.5). Thus, there are strong, equally spaced, interference peaks located by

$$d \sin \theta = m\lambda$$

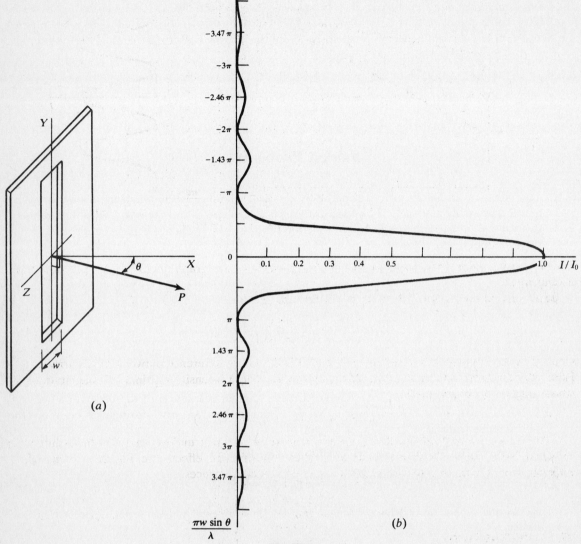

Fig. 37-2

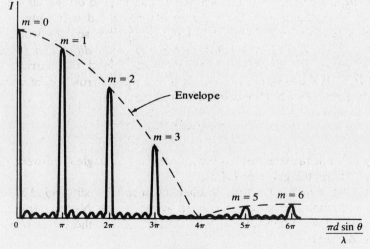

Fig. 37-3

(the *grating equation*), but the heights of these peaks conform to the diffraction envelope, as shown in Fig. 37-3. The value of m $(0, 1, 2, \ldots$; we need only consider $\theta \geq 0)$ corresponding to a given peak is called its *order*. Certain orders may be absent from the grating pattern, depending on the relation between the slit width w and the slit spacing d.

Solved Problems

37.1. Two identical radiators have a separation $d = \lambda/8$, where λ is the wavelength of the waves emitted by either source. The phase difference $\Delta\phi'$ of the sources is $\pi/4$. Find the intensity distribution in the radiation field as a function of the angle θ which specifies the direction from the radiators to the observation point P [see Fig. 37-1(a)].

Since $\Delta r = (\lambda/8) \sin \theta$, the phase difference at P is

$$\phi = \frac{2\pi}{\lambda} \Delta r + \Delta\phi' = \frac{\pi}{4} (\sin \theta + 1)$$

and

$$I(\theta) = 4I_0 \cos^2 \frac{\phi}{2} = 4I_0 \cos^2 \left[\frac{\pi}{8} (\sin \theta + 1) \right]$$

37.2. A Young's double-slit experiment is set up using a light source of wavelength 500 nm and placing slits 2 m from a screen. Given that the angular resolution of the observer's eye is 1′ (0.000291 rad), how far apart are the two slits if he can just distinguish between the interference fringes?

Refer to Fig. 37-1(a). Since the bright fringes are located by

$$\sin \theta_m \approx \theta_m = \frac{m\lambda}{d}$$

the angular separation of adjacent fringes is λ/d. Thus the condition for seeing adjacent fringes as distinct is

$$\frac{\lambda}{d_{max}} = 0.000291 \qquad \text{or} \qquad d_{max} = \frac{500 \times 10^{-9}}{0.000291} = 1.72 \text{ mm}$$

37.3. In a Young's double-slit experiment, the slits are 2 mm apart and are illuminated with a mixture of two wavelengths, $\lambda = 750$ nm and $\lambda' = 900$ nm. At what minimum distance from the common central bright fringe on a screen 2 m from the slits will a bright fringe from one interference pattern coincide with a bright fringe from the other?

Refer to Fig. 37-1(a). The mth bright fringe of the λ-pattern and the m'th bright fringe of the λ'-pattern are located at

$$y_m = \frac{mD\lambda}{d} \qquad \text{and} \qquad y'_{m'} = \frac{m'D\lambda'}{d}$$

Equating these distances gives

$$\frac{m}{m'} = \frac{\lambda'}{\lambda} = \frac{900}{750} = \frac{6}{5}$$

Hence, the first position at which overlapping occurs is

$$y_6 = y'_5 = \frac{(6)(2)(750 \times 10^{-9})}{2 \times 10^{-3}} = 23 \text{ mm}$$

37.4. Consider a thin slab of thickness t, as shown in Fig. 37-4. A beam of light, a, in medium 1 at almost-normal incidence to the interface with the slab, medium 2, is split into a reflected ray, b, and refracted ray, d. The refracted ray is (partially) reflected on the bottom interface and finds its way back into medium 1 as a ray parallel to b. Find the interference conditions for rays b and d.

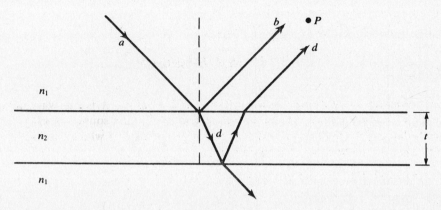

Fig. 37-4

We compute the difference in phase, ϕ, between the two rays at P. Ray d has traveled an extra distance $2t$ (assuming nearly-normal incidence), in a medium where the wavelength is $\lambda_2 = (n_1/n_2)\lambda_1$. Moreover, one of the rays (b, if $n_2 > n_1$) has suffered a 180° phase change in reflection. Consequently,

$$\phi = \frac{2\pi}{\lambda_2}(2t) - \pi = 2\pi\left(\frac{2n_2 t}{n_1 \lambda_1} - \frac{1}{2}\right)$$

The condition for maximum constructive interference is then

$$\phi = 2\pi m \qquad \text{or} \qquad \frac{2n_2 t}{n_1 \lambda_1} = m + \tfrac{1}{2} \quad (m = 0, 1, 2, \ldots)$$

whereas the condition for maximum destructive interference is

$$\phi = 2\pi(m + \tfrac{1}{2}) \qquad \text{or} \qquad \frac{2n_2 t}{n_1 \lambda_1} = m + 1 \quad (m = 0, 1, 2, \ldots)$$

37.5. By an anodizing process, a transparent film of aluminum oxide, of thickness $t = 250$ nm and index of refraction $n_2 = 1.80$, is deposited on a sheet of polished aluminum. What is the color of utensils made from this sheet when observed in white light? Assume normal incidence of the light.

We must find which colors in the visible region, having vacuum wavelengths from 400 nm (violet) to 700 nm (red), will interfere constructively, and which destructively. From Problem 37.4, with $n_1 = 1$ (air), maximum constructive interference occurs for

$$\lambda_1 = \frac{2(n_2/n_1)t}{m + \tfrac{1}{2}} = \frac{(900 \text{ nm})}{m + \tfrac{1}{2}} \quad (m = 0, 1, 2, \ldots)$$

Only the value corresponding to $m = 1$, i.e. $\lambda_1 = 600$ nm (orange), falls in the visible range.
Maximum destructive interference occurs for

$$\lambda_1 = \frac{2(n_2/n_1)t}{m + 1} = \frac{(900 \text{ nm})}{m + 1} \quad (m = 0, 1, 2, \ldots)$$

and of these values, only $\lambda_1 = 450$ nm (violet) is in the visible range.

We infer that the red-orange-yellow end of the spectrum will be strongly reflected, while the violet-blue end will be greatly diminished in intensity as compared with the illuminating white light.

37.6. A radar antenna is located atop a high cliff on the edge of a lake. This antenna operates on a wavelength of 400 m. Venus rises above the horizon and is tracked by the antenna. The first minimum in the signal reflected off the surface of Venus is recorded when Venus is 35° above the horizon. Find the height of the cliff.

The radar antenna receives signals directly from Venus and by reflection from the surface of the lake (Fig. 37-5). Venus may be considered infinitely far away, so that the disturbances at B and D have the same phases at every instant. The path difference between the reflected and direct rays is

$$\overline{BE} - \overline{DE} = \frac{y}{\sin 35°}(1 - \sin 20°)$$

and there is a phase shift of π due to the reflection at B. Thus the phase difference at E is

$$\phi = \frac{2\pi y}{\lambda \sin 35°}(1 - \sin 20°) + \pi$$

and for an intensity minimum this must equal $\pi, 3\pi, 5\pi, \ldots$. Ruling out $\phi = \pi$, which would imply $y = 0$, we have $\phi = 3\pi$, whence

$$y = \frac{\lambda \sin 35°}{1 - \sin 20°} = \frac{(400)(0.5736)}{1 - 0.3420} = 349 \text{ m}$$

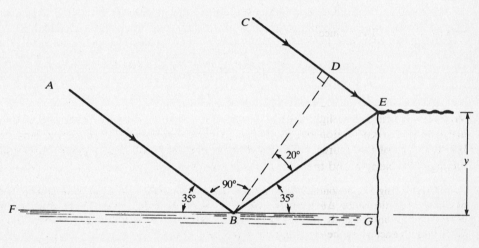

Fig. 37-5

37.7. A slit is located "at infinity" in front of a lens of focal length 1 m and is illuminated normally with light of wavelength 600 nm. The first minima on either side of the central maximum of the diffraction pattern observed in the focal plane of the lens are separated by 4 mm. What is the width of the slit?

A fundamental property of any focusing system, e.g. a lens, is that all rays traverse the same optical path length through it. Thus the lens does not alter the diffraction pattern of the slit. In fact, since rays through the center of the lens are undeviated, the net effect of the lens is to put the slit a distance f in front of the viewing screen.

From (37.5), or Fig. 37-2(b), the angle of the first minimum on the positive side of the central maximum is given by

$$\frac{\pi w \sin \theta}{\lambda} = \pi \qquad \text{or} \qquad \sin \theta = \frac{\lambda}{w}$$

where w is the slit width.   But

$$\sin \theta \approx \tan \theta = \frac{2 \text{ mm}}{1 \text{ m}} = 0.002$$

Therefore,

$$w = \frac{\lambda}{0.002} = \frac{600 \times 10^{-9}}{0.002} = 0.3 \text{ mm}$$

37.8. Figure 37-2(b) indicates that the first secondary peak in the single-slit Fraunhofer diffraction pattern is 0.047 times as high as the central peak.   Verify this value.

By (37.5) the first two intensity minima (zeros) are located at $\phi = 2\pi$ and $\phi = 4\pi$.   Supposing that the ϕ-value for the first secondary peak is midway between these two values, i.e. $\phi = 3\pi$, we have

$$I = I_0 \frac{\sin^2 (3\pi/2)}{(3\pi/2)^2} = 0.045 \, I_0$$

The error resulting from our assumption is only about 4%.

37.9. A light source emits a mixture of wavelengths from 450 to 600 nm.   When a diffraction grating is illuminated normally by this source, it is noted that two adjacent spectra barely overlap at an angle of 30°.   How many lines per meter are ruled on the grating?

According to the grating equation, the long-wavelength limit of the mth-order spectrum and the short-wavelength limit of the $(m + 1)$th-order spectrum will just coincide if

$$d \sin \theta = m (600 \times 10^{-9}) = (m + 1)(450 \times 10^{-9})$$

whence $m = 3$.   Then, since $\theta = 30°$,

$$\frac{1}{d} = \frac{\sin 30°}{3(600 \times 10^{-9})} = 116\,700 \text{ lines/m}$$

37.10. When monochromatic light from a long, very narrow source—say, a slit in a screen—falls normally on a diffraction grating, each principal maximum in the Fraunhofer pattern is a bright image of the source.   Express the angular width of such a *spectral line* in terms of the incident wavelength and the width of the grating.

Disregarding the secondary maxima and minima (see Fig. 37-3), we can assume that a principal maximum fills the angle $\Delta \theta$ between two consecutive minima.   From (37.4), then, we infer that the central maximum extends from $N\phi/2 = -\pi$ to $N\phi/2 = +\pi$; that is, $\Delta \phi = 4\pi/N$.   This same phase difference characterizes the other principal maxima.   But

$$\phi = \frac{2\pi d}{\lambda} \sin \theta$$

and so

$$\Delta \phi = \frac{2\pi d}{\lambda} \Delta(\sin \theta) = \frac{2\pi d}{\lambda} \cos \theta \, \Delta \theta = \frac{4\pi}{N}$$

or

$$\Delta \theta = \frac{2\lambda}{Nd \cos \theta} \tag{1}$$

Equation (1) applies to the mth-order peak when $\cos \theta$ is evaluated from the grating equation.   Notice that $\Delta \theta$ is inversely proportional to the width, Nd, of the grating.   Hence, the wider the grating, the sharper the spectral lines.

37.11. Reconsider Problem 37.10 when the source emits a continuous mixture of wavelengths.   The *resolving power*, at wavelength λ, of the grating is defined as $R \equiv \lambda/\delta$, where δ is the smallest wavelength difference for which the spectral lines $\lambda - \frac{1}{2}\delta$ and $\lambda + \frac{1}{2}\delta$ are resolvable.   Find R_m, the resolving power in the mth order.

According to *Rayleigh's criterion*, two peaks are just resolvable when their angular separation is half the angular width of either peak. This minimal separation is given by (*1*) of Problem 37.10 as

$$(\Delta\theta)_{min} = \frac{\lambda}{Nd\cos\theta} \qquad \text{or} \qquad d\cos\theta\,(\Delta\theta)_{min} = \frac{\lambda}{N}$$

Differentiation of the grating equation, $d\sin\theta = m\lambda$, gives

$$d\cos\theta\,\Delta\theta = m\,\Delta\lambda \qquad \text{whence} \qquad d\cos\theta\,(\Delta\theta)_{min} = m\delta$$

Comparing the two expressions for $d\cos\theta\,(\Delta\theta)_{min}$ gives $R_m = mN$.

Note that the resolving power is the same for all wavelengths.

Supplementary Problems

37.12. Light consisting of two wavelengths, 559.0 nm and 559.5 nm, falls normally on a plane diffraction grating with 200 000 rulings per meter. What is the angular separation of the two lines observed in the second-order spectrum?   *Ans.* $d\theta = 2.05 \times 10^{-4}$ rad

37.13. In a Young's double-slit experiment employing 600 nm light, the zero-order bright fringe of the interference pattern is centrally placed on the viewing screen. When a thin film of transparent material of refractive index 1.50 is placed over one of the slits, the central position is now found to be occupied by the fifth-order bright fringe. What is the thickness of the inserted film?   *Ans.* 6 μm

37.14. In a double-slit experiment, light from a sodium vapor lamp falls normally upon two parallel slits which are 2.0 mm apart. The distance between adjacent fringes is 9.4 mm when the screen is placed 3.20 m from the slits. Find the wavelength of the yellow light.   *Ans.* 590 nm

37.15. Show that in a Newton's rings experiment (Fig. 37-6) in which the fringe pattern is viewed in reflected light which has struck the plane surface nearly normally, $r_{m+s}^2 - r_m^2 = Rs\lambda$, where λ is the wavelength of the light used, R is the radius of curvature of the bottom face of the lens, and r_{m+s} and r_m are the radii of the $(m+s)$th and mth rings of the pattern.

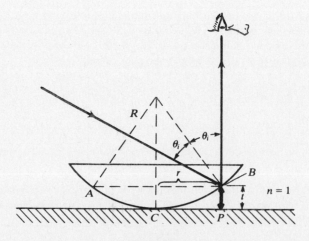

Fig. 37-6

37.16. A long narrow slit, 2 μm wide, is illuminated by waves of light of wavelength 1000 nm. What is the angle θ corresponding to the first minimum of intensity in the Fraunhofer diffraction pattern?   *Ans.* 30°

37.17. A diffraction grating having 100 000 lines/m is illuminated at normal incidence.   An intensity maximum is seen at an angle of 20° to the normal.   What is the wavelength of the light, if this is a first-order maximum?   *Ans.*   384 nm

37.18. How many grooves must a diffraction grating have if it is just to resolve the sodium doublet (589.592 nm and 588.995 nm) in the second-order spectrum?   *Ans.*   494

Chapter 38

Special Relativity

38.1 THE TWO BASIC POSTULATES

Recall, from Section 4.3, the definition of an *inertial frame* as a coordinate system with respect to which Newton's first law of motion is valid.

Postulate I. All laws of physics, expressed in some consistent system of units of length, mass, time, etc., have exactly the same mathematical forms and involve the same numerical constants in all inertial frames. That is, so far as the laws of physics (mechanics, as well as electricity and magnetism) are concerned, *all inertial frames are equivalent.*

Postulate II. The speed of light in empty space has the unique value $c = 2.997925 \times 10^8$ m/s when measured in any inertial frame. This is true regardless of the motion of the source of light.

38.2 CONSEQUENCES OF THE POSTULATES

The Lorentz Transformations

Let two inertial frames, $\mathscr{S}$ and $\mathscr{S}'$, be in motion at constant relative velocity v along the common X, X' axis (Fig. 38-1). It is supposed that a clock is fixed to *each point* throughout either inertial frame and that all clocks of a particular frame are synchronized. Also, for convenience, it is assumed that when the origins O and O' coincide, $t = t' = 0$.

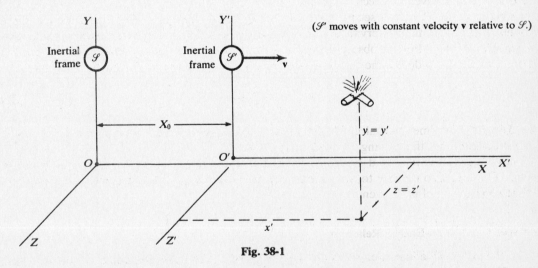

Fig. 38-1

Let an *event* be specified in $\mathscr{S}$ by space coordinates x, y, z and time coordinate t, and the same event in $\mathscr{S}'$ by space coordinates x', y', z' and time coordinate t'. Then the two sets of coordinates are related by

$$x = \frac{x' + vt'}{\sqrt{1 - (v^2/c^2)}} \qquad y = y' \qquad z = z' \qquad t = \frac{t' + (vx'/c^2)}{\sqrt{1 - (v^2/c^2)}} \qquad (38.1)$$

or by

$$x' = \frac{x - vt}{\sqrt{1 - (v^2/c^2)}} \qquad y' = y \qquad z' = z \qquad t' = \frac{t - (vx/c^2)}{\sqrt{1 - (v^2/c^2)}} \qquad (38.2)$$

Note that, in accordance with Postulate I, one obtains (38.2) from (38.1), or vice versa, by interchanging primed and unprimed coordinates and changing the sign of v.

The Lorentz coordinate transformations imply the following transformation between $\mathbf{u}' = (u'_x, u'_y, u'_z)$, the velocity of a particle as measured in $\mathscr{S}'$, and $\mathbf{u} = (u_x, u_y, u_z)$, its velocity in $\mathscr{S}$:

$$u_x = \frac{u'_x + v}{1 + (v/c^2)u'_x} \qquad u_y = \frac{u'_y\sqrt{1 - (v^2/c^2)}}{1 + (v/c^2)u'_x} \qquad u_z = \frac{u'_z\sqrt{1 - (v^2/c^2)}}{1 + (v/c^2)u'_x} \qquad (38.3)$$

or, inversely,

$$u'_x = \frac{u_x - v}{1 - (v/c^2)u_x} \qquad u'_y = \frac{u_y\sqrt{1 - (v^2/c^2)}}{1 - (v/c^2)u_x} \qquad u'_z = \frac{u_z\sqrt{1 - (v^2/c^2)}}{1 - (v/c^2)u_x} \qquad (38.4)$$

Length Contraction

Imagine a rigid rod fixed along the X' axis in Fig. 38-1. An observer in $\mathscr{S}'$, at rest with respect to the rod, measures its length as ℓ', the *rest length*. However, an observer in $\mathscr{S}$, for whom the rod is moving with velocity $\mathbf{v}$, measures its length as

$$\ell = \ell'\sqrt{1 - (v^2/c^2)} \qquad (38.5)$$

Since $\ell < \ell'$, the rod is contracted (relative to its rest length) for the $\mathscr{S}$-observer. If the rod has an arbitrary fixed orientation in $\mathscr{S}'$, then (38.5) would apply only to the X- and X'-projections of the rod; both observers would find the same lengths for the Y- and Y'-projections and for the Z- and Z'-projections.

Time Dilation

Suppose that an event E_1 occurs at point $P(x', y', z')$ in $\mathscr{S}'$ at time t'_1, as measured by the clock in $\mathscr{S}'$ attached to point P. Let a second event, E_2, occur *at the same point P* at time t'_2, as measured by the same clock. The time interval $\Delta t' \equiv t'_2 - t'_1$ measured in $\mathscr{S}'$ is known as the *proper time interval* between E_1 and E_2. For an observer in $\mathscr{S}$ the two events occur *at different locations*, so that he requires two clocks to determine the time interval $\Delta t \equiv t_2 - t_1$ between E_1 and E_2. We have:

$$\Delta t = \frac{\Delta t'}{\sqrt{1 - (v^2/c^2)}} \qquad (38.6)$$

Since $\Delta t > \Delta t'$, the time interval is dilated (relative to the proper time interval) for the $\mathscr{S}$-observer; to him it appears that the moving $\mathscr{S}'$-clock runs slow.

Caution: (38.6) is *not* valid for two events that occur at different locations in both reference frames. One must go directly to the Lorentz transformations (38.1) to obtain the relation between the time separations of such events.

Mass-Speed and Mass-Energy Relations

Let the mass of a particle, measured while at rest in an inertial frame $\mathscr{S}$, be m_0, the *rest mass*. Then if the particle moves with velocity $\mathbf{u}$ relative to $\mathscr{S}$, its mass (measured in $\mathscr{S}$) is

$$m = \frac{m_0}{\sqrt{1 - (u^2/c^2)}} \qquad (38.7)$$

and its total energy (measured in $\mathscr{S}$) is

$$E = mc^2 = \frac{E_0}{\sqrt{1 - (u^2/c^2)}} \qquad (38.8)$$

where $E_0 \equiv m_0 c^2$ is the *rest energy* of the particle. The rest mass, and hence the rest energy, includes the contributions of all the forms of energy—gravitational, electromagnetic, thermal, etc.—possessed by the particle when at rest in $\mathscr{S}$. The difference $E - E_0$ is defined as the *kinetic energy*, K, of the particle; i.e.

$$E = m_0 c^2 + K$$

Total energy E of an isolated system is conserved under Special Relativity, but rest energy E_0—and hence rest mass E_0/c^2—is not separately conserved. (See Problem 38.11.)

Solved Problems

In all problems, $\mathscr{S}$ and $\mathscr{S}'$ are inertial frames whose relative motion is as indicated in Fig. 38-1. The speed of light is taken as $c = 2.997925 \times 10^8$ m/s.

38.1. At time $t' = 4 \times 10^{-4}$ s, as measured in $\mathscr{S}'$, a particle is at the point $x' = 10$ m, $y' = 4$ m, $z' = 6$ m. (Note that this constitutes an event.) Compute the corresponding values of x, y, z, t, as measured in $\mathscr{S}$, for (a) $v = +500$ m/s, (b) $v = -500$ m/s, (c) $v = 2 \times 10^8$ m/s.

Apply (*38.1*).

(a)
$$x = \frac{10 + (500)(4 \times 10^{-4})}{\sqrt{1 - (500^2/c^2)}} \approx 10 + (500)(4 \times 10^{-4}) = 10.2 \text{ m}$$

$$y = 4 \text{ m}$$

$$z = 6 \text{ m}$$

$$t = \frac{(4 \times 10^{-4}) + (500)(10)/c^2}{\sqrt{1 + (500^2/c^2)}} \approx 4 \times 10^{-4} \text{ s}$$

(b)
$$x = \frac{10 - (500)(4 \times 10^{-4})}{\sqrt{1 - (500^2/c^2)}} \approx 10 - (500)(4 \times 10^{-4}) = 9.8 \text{ m}$$

$$y = 4 \text{ m}$$

$$z = 6 \text{ m}$$

$$t = \frac{(4 \times 10^{-4}) - (500)(10)/c^2}{\sqrt{1 - (500^2/c^2)}} \approx 4 \times 10^{-4} \text{ s}$$

(c)
$$x = \frac{10 + (2 \times 10^8)(4 \times 10^{-4})}{\sqrt{1 - (2/2.997925)^2}} = \frac{8.001 \times 10^4}{0.744943} = 1.0740 \times 10^5 \text{ m}$$

$$y = 4 \text{ m}$$

$$z = 6 \text{ m}$$

$$t = \frac{(4 \times 10^{-4}) + (2 + 10^8)(10)/c^2}{\sqrt{1 - (v^2/c^2)}} = 5.36984 \times 10^{-4} \text{ s}$$

38.2. Suppose that at $t = 6 \times 10^{-4}$ s, the space coordinates of a particle are $x = 1100$ m, $y = 100$ m, $z = 300$ m, all values measured in the "stationary" $\mathscr{S}$-frame. Compute corresponding values as measured in the "moving" $\mathscr{S}'$-frame, if $v = 2 \times 10^8$ m/s.

Apply (38.2):

$$x' = \frac{1100 - (2 \times 10^8)(6 \times 10^{-4})}{\sqrt{1 - (2/2.997925)^2}} = -1.5961 \times 10^5 \text{ m}$$

$$y' = 100 \text{ m}$$

$$z' = 300 \text{ m}$$

$$t' = \frac{(6 \times 10^{-4}) - (2 \times 10^8)(1100/c^2)}{\sqrt{1 - (v^2/c^2)}} = 8.02145 \times 10^{-4} \text{ s}$$

38.3. At $t = 10^{-3}$ s (in $\mathscr{S}$), an explosion occurs at $x = 5$ km. What is the time of the event for the $\mathscr{S}'$-observer, if for him it occurs at $x' = 35.354$ km?

From the first relation (38.2),

$$35.354 \times 10^3 = \frac{(5 \times 10^3) - v(10^{-3})}{\sqrt{1 - (v^2/c^2)}} \qquad \text{or} \qquad v = -3 \times 10^7 \text{ m/s}$$

Then, from the fourth relation (38.2),

$$t' = \frac{10^{-3} + (3 \times 10^7)(5 \times 10^3)/c^2}{\sqrt{1 - (v^2/c^2)}} = 1.0067 \times 10^{-3} \text{ s}$$

38.4. Suppose that at the instant origin O' coincides with O (at $t = t' = 0$), a flashbulb is exploded at this common origin. According to observers in $\mathscr{S}$, a spherical wavefront expands outward from O at speed c. Show that, even though $\mathscr{S}'$ is moving relative to $\mathscr{S}$ with velocity v, observers in $\mathscr{S}'$ note an exactly similar wavefront expanding outward from O'.

The equation of the wavefront in $\mathscr{S}$ is

$$x^2 + y^2 + z^2 = c^2 t^2 \tag{1}$$

The Lorentz transformation of (1) is

$$\left[\frac{x' + vt}{\sqrt{1 - (v^2/c^2)}} \right]^2 + (y')^2 + (z')^2 = c^2 \left[\frac{t' + (vx'/c^2)}{\sqrt{1 - (v^2/c^2)}} \right]^2$$

which reduces to
$$(x')^2 + (y')^2 + (z')^2 = c^2(t')^2 \tag{2}$$

Equation (2) represents a spherical wavefront expanding outward from O' at speed c.

Actually, the reasoning goes the other way round. The $\mathscr{S}$- and $\mathscr{S}'$-observers must see exactly the same kind of wave, by Postulates I and II. This (almost) forces the relations between the coordinates in the two frames to have a particular form—that of the Lorentz transformations. See Problem 38.19.

38.5. A spaceship of (rest) length 100 m takes 4 μs to pass an observer on earth. What is its velocity relative to the earth?

The observer measures the length of the spaceship to be

$$\ell = 100\sqrt{1 - (v^2/c^2)} \quad \text{(m)}$$

where v is the velocity of the ship relative to him. He must then find the flyby time to be ℓ/v. Thus,

$$100\sqrt{\frac{1}{v^2} - \frac{1}{c^2}} = 4 \times 10^{-6}$$

from which $v = 0.083 c$.

Alternatively, from the viewpoint of the spaceship, it takes a time

$$\Delta t = \frac{4 \times 10^{-6}}{\sqrt{1 - (v^2/c^2)}} \quad \text{(s)}$$

for the earth observer to pass by. Equating this to $100/v$ (s) yields the same value for v.

38.6. A clock is fastened to the $\mathscr{S}'$-frame at some point (x', y', z'). At a certain moment it reads exactly 200 seconds. Later the same clock reads 300 seconds. (Two events at the same location in $\mathscr{S}'$). Find the time interval between these two events, as measured in the "stationary" frame $\mathscr{S}$, if (a) $v = 2 \times 10^5$ m/s, (b) $v = 2 \times 10^8$ m/s.

(a)
$$\Delta t = \frac{100}{\sqrt{1 - (v^2/c^2)}} = 100.000\,022\,3 \text{ s}$$

(b)
$$\Delta t = 134.2385 \text{ s}$$

In (a) the time intervals are practically equal, but in (b) the result indicates that the $\mathscr{S}'$-clock runs quite slowly as compared with the one in $\mathscr{S}$.

38.7. At a certain location in $\mathscr{S}'$ it requires 45 minutes to bake a potato. What time is required for this process according to measurements made in $\mathscr{S}$? Let $v = 2 \times 10^8$ m/s.

$$\Delta t = \frac{45}{\sqrt{1 - (v^2/c^2)}} = 60.4 \text{ min}$$

38.8. A man drives a nail at point $P_1(4000 \text{ m}, -100 \text{ m}, 200 \text{ m})$ of $\mathscr{S}$; the clock at P_1 records $t = 15\,\mu$s. At the same moment—as read on another, synchronized clock located at $P_2(3000 \text{ m}, 500 \text{ m}, 150 \text{ m})$—his brother drives a nail at P_2. (Two events occur simultaneously in $\mathscr{S}$.) Compute the time of each event as determined in $\mathscr{S}'$, if $v = 2 \times 10^8$ m/s.

Applying the fourth relation (38.2),

$$t_1' = \frac{(15 \times 10^{-6}) - (2 \times 10^8)(4000)/c^2}{\sqrt{1 - (v^2/c^2)}} = 8.186940\,\mu\text{s}$$

$$t_2' = \frac{(15 \times 10^{-6}) - (2 \times 10^8)(3000)/c^2}{\sqrt{1 - (v^2/c^2)}} = 11.1741490\,\mu\text{s}$$

The events do not occur simultaneously in $\mathscr{S}'$. Hence, from the point of view of $\mathscr{S}$, the $\mathscr{S}'$-clocks are not synchronized, even though the $\mathscr{S}'$-clocks have been synchronized among themselves.

38.9. An electron has velocity $\mathbf{u}'$ relative to $\mathscr{S}'$, the components of which are

$$u_x' = 6 \times 10^7 \text{ m/s} \qquad u_y' = 4 \times 10^7 \text{ m/s} \qquad u_z' = 3 \times 10^7 \text{ m/s}$$

Find the velocity components as measured in $\mathscr{S}$, as well as the magnitude of $\mathbf{u}$.

From (38.3),

$$u_x = \frac{(6 \times 10^7) + (2 \times 10^8)}{1 + (6 \times 10^7)(2 \times 10^8)/c^2} = 2.293744 \times 10^8 \text{ m/s}$$

$$u_y = \frac{(4 \times 10^7)\sqrt{1 - (v^2/c^2)}}{1 + (6 \times 10^7)(2 \times 10^8)/c^2} = 2.62878 \times 10^7 \text{ m/s}$$

Likewise, $u_z = 1.97159 \times 10^7$ m/s. Then,

$$u = (u_x^2 + u_y^2 + u_z^2)^{1/2} = 2.3171616 \times 10^8 \text{ m/s}$$

38.10. Conservation of linear momentum of an isolated system holds in Special Relativity; indeed, it is this principle that leads to the nonconstancy of mass, (38.7). (a) Derive the relation between momentum and total energy. (b) Calculate the momentum of a 1 MeV electron.

(a) Square (38.7) and then multiply both sides by $c^4[1 - (u^2/c^2)]$, obtaining

$$m^2 c^4 - m^2 u^2 c^2 = m_0^2 c^4 \qquad \text{or} \qquad (mc^2)^2 = (m_0 c^2)^2 + (muc)^2$$

But $mc^2 = E$, the total energy; $m_0c^2 = E_0$, the rest energy; $mu = p$, the magnitude of the momentum vector.   Thus the desired relation is

$$E^2 = E_0^2 + (pc)^2 \qquad\qquad (38.9)$$

Because of the form of (38.9), momenta are often specified in MeV/c, on the atomic level, where energies are given in eV or MeV.   (See Problem 38.20.)

(b)   The total energy of the electron is its rest energy, 0.511 MeV, plus its kinetic energy, 1 MeV.   Then (38.9) gives

$$(1.511)^2 = (0.511)^2 + (pc)^2$$

from which $pc = 1.42$ MeV, or $p = 1.42$ MeV/c.

38.11. Show that rest mass is not conserved in a symmetrical, perfectly inelastic collision.

Let two identical bodies, each with rest mass m_0, approach each other at equal speeds u, collide, and stick together.   By momentum conservation, the conglomerate body must be at rest, with energy M_0c^2.   The initial energy of the system was $2[m_0c^2/\sqrt{1-(u^2/c^2)}]$.   Hence,

$$\frac{2m_0c^2}{\sqrt{1-(u^2/c^2)}} = M_0c^2 \qquad \text{or} \qquad M_0 = \frac{2m_0}{\sqrt{1-(u^2/c^2)}}$$

It is seen that the final rest mass, M_0, exceeds the initial rest mass, $2m_0$.

In the "collision" of elementary particles to form the nucleus, the rest mass decreases (see Problem 38.12).

38.12. The rest masses of the proton, neutron, and deuteron are:

$$m_p = 1.67261 \times 10^{-27} \text{ kg} \qquad m_n = 1.67492 = 10^{-27} \text{ kg} \qquad m_d = 3.34357 \times 10^{-27} \text{ kg}$$

The deuteron (the nucleus of heavy hydrogen) consists of a proton and a neutron.   How much energy should be liberated in the formation of a deuteron from a free proton and a free neutron, initially at rest "at infinity"?

The rest mass lost in the formation of a deuteron is

$$(m_p + m_n) - m_d = (1.67261 + 1.67492 - 3.34357) \times 10^{-27} = 3.96 \times 10^{-30} \text{ kg}$$

which is equivalent to a rest-energy loss

$$\Delta E_0 = 3.96 \times 10^{-30} c^2 = 3.56 \times 10^{-13} \text{ J} = 2.22 \text{ MeV}$$

Hence, by energy conservation, the surroundings must gain exactly 2.22 MeV of energy.   This same amount of energy, the *binding energy* of the deuteron, would have to be supplied to the deuteron to tear it apart into an infinitely separated proton and neutron.

38.13. At rest on the earth a spaceship has mass $m_0 = 10\,000$ kg.   In flight its mass, as determined from earth, is $m = 10\,100$ kg.   What is its speed u relative to the earth?

Regarding an earth frame as inertial and applying (38.7),

$$10\,100 = \frac{10\,000}{\sqrt{1-(u^2/c^2)}}$$

from which $u = 4.2082 \times 10^7$ m/s.

38.14. The rest mass of a proton is $m_0 = 1.672614 \times 10^{-27}$ kg.   Its laboratory speed after having fallen through a high difference in potential, ΔV, is $u = 2 \times 10^8$ m/s.   (a) Evaluate ΔV.   (b) Find the total energy of the proton.

(a) The kinetic energy of the proton is

$$K = (m - m_0)c^2 = \left[\frac{1}{\sqrt{1-(u^2/c^2)}} - 1\right]m_0c^2 = e\,\Delta V$$

from which

$$\Delta V = \frac{[(1/0.744943) - 1](1.672614 \times 10^{-27})(2.997925 \times 10^8)^2}{1.602 \times 10^{-19}} = 321\ \text{MV}$$

(b) The rest energy of the proton is $m_0c^2 = 938.3$ MeV, and so the total energy is

$$E = 938.3 + 321 = 1259.3\ \text{MeV}$$

38.15. Find the increase in mass of 100 kg of copper ($C = 0.389$ kJ/kg $\cdot$ K) if its temperature is increased 100 °C.

$$\Delta E = mC\,\Delta T = (100)(0.389)(100) = 3890\ \text{kJ}$$

$$\Delta m = \frac{\Delta E}{c^2} = 4.33 \times 10^{-11}\ \text{kg}$$

Supplementary Problems

38.16. As determined in $\mathscr{S}$, the origin O' of $\mathscr{S}'$ is struck by a meteor at time t. What is the time of the strike as read on the $\mathscr{S}'$-clock at O'? *Ans.* $t' = t\sqrt{1 - (v^2/c^2)}$

38.17. An event happens at $x' = 1000$ m on the X' axis; at that moment the clock attached to this point reads $t_1' = 3 \times 10^{-4}$ s. Later, when another event occurs at the same point, the clock reads $t_2' = 8 \times 10^{-4}$ s. What is the time interval between the events (a) as measured in $\mathscr{S}'$? (b) as measured in $\mathscr{S}$? Let $v = 2 \times 10^8$ m/s. *Ans.* (a) 5×10^{-4} s; (b) 6.712×10^{-4} s

38.18. An electron leaves an electron gun at O' with velocity components

$$u_x' = 12\,000\ \text{km/s} \qquad u_y' = 8000\ \text{km/s} \qquad u_z' = 9000\ \text{km/s}$$

If $v = 2 \times 10^5$ km/s, what is the electron's speed (a) in $\mathscr{S}'$? (b) in $\mathscr{S}$?
Ans. (a) 17 000 km/s; (b) 206 670 km/s

38.19. Verify that under the Lorentz transformation the wave equation in $\mathscr{S}$,

$$\frac{\partial^2\phi}{\partial x^2} + \frac{\partial^2\phi}{\partial y^2} + \frac{\partial^2\phi}{\partial z^2} = \frac{1}{c^2}\frac{\partial^2\phi}{\partial t^2}$$

goes over into the wave equation in $\mathscr{S}'$,

$$\frac{\partial^2\phi}{\partial x'^2} + \frac{\partial^2\phi}{\partial y'^2} + \frac{\partial^2\phi}{\partial z'^2} = \frac{1}{c^2}\frac{\partial^2\phi}{\partial t'^2}$$

38.20. Compute the conversion factor between the ordinary units of momentum and MeV/c.
Ans. 1 MeV/c = 0.534×10^{-21} kg $\cdot$ m/s

38.21. The rest mass of a lithium atom is $m_0 = 1.15224 \times 10^{-26}$ kg. At what speed (relative to the laboratory) would its mass be double this value? *Ans.* 2.59628×10^8 m/s

38.22. A rectangular tank is fixed in $\mathscr{S}'$, with its edges parallel to the coordinate axes.   If the tank is filled with liquid of density ρ, as measured in $\mathscr{S}'$, what is the density as measured in $\mathscr{S}$?

Ans. $\dfrac{\rho}{1-(v^2/c^2)}$

38.23. Refer to Problem 38.22.   Show that the $\mathscr{S}'$-observer and the $\mathscr{S}$-observer agree as to the rate at which mass is being lost from the tank by evaporation.

Chapter 39

Photons

39.1 DUAL NATURE OF LIGHT

Electromagnetic radiation, considered as a wave, is characterized by its frequency ν or its wavelength λ, which are related by

$$\lambda\nu = c \tag{39.1}$$

The same radiation, considered as a stream of quanta, called *photons*, is characterized by the energy E or the momentum p of the individual photons. Photons are identical, uncharged, zero-rest-mass particles that travel at the single speed c (in any inertial frame). If the radiation, in its wave aspect, has frequency ν, each constituent photon is of energy

$$E = h\nu \tag{39.2}$$

where h is Planck's constant [Problem 32.5(b)]. The momentum of each photon is, from (38.9) with $E_0 = 0$,

$$p = \frac{E}{c} = \frac{h\nu}{c} = \frac{h}{\lambda} \tag{39.3}$$

On the wave picture, the intensity of a beam of electromagnetic radiation is directly proportional to the square of the amplitude [see (20.9)]; the proportionality factor turns out to be $c\epsilon_0/2 \approx 1.33 \times 10^{-3}$ S. On the particle picture, the intensity is just NE, where N is the number of photons per unit time crossing a unit area perpendicular to the beam and E is the energy of each photon. Hence,

$$I = NE = \frac{c\epsilon_0 E_0^2}{2} \quad (\text{W/m}^2) \tag{39.4}$$

where E_0 is the amplitude of the electric field.

39.2 PHOTOELECTRIC EFFECT

The release of electrons from a material under the action of radiation is called the *photoelectric effect*. Einstein's *photoelectric equation* is

$$K_{max} = h\nu - W_{min} \tag{39.5}$$

Here, K_{max} is the maximum kinetic energy (nonrelativistic) of the electrons ejected from a metal surface, $h\nu$ is the energy of the photons striking the surface, and W_{min}, the *work function* of the metal, is the binding energy of the least tightly bound electron in the metal atom.

The *stopping potential, V_s*, is the value of the retarding potential difference that is just sufficient to halt the most energetic photoelectrons; i.e. $eV_s = K_{max}$. The *threshold frequency, ν_c*, is the frequency of the incident electromagnetic wave below which there is no photoelectric current, regardless of the incident intensity; i.e. $h\nu_c = W_{min}$. Einstein's equation may then be rewritten as

$$eV_s = h(\nu - \nu_c) \tag{39.6}$$

334

39.3 COMPTON SCATTERING

When a photon of wavelength λ collides with a free electron of rest mass m_e, the electron recoils and the scattered photon has less energy than the incident photon, and therefore a longer wavelength, λ'. *Compton's equation* is

$$\lambda' - \lambda = \frac{h}{m_e c}(1 - \cos\theta) \tag{39.7}$$

where θ is the scattering angle, shown in Fig. 39-1. The term $h/m_e c$, called the *Compton wavelength*, has the value $0.0243\,\text{Å} = 2430\,\text{fm}$. For the Compton effect to be significant, the incident wavelength should be comparable to the Compton wavelength (i.e., in the X-ray or γ-ray region).

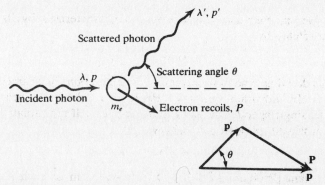

Fig. 39-1. Conservation of Momentum in Compton Scattering

39.4 PAIR ANNIHILATION, PAIR PRODUCTION

The electron and the positron constitute a pair of *antiparticles*. They have equal masses m_e but opposite charges, and they can annihilate each other when they come close enough to interact significantly. In the usual case, two photons remain after the annihilation, so that conservation of energy for the process reads

$$E_{\text{electron}} + E_{\text{positron}} = E_{\text{photon}} + E_{\text{photon}}$$

If the two antiparticles annihilate at rest, the law of conservation of momentum requires that the two photons have momentum vectors in opposite directions and of equal magnitudes (which means equal photon energies).

If a photon has enough energy, it can produce a pair of antiparticles, an electron and a positron. The law of conservation of momentum requires the presence of a third particle, ordinarily the nucleus of a massive atom, to carry off some momentum. The kinetic energy obtained by this nucleus is negligible compared to the photon energy. The law of conservation of energy for this pair-production process is

$$E_{\text{photon}} = E_{\text{electron}} + E_{\text{positron}} = 2m_e c^2 + K_{\text{electron}} + K_{\text{positron}}$$

Solved Problems

39.1. What is the energy in a quantum of radiation of wavelength 700 nm?

Using the convenient conversion $hc = 19.865 \times 10^{-26}\,\text{J}\cdot\text{m} = 1241\,\text{eV}\cdot\text{nm}$, we have

$$E = \frac{hc}{\lambda} = \frac{1241}{700} = 1.77\,\text{eV}$$

39.2. A sensor is exposed for 0.1 s to a 200-W lamp 10 m away. The sensor has an opening 20 mm in diameter. How many photons enter the sensor, if the wavelength of the light is 600 nm? Assume all the energy of the lamp is given off as light.

The energy of a photon of the light is

$$E = \frac{hc}{\lambda} = \frac{(6.63 \times 10^{-34})(3 \times 10^8)}{600 \times 10^{-9}} = 3.3 \times 10^{-19} \text{ J}$$

The lamp uses 200 W of power. The number of photons emitted per second is therefore

$$n = \frac{200}{3.3 \times 10^{-19}} = 6.1 \times 10^{20} \text{ photons/s}$$

Since the radiation is spherically symmetrical, the number of photons entering the sensor per second is n multiplied by the ratio of the aperture area to the area of a sphere of radius 10 m:

$$(6.1 \times 10^{20}) \frac{\pi (0.010)^2}{4\pi (10)^2} = 1.53 \times 10^{14} \text{ photons/s}$$

and the number of photons that enter the sensor in 0.1 s is $(0.1)(1.53 \times 10^{14}) = 1.53 \times 10^{13}$.

39.3. A desk lamp illuminates a desk top with violet light of wavelength 412 nm. The amplitude of this electromagnetic wave is 63.2 V/m. Find the number N of photons striking the desk per second per unit area, assuming the illumination is normal.

From $E = hc/\lambda$ and (39.4),

$$N = \frac{\lambda \epsilon_0 E_0^2}{2h} = \frac{(412 \times 10^{-9})(8.85 \times 10^{-12})(63.2)^2}{2(6.63 \times 10^{-34})} = 1.10 \times 10^{19} \text{ photons/s} \cdot \text{m}^2$$

39.4. Light of wavelength 600 nm falls on a metal having photoelectric work function 2 eV. Find (a) the energy of a photon, (b) the kinetic energy of the most energetic photoelectron, and (c) the stopping potential.

(a)
$$E = \frac{hc}{\lambda} = \frac{1241 \text{ eV} \cdot \text{nm}}{600 \text{ nm}} = 2.07 \text{ eV}$$

(b)
$$K_{max} = E - W_{min} = 2.07 - 2 = 0.07 \text{ eV}$$

(c)
$$eV_s = K_{max} = 0.07 \text{ eV} \qquad \text{or} \qquad V_s = 0.07 \text{ V}$$

39.5. In what amounts to the inverse of the photoelectric effect, X-ray photons are produced when a tungsten target is bombarded by accelerated electrons. If an X-ray machine has an accelerating potential of 60 000 V, what is the shortest wavelength present in its radiation?

Since the work function of tungsten is very much smaller than the accelerating potential, we may suppose that the entire kinetic energy of an electron is lost to create a single photon of maximum energy:

$$eV = \frac{hc}{\lambda_{min}} \qquad \lambda_{min} = \frac{hc}{eV} = \frac{1241 \text{ eV} \cdot \text{nm}}{60\,000 \text{ eV}} = 0.0207 \text{ nm}$$

39.6. A surface has light of wavelength $\lambda_1 = 550$ nm incident on it, causing the ejection of photo-electrons for which the stopping potential is $V_{s1} = 0.19$ V. Suppose that radiation of wavelength $\lambda_2 = 190$ nm were incident on the surface. Calculate (a) the stopping potential V_{s2}, (b) the work function of the surface, (c) the threshold frequency for the surface.

(a) From (39.6), $e(V_{s2} - V_{s1}) = h(\nu_2 - \nu_1)$ or

$$V_{s2} = V_{s1} + \frac{h}{e}(\nu_2 - \nu_1) = V_{s1} + \frac{hc}{e}\left(\frac{1}{\lambda_2} - \frac{1}{\lambda_1}\right) = 0.19 + 1241\left(\frac{1}{190} - \frac{1}{550}\right) = 4.47 \text{ V}$$

(b)
$$W_{\min} = \frac{hc}{\lambda_1} - eV_{s1} = \frac{1241}{550} - 0.19 = 2.07 \text{ eV}$$

(c)
$$h\nu_c = W_{\min} \qquad \text{or} \qquad \nu_c = \frac{W_{\min}}{h} = \frac{(2.07)(1.602 \times 10^{-19})}{6.63 \times 10^{-34}} = 498 \text{ THz}$$

39.7. Verify Compton's equation, (39.7), when the photon is back-scattered ($\theta = 180°$).

See Fig. 39-1. Assuming the electron to be at rest initially, we have from conservation of momentum,

$$\frac{h}{\lambda} = P - \frac{h}{\lambda'} \qquad \text{or} \qquad P = h\left(\frac{1}{\lambda} + \frac{1}{\lambda'}\right)$$

and from conservation of relativistic energy,

$$\frac{hc}{\lambda} + E_0 = \frac{hc}{\lambda'} + E \qquad \text{or} \qquad E = E_0 + hc\left(\frac{1}{\lambda} - \frac{1}{\lambda'}\right)$$

Substituting these expressions for the electron's final momentum and final energy into the momentum-energy relation, (38.9), we obtain

$$\left[E_0 + hc\left(\frac{1}{\lambda} - \frac{1}{\lambda'}\right)\right]^2 = E_0^2 + \left[hc\left(\frac{1}{\lambda} + \frac{1}{\lambda'}\right)\right]^2$$

$$2E_0 hc\left(\frac{1}{\lambda} - \frac{1}{\lambda'}\right) - 2h^2 c^2 \frac{1}{\lambda\lambda'} = 2h^2 c^2 \frac{1}{\lambda\lambda'}$$

$$E_0(\lambda' - \lambda) = 2hc$$

whence $\lambda' - \lambda = 2hc/E_0 = 2h/m_e c$, which is Compton's equation with $\cos\theta = -1$.

39.8. Suppose that a beam of 0.2-MeV photons is scattered by the electrons in a carbon target. (a) What is the wavelength associated with these photons? (b) What is the wavelength of those photons scattered through an angle of 90°? (c) What is the energy of the scattered photons that emerge at an angle of 60° relative to the incident direction?

(a)
$$\lambda = \frac{hc}{E} = \frac{1241 \text{ MeV} \cdot \text{fm}}{0.2 \text{ MeV}} = 6200 \text{ fm}$$

(b)
$$\lambda' = \lambda + \frac{h}{m_e c}(1 - \cos\theta) = 6200 + (2430)(1 - 0) = 8630 \text{ fm}$$

(c)
$$\frac{hc}{\lambda'} = \frac{hc}{\lambda + (h/m_e c)(1 - \cos\theta)} = \frac{1241}{6200 + (2430)(1 - \frac{1}{2})} = 0.168 \text{ MeV}$$

39.9. Show that when a positron and an electron (both essentially at rest) annihilate, creating two photons, either photon has the Compton wavelength.

Total energy before annihilation is $2m_e c^2$; and after, $2hc/\lambda$ (momentum conservation requires that the photon energies be equal). Then, by conservation of energy,

$$2m_e c^2 = 2\frac{hc}{\lambda} \qquad \text{or} \qquad \lambda = \frac{h}{m_e c} = 2430 \text{ fm}$$

39.10. Show that the threshold wavelength for the production of a positron-electron pair is half the Compton wavelength.

The incident photon must have at least enough energy to create a pair having zero kinetic energy (we ignore any kinetic energy of the recoil nucleus). Thus,

$$\frac{hc}{\lambda} \geq 2m_e c^2 \qquad \text{or} \qquad \lambda \leq \frac{hc}{2m_e c} = 1215 \text{ fm}$$

Supplementary Problems

39.11. Find the energy of the photons in a beam whose wavelength is 526 nm.   *Ans.*   2.36 eV

39.12. In the photoionization of atomic hydrogen, what will be the maximum kinetic energy of the ejected electron when a 60-nm photon is absorbed by the atom?   The ionization energy of H is 13.6 eV. *Ans.*   7.0 eV

39.13. A tungsten surface is illuminated with ultraviolet light of wavelength 250 nm.   Electrons are ejected from the tungsten, and a potential of 2.4 V applied between the tungsten and another electrode is just enough to prevent any electrons from reaching the second electrode.   What is the work function W_{min} for this surface?   *Ans.*   2.62 eV

39.14. A potential drop of 100 kV is maintained between the anode and the cathode of an X-ray tube.   Find the shortest wavelength in the spectrum of the emitted X-rays.   *Ans.*   0.0125 nm

39.15. The photoelectric work function for a sheet of copper is 3.0 eV.   (*a*) What is the threshold frequency for the sample?   (*b*) If light of wavelength 310 nm is incident on the sheet, what is the maximum kinetic energy of the ejected electrons?   *Ans.*   (*a*) 720 THz;   (*b*) 1.06 eV

39.16. After pair annihilation, two 1-MeV photons move off in opposite directions.   If the electron and positron had the same kinetic energy, find its value.   *Ans.*   0.49 MeV

Chapter 40

The Bohr Atom

40.1 INTRODUCTION

An atom is specified by giving the composition of its nucleus, using the formalism $^A_Z X$. Here, X is the chemical symbol for the element, Z is the number of protons in the nucleus (which is equal to the number of electrons in the neutral atom), and A is the total number of nucleons (protons and neutrons).

EXAMPLE 40.1. The most abundant species of carbon is $^{12}_6 C$, which has 6 protons and 6 neutrons in the nucleus. The *atomic mass unit* (u) is chosen so as to make the mass of the $^{12}_6 C$ atom exactly 12 u. Since, by definition, 12 grams of $^{12}_6 C$ constitutes 1 mol, we have

$$1 \text{ u} = \frac{0.001 \text{ kg}}{N_0} = 1.660 \times 10^{-27} \text{ kg}$$

where N_0 is Avogadro's number (Section 17.1).

The proton and the neutron are almost equal in mass, and are far more massive than the electron. Consequently, the mass of the neutral atom $^A X$ will be very nearly A u.

The Bohr model applies to atomic hydrogen, $^1_1 H$; to singly ionized helium, $^3_2 He^+$ or $^4_2 He^+$; and, in general, to any *hydrogenlike atom*, by which is meant a neutral atom that has been stripped of all but one of its electrons. In the light of quantum mechanics, which forbids precise circular or elliptical orbits for the electron, the Bohr theory is incorrect. Nevertheless, it predicts very well the ionization energies and spectra of hydrogenlike atoms.

40.2 CLASSICAL ENERGY OF THE ATOM

Let a single electron revolve in a circular orbit of radius r about a stationary nucleus, of charge Ze. Then the total energy of the electron, which is the energy of the atom, is given by

$$E = -\frac{Ze^2}{8\pi\epsilon_0 r} \quad \text{(J)} \tag{40.1}$$

(See Problem 40.1.) Recall the approximate values

$$e \approx 1.602 \times 10^{-19} \text{ C} \qquad \epsilon_0 \approx \frac{10^{-9}}{36\pi} \text{ F/m}$$

40.3 POSTULATES OF THE BOHR MODEL

(1) The electron cannot occupy an orbit of arbitrary radius, but only one in which its angular momentum, mvr, is an integral multiple of $h/2\pi$:

$$mvr = \frac{nh}{2\pi} \quad (n = 1, 2, 3, \ldots) \tag{40.2}$$

Recall the values $h = 6.6262 \times 10^{-34} \text{ J} \cdot \text{s}$ and $m = 9.109 \times 10^{-31} \text{ kg}$ for Planck's constant and the electronic (rest) mass. The integer n is called the *quantum number* of the orbit.

339

(2) The electron does not radiate energy while revolving in one of the allowed orbits, although it should according to classical laws of electricity and magnetism.

(3) The electron can "fall" from an outer (higher-energy) orbit, in which the energy is E_h, into an inner (lower-energy) orbit of energy E_ℓ, and in so doing radiate a photon of energy

$$h\nu = E_h - E_\ell \tag{40.3}$$

(4) The electron can be "raised" from an inner orbit to an outer orbit by absorbing a quantum of energy as given by (40.3). This can be supplied by a bombarding electron or as radiant energy.

40.4 ENERGY LEVELS

Setting the Coulomb force between electron and nucleus equal to centripetal force gives

$$\frac{Ze^2}{4\pi\epsilon_0 r^2} = \frac{mv^2}{r} \tag{40.4}$$

Now solving (40.1), (40.2), (40.4) for r, v, and E as functions of n, we obtain

$$r_n = \frac{n^2 r_1^\circ}{Z} \quad \left(r_1^\circ \equiv \frac{\epsilon_0 h^2}{\pi m e^2} = 0.529 \text{ Å} = 0.0529 \text{ nm}\right) \tag{40.5}$$

$$v_n = \frac{Z v_1^\circ}{n} \quad \left(v_1^\circ \equiv \frac{e^2}{2\epsilon_0 h} = \frac{c}{137.0} \quad (\text{m/s})\right) \tag{40.6}$$

$$E_n = -\frac{Z^2 E_1^\circ}{n^2} \quad \left(E_1^\circ \equiv \frac{me^4}{8\epsilon_0^2 h^2} = 13.60 \text{ eV}\right) \tag{40.7}$$

The quantized energy values E_n are called the *energy levels* of the hydrogenlike atom. When $n = 1$, the atom has minimum energy; it is then said to be in the *ground level* or *ground state*. Note that r_1° and E_1° respectively represent the radius and ionization energy of $_1^1\text{H}$ (in the ground state); the values 0.529 Å and 13.60 eV are in good agreement with experiment.

40.5 ATOMIC SPECTRA

When the allowed energies, (40.7), and $\nu = c/\lambda$ are substituted in (40.3), one obtains for the wavelength of the emitted or absorbed photon

$$\frac{1}{\lambda} = Z^2 R_\infty \left(\frac{1}{n_\ell^2} - \frac{1}{n_h^2}\right) \tag{40.8}$$

where the *Rydberg constant* is given by

$$R_\infty \equiv \frac{me^4}{8\epsilon_0^2 h^3 c} = 1.09737 \times 10^{-3} \text{ Å}^{-1} = 1.09737 \times 10^{-2} \text{ nm}^{-1} \tag{40.9}$$

The above value for the Rydberg constant reflects our assumption of an infinitely massive, and therefore stationary, nucleus. The necessary correction of the Bohr model for a finite nuclear mass, M, amounts to replacing the electronic mass m in (40.5), (40.7), and (40.9) by

$$m' = \frac{mM}{m + M}$$

the *reduced mass* of the atom. The corrected theoretical value of the Rydberg constant for $_1^1\text{H}$ is then

$$R_H = \frac{R_\infty}{1 + (m/M)} = \frac{R_\infty}{1 + (1/1836)} = 1.096773 \times 10^{-3} \text{ Å}^{-1}$$

(The corrected value of E_1° is 13.598 eV.)

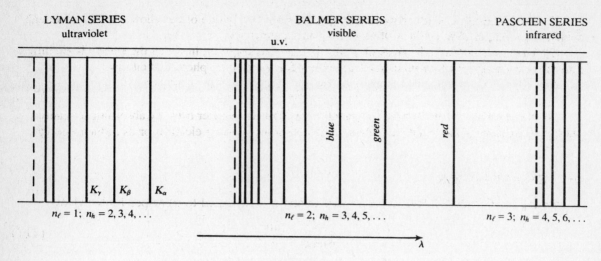

Fig. 40-1

Now the observed emission spectrum of atomic hydrogen is in fact described by an expression of exactly the type *(40.8)*, with an empirical constant of $1.096776 \times 10^{-3}\ \mathring{A}^{-1}$! The spectral "lines" (Problem 37.10) are grouped into series answering to a common value of n_ℓ in *(40.8)*. The three most prominent series for $^1_1$H are indicated (not to scale) in Fig. 40-1, where the dashed lines represent the *series limits* corresponding to $n_h = \infty$. The transitions responsible for these series are indicated in the energy-level diagram for $^1_1$H, Fig. 40-2.

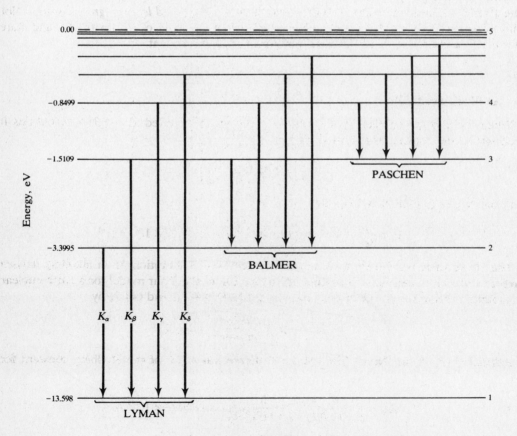

Fig. 40-2

Solved Problems

40.1. Derive *(40.1)*.

By *(25.6)*, the potential energy of the electron in the field of the nucleus is

$$U = (-e)\left(\frac{+Ze}{4\pi\epsilon_0 r}\right) = -\frac{Ze^2}{4\pi\epsilon_0 r}$$

Then, from Problem 13.21, the total energy of the electron is

$$E = \frac{1}{2}U = -\frac{Ze^2}{8\pi\epsilon_0 r}$$

40.2. Using the reduced mass of $^1_1$H, compute the radius of (a) the smallest Bohr orbit, (b) the fifth Bohr orbit.

(a)
$$r_{1H} = \left(1 + \frac{m}{M}\right)r_1^\circ = \left(1 + \frac{1}{1836}\right)(0.529) = 0.5293 \text{ Å}$$

(b)
$$r_5 = 25r_{1H} = 13.232 \text{ Å}$$

40.3. Compute the electron's speed and angular momentum in the two orbits of Problem 40.2.

(a) Total angular momentum, L_n, depends only on the quantum number, n. Thus

$$L_1 = 1\left(\frac{h}{2\pi}\right) = 1.055 \times 10^{-34} \text{ J} \cdot \text{s}$$

But, from Fig. 40-3, and the definition of the center of mass,

$$L_1 = mv_1 r_1^\circ + MV_1(r_{1H} - r_1^\circ) = mv_1 r_{1H} = mv_1\left(1 + \frac{m}{M}\right)r_1^\circ$$

Hence,
$$v_1 = \frac{L_1}{m\left(1 + \frac{m}{M}\right)r_1^\circ} = \frac{h/2\pi}{m\left(1 + \frac{m}{M}\right)(\epsilon_0 h^2/\pi m e^2)}$$

$$= \frac{v_1^\circ}{1 + (m/M)} = \frac{c/137.0}{1 + (1/1836)} = 2189 \text{ km/s}$$

(b)
$$L_5 = 5L_1 = 5.277 \times 10^{-34} \text{ J} \cdot \text{s} \qquad v_5 = \frac{v_1}{5} = 437.8 \text{ km/s}$$

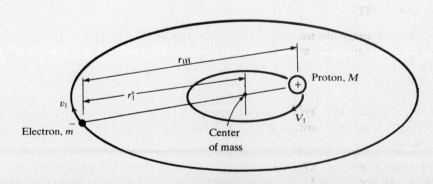

Fig. 40-3

40.4. Verify the values given in Fig. 40-2 for the first four energy levels of $^1_1$H.

Making the reduced-mass correction in (40.7), we have:

$$E_1 = -\frac{13.60}{1 + (1/1836)} = -13.598 \text{ eV} \qquad E_3 = \frac{E_1}{9} = -1.5109 \text{ eV}$$

$$E_2 = \frac{E_1}{4} = -3.3995 \text{ eV} \qquad\qquad E_4 = \frac{E_1}{16} = -0.8499 \text{ eV}$$

40.5. Calculate the wavelengths of the first three spectral lines of $^1_1$H in (a) the Lyman series, (b) the Balmer series.   Also find the series limits.

Use the Rydberg formula,

$$\frac{1}{\lambda} = 1.096776 \times 10^{-3} \left(\frac{1}{n_\ell^2} - \frac{1}{n_h^2} \right) \quad (\text{Å}^{-1})$$

(a)   For $n_\ell = 1$ and $n_h = 2, 3, 4, \infty$:

$$\lambda_1 = 1215.68 \text{ Å} \qquad \lambda_3 = 972.55 \text{ Å}$$

$$\lambda_2 = 1025.73 \text{ Å} \qquad \lambda_\infty = 911.76 \text{ Å}$$

(b)   For $n_\ell = 2$ and $n_h = 3, 4, 5, \infty$:

$$\lambda_1 = 6564.69 \text{ Å} \qquad \lambda_3 = 4341.72 \text{ Å}$$

$$\lambda_2 = 4862.74 \text{ Å} \qquad \lambda_\infty = 3647 \text{ Å}$$

40.6. The combination of fundamental quantities [see (40.6)]

$$\frac{e^2}{2\epsilon_0 hc} \equiv \alpha$$

is known as the *fine structure constant*.   Show that α is a pure number, of magnitude 1/137.

Unit-dimensionally,

$$\alpha = \left| \frac{C^2}{(F/m)(J \cdot s)(m/s)} \right| = \left| \frac{C^2}{F \cdot J} \right| = \left| \frac{C/F}{J/C} \right| = \left| \frac{V}{V} \right| = |1|$$

Numerically,

$$\frac{1}{\alpha} = \frac{2(8.85432 \times 10^{-12})(6.6262 \times 10^{-34})(2.997925 \times 10^8)}{(1.6021 \times 10^{-19})^2} = 137.05$$

40.7. What is the maximum wavelength of light that could accomplish the photoionization of ground-state $^1_1$H?

The bombarding photon must have enough energy to lift the electron from $n_\ell = 1$ to $n_h = \infty$. Thus its wavelength could not exceed that of the Lyman series limit, found in Problem 40.5(a) to be 911.76 Å.

40.8. A gas of monatomic hydrogen is bombarded with a stream of electrons that have been accelerated from rest through a potential difference of 12.75 V.   Which spectral lines should be emitted?

If an atom absorbs 12.75 eV of energy, it will be lifted from the ground level ($n = 1$) to the fourth level ($n = 4$); see Fig. 40-2.   Hence, in falling back to the ground level there is the possibility of exciting the first three Lyman, the first two Balmer, and the first Paschen lines.

40.9. (*a*) Compute the first three energy levels of doubly ionized lithium. What is its ionization potential? (*b*) Find the wavelengths of the first three lines of the *K*-series ($n_\ell = 1$; $n_h = 2, 3, 4$).

(*a*) With $Z = 3$ and $m' \approx m$, (*40.7*) becomes

$$E_n = -\frac{122.40 \text{ eV}}{n^2}$$

Hence $E_1 = -122.40$ eV (ionization potential is 122.40 V) and

$$E_2 = -30.60 \text{ eV} \qquad E_3 = -13.60 \text{ eV} \qquad E_4 = -7.65 \text{ eV}$$

(*b*) According to (*40.8*), the wavelengths for doubly ionized lithium ($Z = 3$) will be 1/9 the corresponding wavelengths for hydrogen. Thus, using the results of Problem 40.5(*a*),

$$\lambda_1 = 135.08 \text{ Å} \qquad \lambda_2 = 113.97 \text{ Å} \qquad \lambda_3 = 108.06 \text{ Å}$$

(The above values are slightly too large, since Problem 40.5 uses the empirical Rydberg constant for $^1_1$H instead of R_∞.)

40.10. Suppose that means were available for stripping 28 electrons from $_{29}$Cu in a vapor of this metal. (*a*) Compute the first three energy levels for the remaining electron. (*b*) Find the wavelengths of the spectral lines of the series for which $n_\ell = 1$; $n_h = 2, 3, 4$. What is the ionization potential for the last electron?

(*a*) Here $Z = 29$ and $m' \approx m$, and so the allowed energies will be $29^2 = 841$ times the corresponding energies for $_1$H:

$$E_1 = 841(-13.60) = -11\,438 \text{ eV} \qquad E_2 = \frac{E_1}{4} = -2859 \text{ eV} \qquad E_3 = \frac{E_1}{9} = -1271 \text{ eV}$$

(*b*) Dividing the wavelengths found in Problem 40.5(*a*) by 841,

$$\lambda_1 = 1.44 \text{ Å} \qquad \lambda_3 = 1.15 \text{ Å}$$
$$\lambda_2 = 1.22 \text{ Å} \qquad \lambda_\infty = 1.08 \text{ Å}$$

These wavelengths are in the X-ray region. The ionization potential is the voltage corresponding to λ_∞.

$$\frac{hc}{\lambda_\infty} = \frac{12.41 \text{ keV} \cdot \text{Å}}{1.08 \text{ Å}} = 11.5 \text{ keV}$$

i.e. an ionization potential of 11.5 kV.

Supplementary Problems

40.11. The first Balmer line of the hydrogen isotope deuterium lies at 6561.01 Å, just below the corresponding line for ordinary hydrogen, at 6562.80 Å. From this infer the formula $^2_1$H for deuterium. (*Hint*: The mass of the nucleus enters the Rydberg constant via the reduced mass.)

40.12. An electron and its antiparticle (Section 39.3), the positron, can form a bound system, called *positronium*. What should be the ionization potential of positronium? *Ans.* 6.80 V ($R_P = R_\infty/2$)

40.13. Compute, in μm, the wavelengths of the first three lines in the Paschen series (infrared) for $^1_1$H. *Ans.* 1.87563 μm; 1.28217 μm; 1.09412 μm

40.14. (*a*) Show that the frequency of revolution of the electron in the *n*th circular Bohr orbit of a hydrogenlike atom is given by

$$f = \frac{2Z^2 R_\infty c}{n^3}$$

(*b*) Verify the units s^{-1} or Hz for the right-hand side of the above relation.   (*c*) The mean lifetime of $_1^1$H in its first excited state is 10 ns.   How many revolutions does the electron make in that time?
Ans.   (*c*) 8.3×10^6

40.15. (*a*) The electron of $_1^1$H drops from the orbit for which $n = 501$ into that for which $n = 500$.   Find the frequency of radiation emitted.   (*b*) Use Problem 40.14(*a*) to calculate the frequency of revolution of the electron in the orbit $n = 500$.   (*c*) Compare the results of (*a*) and (*b*) in the light of Bohr's *correspondence principle*.
Ans.   (*a*) $\nu = 52.6$ MHz.   (*b*) $f = 52.6$ MHz.   (*c*) At high quantum numbers ($n \approx 500$), the photon frequency corresponds to the frequency of electromagnetic radiation predicted from the classical Maxwell's equations; namely, the frequency of the charged electron in its centripetally accelerated circular motion.

40.16. A continuous band of radiation having all wavelengths from about 1000 Å to 10 000 Å is passed through a gas of monatomic hydrogen.   The radiation is then examined with a spectrograph.   What results are found?
Ans.   The lines emitted by the excited hydrogen in the range 1000 Å to 10 000 Å are: 1215 Å and 1025.7 Å in the Lyman series, the entire Balmer series, and all but the first four lines in the Paschen series.

Chapter 41

The Nucleus

41.1 BINDING ENERGY OF STABLE NUCLEI

The binding energy of the deuteron, $^2_1$H, was calculated in Problem 38.12. More generally, the binding energy of the nuclide ^A_ZX, which is composed of Z protons and $N = A - Z$ neutrons, is given by

$$BE_X = (Zm_p + Nm_n - M)c^2 \tag{41.1}$$

where m_p, m_n, and M are the respective rest masses of a proton, a neutron, and the nucleus in question. By adding and subtracting Z electronic rest masses on the right of (41.1), and neglecting the comparatively small binding energies of these electrons, we can rewrite the expression in terms of rest masses of neutral atoms:

$$BE_X = (ZM_H + Nm_n - M_X)c^2 \tag{41.2}$$

Often the factor c^2 is dropped from (41.1) or (41.2), which in effect puts BE_X in mass units, such as u (Example 40.1). Note the values

$$M_H = 1.007825 \text{ u} \qquad m_n = 1.008665 \text{ u}$$

and the mass-energy equivalence $1 \text{ u} \leftrightarrow 931.5 \text{ MeV}$ (Problem 41.1).

41.2 RADIOACTIVE DECAY

Any unstable nucleus will decay into other particles (or photons), provided that the fundamental conservation laws of mass-energy, momentum, nucleons, and electric charge do not preclude such a decay. According to the *law of radioactive decay*, the expected number of undecayed nuclei (N) present in a sample at time t is

$$N = N_0 e^{-\lambda t} \tag{41.3}$$

where N_0 is the initial (large) number of unstable nuclei in the sample; λ, the *decay constant*, is specific for each radioactive nuclide. The relation (41.3) is a statistical one, independent of the exact mechanism of the decay. Its validity rests on the assumptions (i) that the nuclei decay independently of one another; (ii) that the probability that a given nucleus will decay within the next Δt seconds is $\lambda \, \Delta t$, irrespective of the age of the nucleus. See Problem 41.5.

The *half-life*, $T_{1/2}$, of an unstable species is the time after which any sufficiently large population of undecayed nuclei will have decreased to 1/2 its initial number. From (41.3),

$$\lambda T_{1/2} = \ln 2 \qquad \text{or} \qquad T_{1/2} = \frac{0.693}{\lambda} \tag{41.4}$$

Alternatively, the probability that an undecayed nucleus is still undecayed $T_{1/2}$ seconds from now is 1/2.

The *activity*, A, of a sample is its disintegration rate:

$$A \equiv -\frac{dN}{dt} = \lambda N = N_0 \lambda e^{-\lambda t} = A_0 e^{-\lambda t} \tag{41.5}$$

The SI unit of activity is the *becquerel* (1 Bq = 1 s^{-1}), but the older *curie* (1 Ci = 3.7×10^{10} s^{-1}) is still widely used.

41.3 NUCLEAR REACTIONS

A nuclear reaction in which a particle x strikes a nucleus X and results in the emission of a particle y and a residual nucleus Y is indicated as

$$x + X \rightarrow Y + y \qquad \text{or} \qquad X(x, y)Y$$

x is called the *bullet*; X is the *target*; y is the *product particle*; and Y is the *recoil nucleus*.

The *Q-value* of a nuclear reaction is the energy available from the difference in rest mass between the bullet plus the target and the product particle plus the recoil nucleus:

$$Q \equiv [(M_x + M_X) - (M_y + M_Y)]c^2 \tag{41.6}$$

$$= \text{(rest energy of input particles)} - \text{(rest energy of output particles)}$$

Since total energy $E = E_0 + K$ is conserved, we also have

$$Q = \text{(kinetic energy of output particles)} - \text{(kinetic energy of input particles)} \tag{41.7}$$

Depending on the reaction, Q may be positive, negative, or zero.   Because it depends only on rest masses, the Q-value of a reaction is the same in all inertial frames.

Solved Problems

41.1. If mass 1 u were completely converted into energy, how much energy would be made available?

Using Example 40.1,

$$E = mc^2 = (1.6604 \times 10^{-27} \text{ kg})(2.998 \times 10^8 \text{ m/s})^2$$
$$= 1.4923 \times 10^{-10} \text{ J} = 931.5 \text{ MeV}$$

41.2. Estimate the density of matter in the nucleus.   Take the radius of the nucleus as $R = (1.2 \times 10^{-15})A^{1/3}$ (m), where A is the mass number of the nucleus, and suppose the proton and neutron masses equal at 1.67×10^{-27} kg.

$$\rho = \frac{\text{mass}}{\text{volume}} = \frac{(1.67 \times 10^{-27})A}{(4/3)\pi R^3} = \frac{(1.67 \times 10^{-27})A}{(4/3)\pi (1.2 \times 10^{-15})^3 A} = 2.3 \times 10^{17} \text{ kg/m}^3$$

(Ordinary densities are of the order of 10^3 kg/m^3.)

41.3. Calculate (a) the binding energy of $^{16}_8$O and (b) the binding energy per nucleon.   The mass of neutral $^{16}_8$O is $M_O = 15.994915$ u.

(a) $\text{BE} = (ZM_H + Nm_n) - M_O = 8(1.007825) + 8(1.008665) - 15.994915 = +0.137005$ u

or 127.62 MeV.

(b) $\dfrac{127.62}{16} = 7.98$ MeV/nucleon

41.4. The *separation energy*, SE, is the minimum energy that must be provided to remove the least tightly bound nucleon from the nucleus.   Calculate this energy for $^{16}_8$O, if the least tightly bound nucleon is a proton.   The atomic mass of $^{15}_7$N is 15.000108 u and of $^{16}_8$O is 15.994915 u.

Upon the removal of a proton, $^1_1$H, from the $^{16}_8$O nucleus, a $^{15}_7$N nucleus is left:

$$^{16}_8\text{O} \rightarrow {}^{15}_7\text{N} + {}^1_1\text{H}$$

The SE is therefore, in terms of neutral atomic masses,

$$SE = (\text{atomic mass of } {}^{15}_{7}N + M_H) - (\text{atomic mass of } {}^{16}_{8}O)$$
$$= (15.000108 + 1.007825) - 15.994915 = 0.013018 \text{ u}$$

or 12.13 MeV.

If the nucleus is interpreted according to the *liquid drop model*, the seeming paradox that the SE is larger than the $\overline{BE}$ [calculated as 7.98 MeV in Problem 41.3(b)] can be explained as a "volume effect," whereby the nucleus possesses binding energy $E_v = aA$. The larger the total number of nucleons A, the more difficult it will be to remove the individual proton from the nucleus. It will take a large amount of energy to remove the first proton. But as more and more nucleons are removed, the volume effect and other effects decrease and the BE continues to decrease for the remaining nucleons. Thus the $\overline{BE}$ for ${}^{16}_{8}O$ is smaller than its SE.

41.5. Derive (41.3), the statistical law of radioactive decay.

Starting with a large number, N_0, of a species of unstable nucleus, let $N(t)$ be the expected number of nuclei still undecayed after a time t. [That is, $N(t)$ would be the average number of nuclei observed if the experiment were repeated many times.] Consider what might happen in the infinitesimal time interval $(t, t + \Delta t)$. Since any one of the $N(t)$ nuclei has probability $\lambda \Delta t$ of decaying, the *expected* number that do decay is $N(t) \times (\lambda \Delta t)$. [Compare: "If 1000 fair dice are thrown, the expected number showing a 5 is 1000 $(\frac{1}{6})$."] Consequently, one has

$$N(t + \Delta t) = N(t) - N(t) \times (\lambda \Delta t) \qquad \text{or} \qquad \frac{N(t + \Delta t) - N(t)}{\Delta t} = -\lambda N(t)$$

which becomes, in the limit as $\Delta t \to 0$,

$$\frac{dN}{dt} = -\lambda N$$

Integrating,

$$\int_{N_0}^{N} \frac{dN}{N} = -\lambda \int_{0}^{t} dt$$

$$\ln \frac{N}{N_0} = -\lambda t$$

$$N = N_0 e^{-\lambda t}$$

When N_0 is very large, as it almost always is, the expected number, N, of undecayed nuclei may be identified with the number actually observed.

41.6. Show that the average lifetime of an unstable nucleus is $T_{avg} = 1/\lambda$.

From Problem 41.5, the fraction of nuclei that have lifetime t (i.e. that decay between times t and $t + dt$) is given by

$$\frac{N(t) \times (\lambda \, dt)}{N_0} = e^{-\lambda t} \lambda \, dt$$

Hence

$$T_{avg} = \int_{0}^{\infty} t e^{-\lambda t} \lambda \, dt = \frac{1}{\lambda}$$

41.7. If a radioactive nuclide is produced at the constant rate of n per second (say, by bombarding a target with neutrons), find (a) the (expected) number of nuclei in existence t seconds after the number is N_0, (b) the maximum (expected) number of these radioactive nuclei.

(a) The population N is simultaneously increasing at rate n and decreasing, by decay, at rate λN. Thus, the net rate of increase is

$$\frac{dN}{dt} = n - \lambda N$$

Separating the variables and integrating, we have

$$\int_{N_0}^{N} \frac{dN}{n - \lambda N} = \int_0^t dt$$

$$\frac{1}{\lambda} \ln \frac{n - \lambda N_0}{n - \lambda N} = t$$

$$N = \frac{n}{\lambda} + \left(N_0 - \frac{n}{\lambda}\right) e^{-\lambda t}$$

(b) From the result of (a), it is seen that $N(t)$ starts at N_0 and asymptotically decreases or increases to n/λ, according as $N_0 > n/\lambda$ or $n/\lambda > N_0$. The maximum number of nuclei is thus the larger of N_0 and n/λ.

It is supposed here that n/λ, like N_0, is very large; otherwise the law of radioactive decay, and therefore the result of (a), ceases to be valid at some moment.

41.8. A substance X disintegrates into a radioactive substance Y. Find N_y, given that the initial amounts of X and Y are N_{x_0} and N_{y_0}, respectively.

X decays at the rate

$$\frac{dN_x}{dt} = -\lambda_x N_x = -\lambda_x N_{x_0} e^{-\lambda_x t}$$

For each X-nucleus that disintegrates, one of Y is formed. So Y is formed at the rate $\lambda_x N_x$ nuclei per second. But at the same time Y is being formed, it is disintegrating at the rate $\lambda_y N_y$ nuclei per second. The net increase in Y-nuclei per second is therefore

$$\frac{dN_y}{dt} = \lambda_x N_x - \lambda_y N_y = \lambda_x N_{x_0} e^{-\lambda_x t} - \lambda_y N_y$$

To integrate this, transpose $\lambda_y N_y$ and multiply both sides by $e^{\lambda_y t}\, dt$:

$$e^{\lambda_y t}\, dN_y + \lambda_y N_y e^{\lambda_y t}\, dt = \lambda_x N_{x_0} e^{(\lambda_y - \lambda_x)t}\, dt$$

$$N_y e^{\lambda_y t} = \frac{\lambda_x N_{x_0}}{\lambda_y - \lambda_x} e^{(\lambda_y - \lambda_x)t} + C$$

When $t = 0$, $N_y = N_{y_0}$, which gives

$$C = N_{y_0} - \frac{\lambda_x N_{x_0}}{\lambda_y - \lambda_x}$$

Thus

$$N_y = \left(N_{y_0} - \frac{\lambda_x N_{x_0}}{\lambda_y - \lambda_x}\right) e^{-\lambda_y t} + \left(\frac{\lambda_x N_{x_0}}{\lambda_y - \lambda_x}\right) e^{-\lambda_x t}$$

41.9. The half-life of ^{215}At is $100\ \mu$s. If a sample initially contains 6 mg of the element, what is its activity (a) initially?   (b) after $200\ \mu$s?

(a) The number of radioactive atoms initially present is

$$N_0 = \frac{(6 \times 10^{-3}\ \text{g})}{215\ \text{g/mol}} (6.03 \times 10^{23}\ \text{atoms/mol}) = 1.68 \times 10^{19}\ \text{atoms}$$

and the decay constant of ^{215}At is

$$\lambda = \frac{\ln 2}{T_{1/2}} = \frac{0.693}{100 \times 10^{-6}} = 6930\ \text{s}^{-1}$$

Hence the initial activity is

$$A_0 = \lambda N_0 = (6930)(1.68 \times 10^{19}) = 1.16 \times 10^{23}\ \text{Bq}$$

(b) At $t = 2T_{1/2}$,

$$A = A_0 e^{-\lambda t} = (1.16 \times 10^{23})(\tfrac{1}{2})^2 = 2.9 \times 10^{22}\ \text{Bq}$$

41.10. The table below gives the observed counting rate for a sample of a certain radioactive nuclide. Assuming that counting rate is proportional to activity, find the decay constant of the nuclide.

Time, h	0	2	4	6	8
Decays/min	28 844	15 720	8570	4682	2548

Denoting the proportionality factor by α, we have

$$CR = \alpha A = \alpha A_0 e^{-\lambda t} \equiv \beta e^{-\lambda t}$$

or $\ln CR = \ln \beta - \lambda t$. Thus, if the data are plotted on semi-logarithmic paper, the curve should be a straight line the slope of which is $-\lambda$. Let us suppose, for simplicity, that the "best" straight line through the data runs exactly through the first two data points. Then

$$-\lambda = \frac{\ln 15\,720 - \ln 28\,844}{2 - 0} = \frac{-\ln 1.84}{2} = -0.3049 \text{ h}^{-1}$$

or $\lambda = 0.3049 \text{ h}^{-1} = 8.469 \times 10^{-5} \text{ s}^{-1}$.

Notice that the calculated value of λ is independent of the units chosen for CR.

41.11. It is usually assumed that the production of ^{14}C in the air is balanced by its disintegration with half-life 5568 years. Consequently, living plants absorbing carbon in CO_2 from the air would contain an equilibrium amount of ^{14}C. Suppose a piece of wood of mass 20 grams, taken from an Indian dwelling, has a ^{14}C activity of 2.5 disintegrations per second. (a) Find how long the wood has been dead if plants that are living today have a ^{14}C specific activity of 250 disintegrations per second per kilogram. (b) If the present-day activity is not representative of the past activity of living plants and the specific activity when the tree was chopped down was in fact 170 disintegrations per second per kilogram, what would the true age of the chopped piece be?

(a)   Applying (41.5) to 1 kg of the wood, with $A_0 = 250$ Bq/kg,

$$t = \frac{1}{\lambda} \ln \frac{A_0}{A} = \frac{5568}{\ln 2} \ln \frac{250}{2.5/20 \times 10^{-3}} = 5568 \text{ y}$$

(b)   If, instead, $A_0 = 170$ Bq/kg,

$$t = \frac{5568}{\ln 2} \ln \frac{170}{2.5/20 \times 10^{-3}} = 2471 \text{ y}$$

41.12. Calculate the Q-value of the reaction

$$^2_1H + ^3_1H \rightarrow ^1_0n + ^4_2He$$

if the rest masses of the neutral atoms ^2_1H, ^3_1H, and ^4_2He are 2.014102 u, 3.016049 u, and 4.002603 u, respectively.

Add 2 electronic masses to either side of the reaction so that atomic masses may be used. The neutron rest mass is 1.008665 u, and so

$$Q = (\text{rest mass of reactants}) - (\text{rest mass of products})$$
$$= (2.014102 + 3.016049) - (1.008665 + 4.002603)$$
$$= 0.018883 \text{ u} = 17.6 \text{ MeV}$$

This reaction is *exothermic* $(Q > 0)$; kinetic energy, in the amount 17.6 MeV, is released by the reaction.

41.13. Find the Q-value of the nuclear reaction

$$_2^4\text{He} + {}_7^{14}\text{H} \rightarrow {}_8^{17}\text{O} + {}_1^1\text{H}$$

The rest masses of the neutral atoms $_2^4\text{He}$, $_7^{14}\text{N}$, and $_8^{17}\text{O}$ are 4.002603 u, 14.003074 u, and 16.999133 u, respectively.

After 9 electronic masses are added to either side of the reaction, the given atomic masses may be used, along with the mass of atomic $_1^1\text{H}$, 1.007825 u.

$$Q = (4.002603 + 14.003074) - (16.999133 + 1.007825)$$
$$= -0.001281 \text{ u} = -1.19 \text{ MeV}$$

Unlike the reaction of Problem 41.12, this reaction is *endothermic* ($Q < 0$); 1.19 MeV of kinetic energy is *lost* in the reaction. Clearly, then, the reaction cannot take place unless the bullet (the α-particle, $_2^4\text{He}$) has at least 1.19 MeV of kinetic energy (as measured, so it turns out, in the center-of-mass frame).

41.14. The following *proton-proton cycle* of nuclear reactions has been suggested as a possible source of stellar energy:

$$_1^1\text{H} + {}_1^1\text{H} \rightarrow {}_1^2\text{H} + {}_1^0e + \nu$$

$$_1^2\text{H} + {}_1^1\text{H} \rightarrow {}_2^3\text{He} + \gamma$$

$$_2^3\text{He} + {}_2^3\text{He} \rightarrow {}_2^4\text{He} + 2{}_1^1\text{H}$$

The net result of these reactions is that four protons ($_1^1\text{H}$) are combined to produce an α-particle ($_2^4\text{He}$), two positrons ($_1^0e$), two γ-ray photons, and two neutrinos (ν). Find the Q-value of the cycle. The rest mass of atomic helium is 4.002603 u.

The overall reaction is

$$4{}_1^1\text{H} \rightarrow {}_2^4\text{He} + 2{}_1^0e + 2\gamma + 2\nu$$

in which we may understand neutral atoms, provided $_1^0e$ is interpreted as positronium (Problem 40.12), of mass $2m_e$. Then, recalling that γ and ν have zero rest mass, we have

$$Q = 4M_\text{H} - (M_\text{He} + 4m_e)$$
$$= 4.031300 - (4.002603 + 0.002196)$$
$$= 0.0265 \text{ u} = 24.69 \text{ MeV}$$

Supplementary Problems

41.15. Compute the nuclear binding energy of the deuteron, $_1^2\text{H}$.   *Ans.*   2.22 MeV

41.16. Find the separation energy for a neutron from $_2^4\text{He}$.   *Ans.*   20.58 MeV

41.17. The half-life of radium, $_{88}^{226}\text{Ra}$, is 1620 years. What is the activity of a one-gram sample of radium?
Ans. $3.6 \times 10^{10} \text{ s}^{-1} \approx 1 \text{ Ci}$

41.18. Calculate the mass of a $_{83}^{214}\text{Bi}$ source that has an activity of 1 Ci and a half-life of 19.7 minutes.
Ans.   0.0224 μg

41.19. Calculate the Q-values of the reactions (*a*) $_{62}^{150}\text{Sm}({}_1^1\text{H}, {}_2^4\text{He}){}_{61}^{147}\text{Pm}$, (*b*) $_{41}^{93}\text{Nb}({}_1^1\text{H}, {}_1^2\text{H}){}_{41}^{92}\text{Nb}$. Rest masses of neutral atoms are: $_{62}^{150}\text{Sm}$, 149.917276 u; $_{61}^{147}\text{Pm}$, 146.915108 u; $_2^4\text{He}$, 4.002603 u; $_1^2\text{H}$, 2.014102 u; $_{41}^{93}\text{Nb}$, 92.906382 u; $_{41}^{92}\text{Nb}$, 91.907211 u.   *Ans.*   (*a*) 6.88 MeV;   (*b*) −6.62 MeV

Index

Catalog

If you are interested in a list of SCHAUM'S
OUTLINE SERIES send your name
and address, requesting your free catalog, to:

SCHAUM'S OUTLINE SERIES, Dept. C
McGRAW-HILL BOOK COMPANY
1221 Avenue of Americas
New York, N.Y. 10020